POLARIZED COLLIDER WORKSHOP

CONFERENCE PROCEEDINGS NO. **223**

PARTICLES AND FIELDS SERIES 42

POLARIZED COLLIDER WORKSHOP

UNIVERSITY PARK, PA 1990

EDITORS:
JOHN COLLINS
STEVE F. HEPPELMAN
RICHARD W. ROBINETT
UNIVERSITY PARK

American Institute of Physics   New York

Authorization to photocopy items for internal or personal use, beyond the free copying permitted under the 1978 U.S. Copyright Law (see statement below), is granted by the American Institute of Physics for users registered with the Copyright Clearance Center (CCC) Transactional Reporting Service, provided that the base fee of $2.00 per copy is paid directly to CCC, 27 Congress St., Salem, MA 01970. For those organizations that have been granted a photocopy license by CCC, a separate system of payment has been arranged. The fee code for users of the Transactional Reporting Service is: 0094-243X/87 $2.00.

© 1991 American Institute of Physics.

Individual readers of this volume and nonprofit libraries, acting for them, are permitted to make fair use of the material in it, such as copying an article for use in teaching or research. Permission is granted to quote from this volume in scientific work with the customary acknowledgment of the source. To reprint a figure, table, or other excerpt requires the consent of one of the original authors and notification to AIP. Republication or systematic or multiple reproduction of any material in this volume is permitted only under license from AIP. Address inquiries to Series Editor, AIP Conference Proceedings, AIP, 335 East 45th Street, New York, NY 10017-3483.

L.C. Catalog Card No. 91-71303
ISBN 0-88318-826-0
DOE CONF-9011177

Printed in the United States of America.

Contents

Preface ix

Concluding Talk 1
    J. Collins

PLENARY SESSIONS

Siberian Snake Experiment 13
    A. D. Krisch
Prospects for Polarization at RHIC and SSC 30
    S. Y. Lee
The Spin Dependent Structure Functions of the Nucleon 51
    V. Hughes
Spin Effects at Collider and Supercollider Energies 65
    J. Soffer
Theoretical Interpretation of Polarized Leptoproduction Data 80
    A. V. Efremov
The g_1 Problem: Much Ado About Nothing 90
    A. Manohar
What Have We Learned from Global QCD Analysis of Unpolarized Parton Distributions? 105
    W.-K. Tung
Polarization Prospects for KEK 123
    K.-I. Imai
Recent Results and Future Prospects for the Polarized Beam Program at Fermilab 129
    A. Yokosawa
An Overview of the RHIC Project 139
    T. Ludlam
Remarks on a Polarized RHIC 147
    G. Bunce

WORKSHOP TALKS

Jets in Polarized pp Collisions 155
    J. Ph. Guillet
Direct Photon Production in Polarized pp Collisions 162
    E. L. Berger and J. Qiu
Beyond the Standard Model with Polarized Beams at Future Colliders .. 169
    P. Taxil

Transverse Polarization in Deep Inelastic Collisions 176
    X. Artru
Lepton Pair Production in Polarized Collider Experiments as a Method to Measure the Transverse Polarization of Quarks in the Proton 184
    J. L. Cortes, B. Pire, and J. P. Ralston
Single Spin Asymmetries in Muon Pair Production 191
    R. Carlitz and R. S. Willey
Predictions for Jets in Polarized pp Collisions from EMC Experiment 196
    G. Nardulli
Polarized Protons at RHIC 201
    M. J. Tannenbaum
Polarized pp Production of ψ at Low p_T and the Polarized Gluon Distribution Function 210
    M. A. Doncheski
New Bosons in Polarized Hadronic Collisions 215
    J. Collins and G. Ladinsky
Spin and the Independent Scattering Mechanism 228
    J. Ralston and B. Pire
Asymptotic Perturbative QCD in Elastic Scattering, Color Transparency, and A_{NN} 230
    J. Botts
Hadronic Single Spin Asymmetries at Large p_T 237
    P. G. Ratcliffe
Transverse Spin Observables in Hadron–Hadron and Hadron–Nucleus Collisions 244
    D. Sivers
High Twist Effects in Hadronic Collisions 249
    J. Qiu and G. Sterman
Color Transparency in Nucleus–Nucleus Collisions 255
    L. Frankfurt and M. Strikman
The Proton Spin and the Gluon Anomaly 261
    J. Mandula
Does Parton Model Need Another Gluon Operator? 271
    Z. Ryzak
Flavour Symmetry Breaking and Generalized Goldberger-Treiman Relations 278
    N. A. Törnqvist
Is There a Hard Gluonic Contribution to the First Moment of g_1? 285
    G. T. Bodwin and J. Qui
High-Twist Contributions to $g_2(x, Q^2)$ 292
    X. Ji
The Asymptotic S-Matrix, Mass-Shell Anomalies and Observables 297
    H. F. Contopanagos and M. B. Einhorn
Measurement of the Spin-Dependent Structure Functions of the Proton and Neutron at HERA 305
    R. Milner

Proposal to Measure the Spin-Dependent Neutron Structure Function at SLAC 315
    P. Souder

Advantages of Polarization Experiments at RHIC 321
    D. Underwood

The Stimulated Stern-Gerlach Effect in Charged Particle Storage Rings 327
    Y. S. Derbenev

Parity Violating e^- Deuteron Scattering as a Probe of the Strangeness Content of the Nucleon 335
    S. J. Pollock

Self Polarization of Stored (Anti-) Protons: Status of the Spin Splitter Experiment at IUCF 343
    R. Rossmanith

Feasibility of a Polarized Deuteron Beam in the AGS and RHIC 350
    S. Y. Lee and L. G. Ratner

Efficient Calculation of One-Loop Polarized QCD Amplitudes 358
    Z. Bern and D. Kosower

Is There a Relation Between the Distributions of Transversely and Longitudinally Polarized Quarks? 364
    J. Collins

Spin Asymmetries: From Fixed Targets to Colliders 371
    S. Heppelmann

ROUNDTABLE SUMMARIES

Theoretical Interpretation of the EMC Results: Summary of Roundtable Discussion 377

Roundtable Discussion on Prospects for Polarized Collider Physics 380

Participants 385

NOTE: Underlined names indicate the author who delivered the lecture.

PREFACE

This volume is a record of the Polarized Collider Workshop which was held at Penn State University, University Park Campus, Nov. 15th–Nov. 17th, 1990. It contains versions of the plenary and workshop talks given at the workshop as well as several additional contributed papers and summaries of two roundtable discussions.

The Polarized Collider Workshop was motivated by the fact that recent progress in the field of accelerator design has made the prospects for studying physics with collisions of polarized protons in a collider setting technically feasible. It was this very real possibility which was one of the main focuses of the workshop.

The next generation of high-energy hadron colliders, SSC, LHC, and RHIC will all be scientific projects of immense scope for which much theoretical, experimental, and technological groundwork must be laid. Much effort has already gone into studies of the physics prospects and motivation for such machines. One option for such machines which has been less often discussed is the possibility of polarized proton collisions, since the attainment of polarization in a collider setting has always been a difficult achievement. Recent technical progress in this area (specifically the use of "Siberian snakes" to maintain polarization in the acceleration process) has made it possible to imagine collisions of polarized protons at collider energies (i.e., center-of-mass energies greater than 50 GeV or so) and with high luminosity. (One possibility which had been discussed is the collision of polarized protons at RHIC with luminosity of 10^{32} cm^{-2}s^{-1}, center-of-mass energy in the range $\sqrt{s} = 50$–500 GeV, and 50% polarization, operating for 1–2 months a year.)

Two earlier groundbreaking workshops, organized by Krisch and Chamberlain in 1985, had studied the machine aspects of attaining polarization at the SSC. (See "Polarized Beams at the SSC," American Institute of Physics Conference Proceedings No. 145, edited by A. D. Krisch, A. M. T. Lin, and O. Chamberlain, Ann Arbor-Bodega Bay, AIP, 1985). Since then, the theoretical understanding and experimental testing of such ideas has been extended dramatically.

Moreover, in the meantime, experiments in polarized lepton–proton scattering by the EMC collaboration had raised many questions about the theoretical understanding of the origin of the spin of the proton itself. The theoretical interpretations of such experiments are themselves controversial and an important recent topic in particle physics. This controversy added a new ingredient to the desirability of obtaining collisions of polarized protons at collider energies.

Finally, in addition to being of theoretical importance to resolve many of the basic physics issues suggested by current data, the discussion of polarization options for the various colliders is extremely timely as the problems, both technical and otherwise, of making sure that such options are consistent with the overall design of the machines have to be discussed well in advance of their construction.

The purpose of the meeting, then, was to bring together interested people in the various necessary fields, theoreticians, experimentalists, and machine/accelerator physicists to discuss the physics case that could be made for polarization options at such machines and their technical feasibility. Sessions devoted to the theoretical understanding of the EMC experiment and to resolving any remaining technical disagreements in its interpretation were a large part of the program. Such discussions form a necessary background to assess the current views on the spin structure of the proton. Presentations of future prospects for polarized lepton–nucleon scattering experiments and technical discussions of polarization techniques for hadron accelerators also formed an integral part of the workshop agenda. But the majority of the talks dealt with the physics prospects available at a polarized collider.

We reproduce here the talks given in both the plenary and parallel/workshop sessions of the meeting. Two roundtable discussions, each lasting roughly an hour, were also organized to address specific issues of interest. One, led by John Collins, discussed the present status of theoretical interpretations of the EMC effect and future prospects for experimentation and lattice calculations. The second, led by Jacques Soffer, on the last day of the workshop, featured representatives from all of the different interest groups present and discussed physics prospects for polarized colliders. We have summarized these discussions (not verbatim transcripts), and they appear in this volume. In addition, we have included several other contributed talks which could not be presented in person due to time (and other) constraints.

As with any cooperative enterprise, we have benefited from the advice and help of numerous people and we wish to express our gratitude to the many individuals who contributed to the success of the workshop. In the early planning stages, we benefited greatly from discussions with members of our External Advisory Committee (listed below) who helped shape the program. The workshop itself was supported by the National Science Foundation, the Department of Energy, the International Spin Committee, Brookhaven National laboratory, and the College of Science and Physics Department of the Pennsylvania State University to whom we are grateful. We especially thank Boris Kayser, P. K. Williams, A. Krisch, L. Trueman, G. Geoffroy, and H. Grotch for their personal support and help. We would also like to thank A. Manohar and R. Carlitz for their assistance in writing up the summary of the EMC roundtable discussion. The workshop benefited from the facilities and services provided by the Keller Conference Center, and we thank Norman Lathbury and his fine staff for their support in all aspects of the planning and implementation of the meeting. The local organizers also are grateful for the constant support and advice of the other members of the organizing committee, G. Bunce, J. Soffer, and M. Tannenbaum. Above all, we would like to thank the physicists who attended the workshop for their enthusiastic participation, which we feel made the meeting a success.

J. C. Collins
S. Heppelmann
R. Robinett
University Park
February, 1991

External Advisory Committee

S. Aronson (BNL)
E. Berger (Argonne)
S. Brodsky (SLAC)
R. Carlitz (Pittsburgh)
E. Courant (BNL/SSC)
A. Efremov (Dubna)
M. Einhorn (Michigan/ITP UCSB)

F. Halzen (Wisconsin)
R. Jaffe (MIT)
A. Mueller (Columbia)
J. Ralston (Kansas)
D. Sivers (Argonne)
G. Sterman (Stonybrook)
A. Yokosawa (Argonne)

Organizing Committee

G. Bruce (Brookhaven)
J. Collins (Penn State)
S. Heppelmann (Penn State)
J. Soffer (Marseille)
R. Robinett (Penn State)
M. Tannenbaum (Brookhaven)

CONCLUDING TALK

John C. Collins
Physics Department
Pennsylvania State University
University Park, PA 16802, USA

ABSTRACT

I summarize the physics issues that are important for a polarized hadron—hadron collider.

INTRODUCTION

Our aim in organizing this workshop was to stimulate discussion of the physics case for a polarized hadron collider, given its technical feasibility. Certainly the possibility of polarizing RHIC or the SSC at full luminosity with 70% polarization for both beams opens up some new areas for experimentation that have received relatively little attention.

In the Snowmass studies[1-3] for the SSC, there was very little discussion of polarization. The most notable exception[4] was a calculation of the usefulness of polarization for searching for compositeness of quarks and gluons. One may reasonably suppose that polarization is an option for an upgrade for a collider rather than being important for the first generation of experiments. But equally reasonably, one may argue that after discovery of some new physics at an unpolarized collider, the use of polarization will be important in understanding what it is that one has discovered. Therefore one must be careful to ensure that the possibility of a polarization upgrade does not get designed out by default, by lack of consideration of the physics issues.

A cautionary note is sounded by reading the original proposal for Fermilab.[5] The present physics landscape, populated by quarks, gluons, Ws, Zs, not to mention the Higgs boson, is rather unlike what one finds in the proposal; instead, one finds Regge poles and the like. Of course, the most important change is that we now have a fairly well—established theory of strong, weak and electromagnetic interactions. Thus there is a whole set of phenomena we can predict at a new accelerator, and consequently a limitation on the regimes in which it is reasonable to search for new physics. It is worth remembering that the prediction of the production of W and Z bosons at the $S\bar{p}pS$ is a triumph not only of weak interaction theory but also of QCD; it is a theorem that the Drell—Yan model (plus higher order corrections in $\alpha_S(m_W)$) is true in QCD.

When trying to investigate physics possibilities for a new accelerator which explores regimes previously far from the experimental domain, one should distinguish between what I will call 'rigorous predictions' and 'model prediction' of a theory. Examples of these that received much discussion at this workshop are the Bjorken and the Ellis—Jaffe sum rules for polarized deep inelastic scattering. The Bjorken sum rule is a rigorous prediction. It depends on well established theorems of QCD, plus isospin conservation and neglect of higher order corrections (which in principle can be calculated). Experimental violation of the Bjorken sum role at the level of 10s of per cent is a serious threat to the validity of QCD.

But the Ellis–Jaffe sum rule depends on assuming there are substantially fewer strange quarks than up and down quarks in the nucleon. This assumption is by no means obvious, though somewhat plausible. The violation of this sum-rule is evidence against the assumption, not against the underlying QCD theory. (Needless to say, model predictions may become rigorous predictions or useful approximations thereto, after the development of the theory. The parton model is a case in point. The reverse transition: rigorous prediction to model prediction is evidence of mistakes by theorists.)

PHYSICS AT NEW ACCELERATORS

There are four kinds of physics at a new hadron accelerator; such as RHIC, LHC and the SSC.
1. Standard model physics that is predictable or calculable to a reasonable degree of accuracy. This includes perturbative QCD and the physics of the photon, W and Z, and of leptons.
2. Standard model physics for which we do not have good methods of calculation. This includes soft, low p_T hadron physics.
3. Parts of the standard model that have not yet been discovered — the top quark and the Higgs. This physics lies between the predictions of the standard model and totally new physics.
4. Genuinely new physics. There have been many conjectures in this area, such as supersymmetry, technicolor, compositeness of quarks, leptons etc.

An important category of new physics is "none of the above."

Given that the standard model is valid up to tens or hundreds of GeV, new physics is likeliest to be in a high mass domain. It is therefore to be searched for in processes with a hard collision of quarks or gluons. There the measured cross section is:

$$\sigma = \frac{\text{parton luminosity}}{\text{hadron luminosity}} \times \sigma_{\text{fundamental}} ,$$

where $\sigma_{\text{fundamental}}$ is the short distance cross section at the parton level. It is the parton level cross section that is easily related to the fundamental Hamiltonian of the theory. Given some experimental data, one needs a QCD calculation plus a knowledge of parton densities to interpret a measured cross section in terms of the fundamental physics.

The above argument shows that perturbative QCD is an important subject at a hadron collider, whether one studies it for its own sake, for the exploration of electroweak phenomena, or for the search for new physics. The significant fraction of the workshop devoted to parton densities is a consequence.

On the theoretical side, it is worth systematically determining the generic kinds of new physics that are accessible at current energies, but that have so far remained hidden. The history of physics suggests that, once a theory is well established, it acquires a domain of validity. When the theory is ultimately demonstrated to be invalid, it generally remains a useful approximation to a more complete theory. The main interactions of quarks and leptons with gluons and the photon, W and Z are understood reasonably well. If one therefore regards the standard model (except for the top quark and the Higgs boson) as established, then new physics is to be looked for as an extension to the standard model, or a breakdown at short distances. Possibly there could be something exotic in place of the top and Higgs. Extensions to the standard model with sufficiently high masses or sufficiently weak coupling are commonplace material for theorists. I wish to pose the following question:

Are there any disguises that would permit new physics in a range accessible with current accelerators to have couplings of a size comparable to electroweak or QCD couplings, but that would have prevented it from having been noticed in previous experiments?

WHY POLARIZE?

Spin is a fundamental degree of freedom, and theories make specific predictions for the polarization dependence of whatever processes one might investigate. This alone implies that the study of polarization phenomena is important, at least in principal.

It is quite likely that, many if not almost all, new physics discoveries may be made without the use of polarized beams.[4] The point here is that searches for new physics on high mass scales are often luminosity limited, because of small cross sections. Measurement of polarization asymmetries extracts a substantial extra penalty in required statistics. But after a discovery, one needs to measure couplings, for example. At that point polarization becomes necessary. Perhaps one may get away with something equivalent in the form of a complicated correlation measurement.

A standard and hackneyed, but typical, case is given by a new W or Z. Given sufficient polarization at the parton level, measurements of spin asymmetries determine the relative amounts of left and right handed couplings.

$$A_{LL}^{expt} = A_{beam1} \, A_{beam2} \, A_{parton1/beam1} \, A_{parton2/beam2} \, A_{\hat{\sigma}}.$$

There is a danger of getting a small asymmetry as the product of several not so small numbers. With 10% asymmetries of the partons inside a hadron and a 25% asymmetry of the short-distance cross section, one finds

$$A_{LL} \sim (70\%)^2 \times (10\%)^2 \times (25\%) \sim \frac{1}{5}\% \ .$$

The measurement of such an asymmetry requires good systematic errors and large statistics—millions of events.

4 Concluding Talk

Single spin asymmetries do not suffer from so many factors multiplied together. But they are nonzero in more restricted circumstances. Single helicity asymmetries primarily probe parity violation. Single transverse spin asymmetries are higher twist in all or almost all cases in the standard model, but may be leading twist in new interactions.

PARTON DENSITIES

As part of a program to exploit a polarized collider one must have a program to measure parton densities. Although deep inelastic lepton scattering provides the traditional means for such measurements, it is not very adept at separating out different flavors of quark or at measuring the polarized gluon density, especially in the absence of neutrino scattering on a polarized target.

Fermilab, with its 200 GeV tertiary beam, has the energy, but maybe not the statistics to begin to make useful measurements. The AGS is low enough in energy that we must regard it as marginal for measurements of hard scattering, though it would be nice to see a Drell–Yan experiment. Upgrades to Fermilab and the SpS to use a polarized main ring with a gas jet target could provide useful measurements. The proposed 19 GeV on 19 GeV collider at KEK[15] would be in the same energy range.

Once one gets to RHIC, there is enough energy that processes like jet production are unambiguously in the hard scattering regime, and one could use measurements of various processes to find the helicity dependent parton densities. (Direct photon production is only one of several standard processes. Drell–Yan and jet production are also canonical processes. Next–to–leading QCD corrections will surely be calculated from the theory should the accelerators become reality).

The results of the EMC experiment are suggestive. One must beware, as always, of staking too much on a striking result that only has a significance of two to three standard deviations. But if the results are true at face value, then either sea quarks are relatively highly polarized, or the gluons are, or both. Note that if the gluons are not highly polarized, then the EMC result indicates that strange quarks are surprisingly well polarized. A mechanism for generating polarized $s\bar{s}$ pairs is likely to generate polarized $q\bar{q}$ pairs of all light flavors, so we should expect polarized $\bar{u}$ and $\bar{d}$ quarks. Hence the EMC effect suggests there will be unusually large polarization effects in processes which otherwise would be less interesting.

Consider, jet production at $\sqrt{s}$ = 600 GeV. According to calculations presented here by Guillet[6] and by Nardulli[7], double helicity asymmetries of a few percent occur with conventional distributions, but can be much larger with an EMC–motivated Δg — for example up to 20% at $P_T/\sqrt{s} \sim 1/3$. The rates for jet production can be large: At a $P_T$min = 30 GeV, the cross section at $\sqrt{s}$ = 600 GeV is several tens of nb. In the canonical month of 10^6 sec at a luminosity of 2×10^{32} cm^{-2} sec^{-1}, this gives millions of events — enough to measure a 1% asymmetry with useful precision.

There is an obvious collection of experiments. DIS probes quark distributions but cannot easily separate flavors. The Drell–Yan process is especially good for probing antiquarks; at P_Ts comparable to the dilepton mass, the Drell–Yan process is comparable to direct photon production for probing the gluon distribution, but without the fragmentation problem. Direct photon production usefully probes[16] the gluon density, but there are at present some uncertainties related to fragmentation that have not received sufficient attention in previous studies. Jet production at collider energies will also be valuable in measuring gluon densities.

Bound charm production can also be used to probe the gluon density especially at small x. But the theoretical justification for the application of parton ideas is <u>at present</u> not so sound as for the other processes. Moreover the mass scale of the physics of charmed quarks is low enough that higher power (higher twist) corrections may be substantial. These have the potential to a mess up the interpretation of an asymmetry in data: If the asymmetry is 10% and the higher twist effects are of order 10%, then one is by no means guaranteed at the present of theoretical knowledge that the measured asymmetry can be used to deduce that the leading twist cross section has this asymmetry. There might be a much larger asymmetry in the amplitude for the higher twist term; this might increase or decrease the measured asymmetry. Moreover, whereas the higher twist term in unpolarized scattering is typically of order $1/Q^2$ (times a mass squared) relative to the leading power term, the higher twist term in polarized scattering need only be suppressed by $1/Q$.

None of the issues in the measurement of polarized parton densities is greatly different in principal from those for the unpolarized densities. A measurement of parton densities over a range of processes and kinematic regions also provides a test of QCD: all the measurements must give the same result.

Use of detectors in a forward region — one to several units of rapidity from the central region — should be important. Hard scattering as measured in the forward region is sensitive to configurations in which a valence quark scatters off a low x gluon, quark or antiquark. The distribution of valence quarks is measured well, and the EMC experiment has verified that they are highly polarized. Thus measurements in the forward region particularly directly probe the small x parton.

All the above has concerned the helicity dependence of the parton densities. For the case of transverse polarization, see the next section.

WHAT KIND OF POLARIZATION?

In a circular accelerator, particles polarized transversely to the plane of the accelerator preserve their polarization during their trip round the ring. But the use of a spin rotator — which is what a Siberian snake is — enables both longitudinally and transversely polarized beams to be collided at a detector. Most of the theoretical discussion has been concerned with longitudinal polarization — or helicity. But one should work with all of the possibilities:

6 Concluding Talk

1. Single helicity asymmetries (A_L) are good for probing parity violating phenomena.
2. Double helicity asymmetries (A_{LL}) are the most commonly discussed for QCD hard scattering. Some subprocesses have 100% asymmetry at the parton level.
3. Single transverse spin asymmetries (A_N) are notorious for being higher twist in QCD and in normal electroweak processes.
4. Double transverse spin asymmetries (A_{NN}) are leading twist in some standard processes — e.g. Drell–Yan and jet production.

There appear a number of misconceptions in the literature about the possibility of discussing transverse spin for partons. These are provoked by the fact that the g_2 structure function in deep inelastic lepton scattering is higher twist — that is, it is suppressed by a power mass/Q compared to F_1, F_2 and g_1. This implies that this process is not a good place to measure the transverse spin asymmetry of parton densities. However there are leading twist asymmetries in, for example, the Drell–Yan process and in jet production, provided that one does not perform an azimuthal average over the measured particles. This has been clear since the work of Ralston and Soper[8] on the Drell–Yan process. A recent paper of Artru and Mekhfi[9] presents a fuller and more general discussion in the contest of QCD calculations.

Judging from discussions at this workshop, I see that there is still confusion and controversy. The issues are mathematical, and, I believe, simple, so any remaining controversy should be settled quickly.

According to QCD, single spin asymmetries in hard scattering subprocesses are higher twist — that is, they are suppressed by a power of M/Q, where M is some appropriate mass scale. The power is only unity, since twist-3 operators do enter into single spin asymmetries. Thus the asymmetry in single particle production, e.g. $pp \to \pi + X$, at large transverse momentum, is of order M/p_T times logarithms. There are three caveats in the comparison with experimental data. The first is that the M/p_T can be multiplied by a function of the dimensionless variables $x_T \equiv p_T/\sqrt{s}$ and rapidity y; thus the higher twist nature of the asymmetry cannot be verified without experiments over a range of center-of-mass energy. The second caveat is that the mass scale has absolutely no need to be a quark mass — it can just be the usual confinement scale. The final caveat is that the effect of transverse spin might appear in a shift of the transverse momentum of a jet. As Sivers pointed out,[10] for example, a shift of transverse momentum by a natural amount can give an apparently very large asymmetry in a cross section, since the cross section falls steeply as a function of p_T.

The latest measurement of the asymmetries, at Fermilab at 200 GeV, suggests that the effect is higher twist. But the statistical quality is still poor. One should attempt to do better.

TRANSVERSE SPIN

The discussion of higher twist effects in hadron–hadron scattering is now being put on a sound theoretical basis by Qiu and Sterman.[11] Note also the work on DIS by Chou, Ji and Jaffe.[12] Measurements of the higher twist effects should provide a probe of the amount of the proton spin carried by orbital angular momentum – a subject whose importance is brought out by the EMC results.

Once one has been able to verify the smallness of the single spin asymmetries at high enough energy, as one would hope at a collider, one can use a search for nonzero asymmetry as a test for new physics. To make this quantitative, one must first measure the transverse spin asymmetry of the quarks. Note that this must be done in <u>doubly</u> polarized scattering. (The large, presumably higher twist, effect in single hadron production suggests that investigation[13] of $g_2(x,Q)$ in DIS would be interesting.)

Moreover, the asymmetry of the gluon density is zero, if the parent hadron has spin half. This is a simple consequence of angular momentum conservation, as shown by Artru and Mekhfi.[9] Thus in the many interesting processes where gluon scattering dominates there are predicted to be small asymmetries. Moreover, one also cannot have quarks undergoing Altarelli–Parisi evolution to gluons and then to antiquarks. Thus there is <u>likely</u> to be a small asymmetry in the antiquark and strange quark distributions. Hence the double spin asymmetry Drell–Yan process in proton–proton collisions is unlikely to be a good place to measure the asymmetry of the parton densities. (Polarized antiprotons on polarized protons would not suffer from this problem, but no one has yet shown how to do that in practice). A large higher twist asymmetry would completely mask a small leading twist asymmetry. But a large double spin asymmetry in the Drell–Yan process, once verified to be leading twist, would at the least be interesting. <u>Caution</u>: A higher twist contribution to the double spin asymmetry may also have a corresponding single spin asymmetry.

The most obvious way of measuring the transverse spin asymmetry of the quarks while having a good probability of a large, leading-twist asymmetry in the cross section is to examine jet production with large $E_T/\sqrt{s}$. Given the large single spin asymmetry at fixed target energies, one would need to go to collider energies to avoid being confused by higher twist effects.

It is likely (though not guaranteed) that the transverse spin asymmetry of the valence quarks is large, just as is the helicity asymmetry. Model calculations are needed here. Once one has the QCD under control, transverse spin asymmetries should be a useful probe of certain kinds of new physics.

One should beware of some of the predictions in the literature about transverse spin asymmetries. For example, during the second round-table discussion, Sivers quoted a sum rule relating transverse and longitudinal quark distributions:

$$\int_0^1 g_T(x)\, dx = \int_0^1 g_L(x)\, dx \ .$$

This was asserted to follow from Lorentz invariance. I performed a simple one-loop calculation that showed that this sum rule is false. The calculation used the definitions[8] of the parton densities that are needed to prove the factorization theorem for hard scattering in a collision of polarized beams. The sum-rule quoted was that of Burkhardt and Cottingham.[14] This is actually a sum-rule for g_1 and g_2 in deep-inelastic scattering. These cannot be directly related to the transverse quark density.

CONCLUSION

1. Spin physics is important, because it is a fundamental degree of freedom and probes the chiral structure of couplings.
2. The first question to answer is whether we should design new colliders (RHIC, SSC,...) such that they can be upgraded to use a polarization.
3. The second question is whether it is worthwhile to actually polarize a specific collider, in terms of dollars, effort and (wo)man-years.
4. The answer to the second question depends, to some extent, on the size of the polarization of gluons and quarks inside a proton. In the case of the SSC the answer is much more likely to be 'yes' if new physics is discovered during a first generation experiment.
5. Single spin asymmetries (both longitudinal and transverse) are useful in physics beyond QCD.
6. At colliders with several hundreds of GeV of center-of-mass energy, hard scattering is much cleaner than at fixed target energies, especially for jet production. This is clear in the unpolarized case. Thus one must not be unduly influenced by the results of fixed target experiments.

REFERENCES

1. "Elementary Particle Physics and Future Facilities: Proceedings, 1982 DPF Summer Study, Snowmass, USA, June 28 – July 13, 1982," eds., R. Donaldson, R. Gustafson and F. E. Paige (A.I.P., New York, 1983).

2. "Proceedings of the 1984 Summer Study on the Design and Utilization of the Superconducting Super Collider," eds. R. Donaldson and J. G. Morfin (A.I.P., New York, 1985).

3. "Proceedings of the 1986 Summer Study on the Physics of the Superconducting Supercollider", eds. R. Donaldson and J. Marx (A.I.P., New York, 1987).

4. C. H. Albright et al, p. 31 in Ref. 2.

5. Proposal for Fermilab.

6. J. P. Guillet, these proceedings.

7. G. Nardulli, these proceedings.

8. J. Ralston and D. E. Soper, Nucl. Phys. B152, 109 (1979); J. Ralston, these proceedings.

9. X. Artru and M. Mekhfi, Z. Phys. C45, 669 (1990); X. Artru, these proceedings.

10. D. Sivers, these proceedings.

11. J. -W. Qiu and G. Sterman, Stony Brook preprints, ITP–SB–90–49 and 50; J. -W. Qiu and G. Sterman, these proceedings.

12. X. -D. Ji and C. -H. Chou, Phys. Rev. D42, 3637 (1990); R. L. Jaffe and X. -D. Ji, MIT preprint MIT–CTP–1848; X. -D. Ji, these proceedings.

13. R. L. Jaffe, Comm. Nucl. Part. Phys. 14, 239 (1990).

14. H. Burkhardt and W. H. Cottingham, Ann. Phys. 56, 453 (1970).

15. K. Imai, these proceedings.

16. J. -W. Qiu, these proceedings.

Plenary Sessions

Siberian Snake Experiments‡

A. D. Krisch

Randall Laboratory of Physics, The University of Michigan
Ann Arbor, Michigan 48109-1120 USA

We began studying Siberian Snakes at the 1985 Ann Arbor Workshop on Polarized Protons at the SSC[1]. Fig. 1 is a diagram of how a polarized SSC might appear with Siberian Snakes installed in the various rings. By the end of the Workshop it appeared that it would be practical to have polarized protons at the SSC, but only if the Siberian Snake Concept worked as everyone hoped. The Workshop's first conclusion was that polarized beam acceleration would probably be possible at the SSC. The second conclusion was that one must test the Siberian Snake Concept, which had been invented around 1974 by Derbenev and Kondratenko[2]. While most of the world's distinguished accelerator theorists were quite convinced that the concept would work, we felt that it would be difficult to convince the SSC Director and the funding people to spend tens of millions of dollars on Siberian Snakes, if it was an untested theoretical concept.

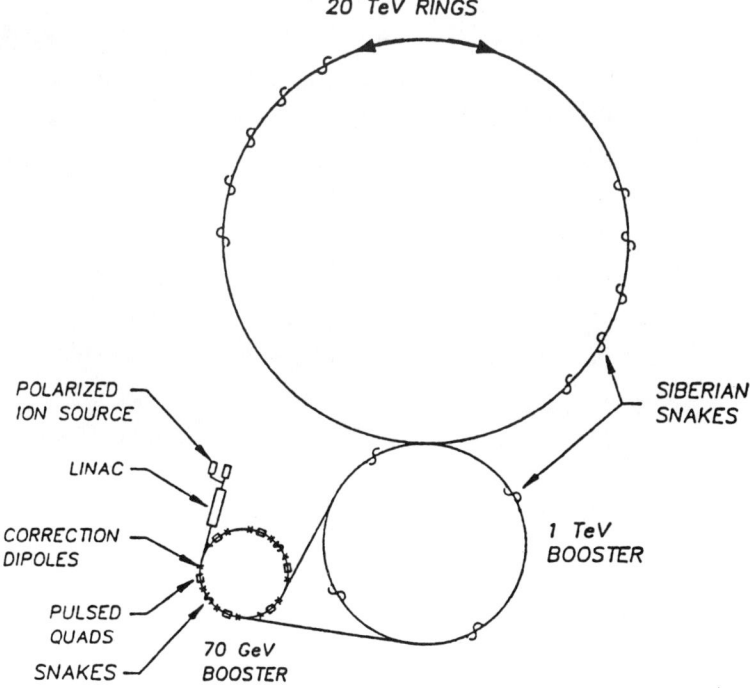

Fig. 1 Polarized proton acceleration at the SSC as seen in 1985.

14 Siberian Snake Experiments

To understand the importance of Siberian Snakes, recall the difficulty of accelerating polarized protons to 22 GeV at the AGS, which was the world's highest energy polarized proton beam. It was necessary to individually overcome each of the many intrinsic and imperfection depolarizing resonances. Fig. 2 shows the oscilloscope trace from the AGS main control room; the bottom curve is the AGS magnet cycle during the acceleration of polarized protons to 16.5 GeV. The correction currents in the 96 correction dipoles for some 25 imperfection resonances are shown in the upper curve; the middle curve shows the pulsed quadrupoles firing three times to overcome three intrinsic resonances. To reach 22 GeV we had to overcome about 45 resonances; this required seven weeks of total dedication of the AGS to these polarized beam studies which cost a million dollars each week. At the SSC each 20 TeV ring will have 36,000 depolarizing resonances; it seems completely impractical to try to overcome them individually. It was clear that one needed a new technique to accelerate polarized protons to TeV energies.

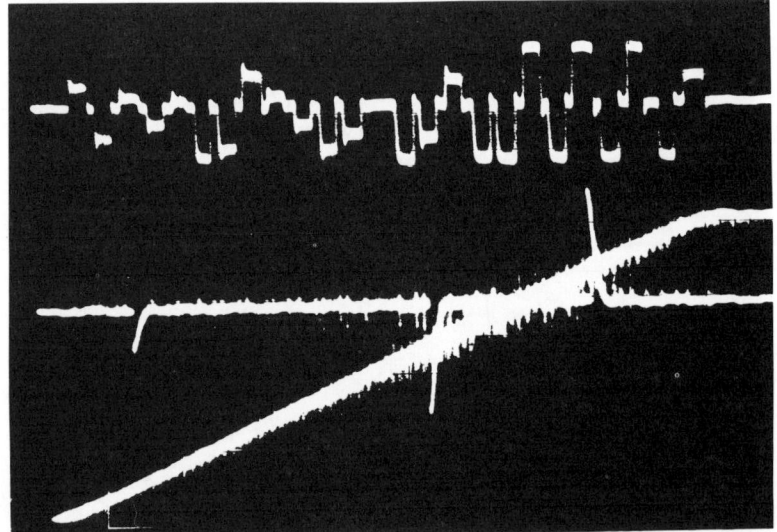

Dipole Correction Currents

Pulsed Quadrupoles

Magnet Cycle

Fig. 2 AGS main control room oscilloscope trace showing the acceleration of polarized protons to 16.5 GeV.

Shortly after the Ann Arbor Workshop, some of the participants began looking around the world for some accelerator where we could test the Siberian Snake concept. As I recall, we looked seriously at the AGS, at Saturne, at KEK, and at Indiana. We settled on the Indiana Cooler Ring mostly because its straight sections were long enough for a Siberian Snake. We then formed a collaboration and began to study Siberian Snakes. The first results were published during the past year in two Physical Review Letters[3,4]. The collaboration includes people from Michigan, Indiana, and Brookhaven who are listed in reference 3.

The Indiana University Cyclotron Facility Cooler Ring is shown in Fig. 3. IUCF had the potential for injecting polarized beam into the Cooler Ring from the two stage cyclotron injector; the maximum energy of the second Cyclotron is 200 MeV. The Cooler Ring itself is a synchrotron storage ring, which can accelerate protons up to 500 MeV. The study of Siberian Snakes at the Cooler Ring required several new hardware items. We used the electron cooling solenoids to create imperfection fields to study the depolarizing resonances. To inject polarized protons into the Cooler Ring, we installed some kicker magnets. The Cooler Ring normally used stripping injection of H$^-$ ions; however the polarized source emitted normal polarized protons, which made stripping injection impossible. Two kicker magnets were built quickly and inexpensively at Michigan using some spare ferrite that remained after building the pulsed quadrupoles for the AGS. These kickers had a risetime of about 100 ns and effectively injected the polarized beam into the Cooler Ring.

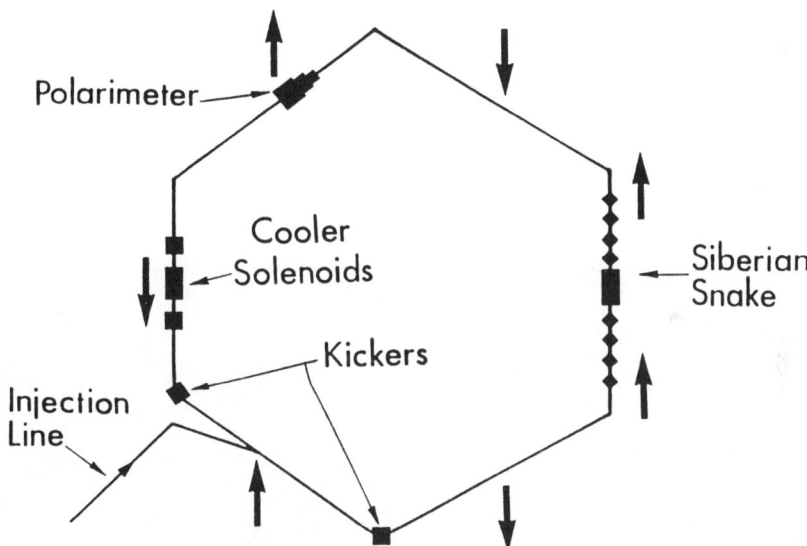

Fig. 3 The IUCF Cooler Ring with the Michigan-Indiana Siberian Snake.

The polarimeter for our experiment was built by Indiana as a spectrometer for various nuclear physics experiments. It had azimuthal symmetry and contained a carbon target, wire chambers, and scintillation counters. We used it as a polarimeter by accepting the $\Delta\theta$ region of about 5° to 17° and splitting it into four $\Delta\phi$ bins of 90° each. The left and right quadrants measured the up-down polarization, while the up-down quadrants measured the sideways polarization. Near the $G\gamma = 2$ depolarizing resonance, which occurs at 108 MeV, the average analyzing power was about 25%. Near the $G\gamma = -3 + \nu_y$ depolarizing resonance at 177 MeV, the cross-section was smaller but the analyzing power was larger.

16   Siberian Snake Experiments

It normally took about 10 minutes to make a 3% measurement of the beam polarization with about 20 nA of stored beam.

The Siberian Snake is shown in Fig. 4; this was of course our most important piece of hardware. This was not a conventional high energy Snake where 8 transverse dipole magnets produced the 180° spin rotation. Such snakes were not very practical at these low energies; instead we used the superconducting solenoid magnet, shown at the center, which easily rotated the spin by 180°. It was a rather strong solenoid magnet, with an $\int B \cdot d\ell$ of about 2 Tesla-meters. Unfortunately, this solenoid caused considerable optical distortion to the low energy beam. This orbit distortion focused the beam and rotated its phase space by about 30° which produced considerable xy-coupling. To compensate for this orbit distortion we used a series of eight quadrupole magnets; the four magnets on the outside were standard quadrupoles, the four on the inside were identical quadrupoles which could be rotated to become skew quadrupoles at various angles. By properly adjusting the currents and the rotation angles of these eight quadrupoles we could eliminate all of the orbit distortions introduced by the solenoid magnet and thus make this straight section optically transparent. The quadrupoles had to be carefully adjusted because they were rather strong; but when all eight were properly adjusted the Snake was optically transparent and there was no beam loss.

Fig. 4 The Michigan-IUCF Siberian Snake.

In our first experiment we studied the beam polarization at 120 MeV, while varying the strength of the $G\gamma = 2$ imperfection resonance at 108 MeV. This energy was some distance from the resonance, but its effect was still strong. With the injection of vertically polarized protons, the measured vertical polarization had a clear peak which looked similar to the imperfection depolarizing resonance correction curves at the ZGS[5] and AGS[6]. The radial polarization was fairly large; it also had a zero crossing and changed sign. We later came to understand that we were rotating the stable spin direction as we varied the longitudinal imperfection field. The imperfection field came from the cooling solenoid which was itself a weak partial Snake that could rotate the stable spin direction.

We repeated this study with the Snake off at 104 MeV as shown in Fig. 5; the curve was now about 10 times sharper than at 120 MeV. We could only maintain full polarization with the Snake off if we exactly corrected the imperfection fields. Any slight imperfection would quickly change the polarization. We next turned the Snake on and injected protons polarized in the radial direction, which was the stable spin direction when the Snake was on. We then found that the polarization remained large and constant as we varied the imperfection field over a wide range. With the Snake on, the polarization appeared totally insensitive to the imperfection fields; the beam remained horizontally polarized even with a strong imperfection field. We considered Fig. 5 to be the first clear indication that the Siberian Snake concept worked.

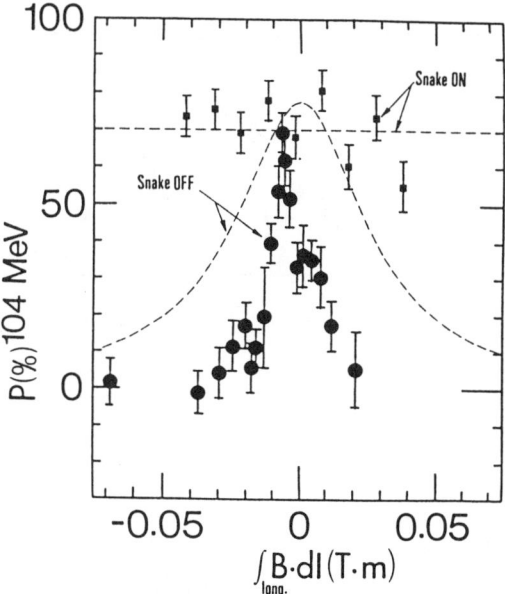

Fig. 5 The 104 MeV polarization in the stable spin directions with the Snake on and the Snake off.

The dashed curve in Fig. 5 shows our calculation of the polarization at 104 MeV. This calculation, which assumed that the resonance was at 108 MeV, did not agree very well with the Snake-off data. At first we were not sure if the calculation was correct, so we did not take this disagreement too seriously. We were quite pleased that the Snake-on calculation agreed with the Snake-on data which showed that the Snake worked.

During 1989 and 1990 we had eight different running periods each lasting about a week. We did many different experiments; I will discuss some of these experiments in an order which is more logical than temporal.

By mid-1990 our measurement ability had improved considerably. We, therefore, remeasured the polarization at 104 MeV in much finer detail as shown in Fig. 6. This curve should have been similar to Fig. 5, but on a much expanded scale. However, with this fine detail, we found a surprise; at first we saw one low data point which we thought was an incorrect measurement. But when we remeasured the curve several more times, there were clear dips on both sides of the central peak. The existence of these dips was further confirmed by identical dips in the radial polarization. Note that the beam was not spin-rotated in these dips; it was fully depolarized because both the vertical and the radial polarizations were zero. We thought that these dips must be due to some type of depolarizing resonance. Eventually we decided that each dip might be a synchrotron depolarizing resonance, which would occur whenever $G\gamma = 2 \pm \nu_s$, where ν_s was the synchrotron tune.

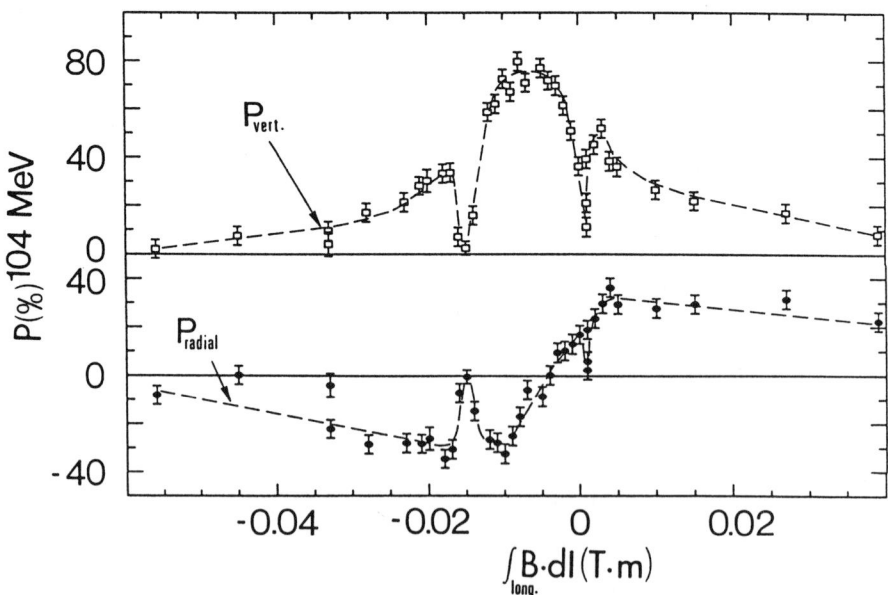

Fig. 6 Observation of synchrotron depolarizing resonances.

To prove that each dip was a synchrotron depolarizing resonance, we measured the polarization while changing the synchrotron tune, ν_s. We thought that the easiest way to vary ν_s might be to vary the RF voltage. Therefore, we set $\int B \cdot d\ell$ in the center of the dip on the left and then measured the polarization while varying the RF voltage. As shown in Fig. 7 this experiment worked very well; there was clearly a strong relationship between the polarization and V_{rf}, the RF voltage. A synchrotron depolarizing resonance seemed to be the only explanation for this relationship. It seems highly probable that each dip in Fig. 6 was due to a synchrotron depolarizing resonance.

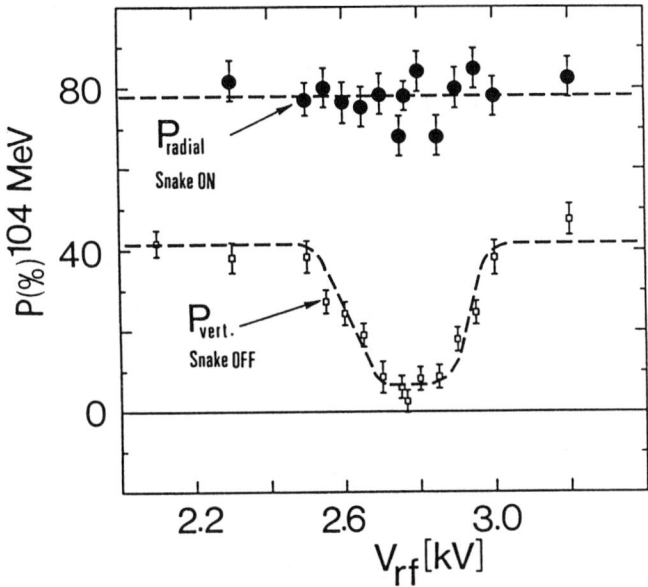

Fig. 7 The Snake overcoming a synchrotron depolarizing resonance

Then with $\int B \cdot d\ell$ still set in the center of the dip, we turned the Snake on and again varied V_{rf} and thus ν_s. As shown in Fig. 7 the Snake now maintained full polarization over the entire range of V_{rf}. We had not originally considered synchrotron depolarizing resonances, but Fig. 7 showed that a Snake can certainly overcome these resonances.

Our next experiment was a study of the effect of the $G\gamma = 2$ resonance on the polarization at some nearby energies. Recall that the calculation did not quite agree with the data in Fig. 5; we thought that perhaps the energy of the Cooler Ring was not properly calibrated. We recalled that at two previous accelerators, the ZGS[5] and AGS[6], the polarized beam studies had discovered an energy miscalibration of about 1%. Our IUCF colleagues were rather concerned about this possibility. Bob Pollock and Hans-Otto Meyer believed that the calibration was correct. Fig. 8 shows a study at 108 MeV which should be near the resonance if the energy calibration was correct; but the width of the resonance curve was quite similar to the width at 104 MeV shown in Fig. 6. This increased the concern about the energy calibration. [This curve was less detailed than Fig. 6, but one could again see the synchrotron resonance dips.]

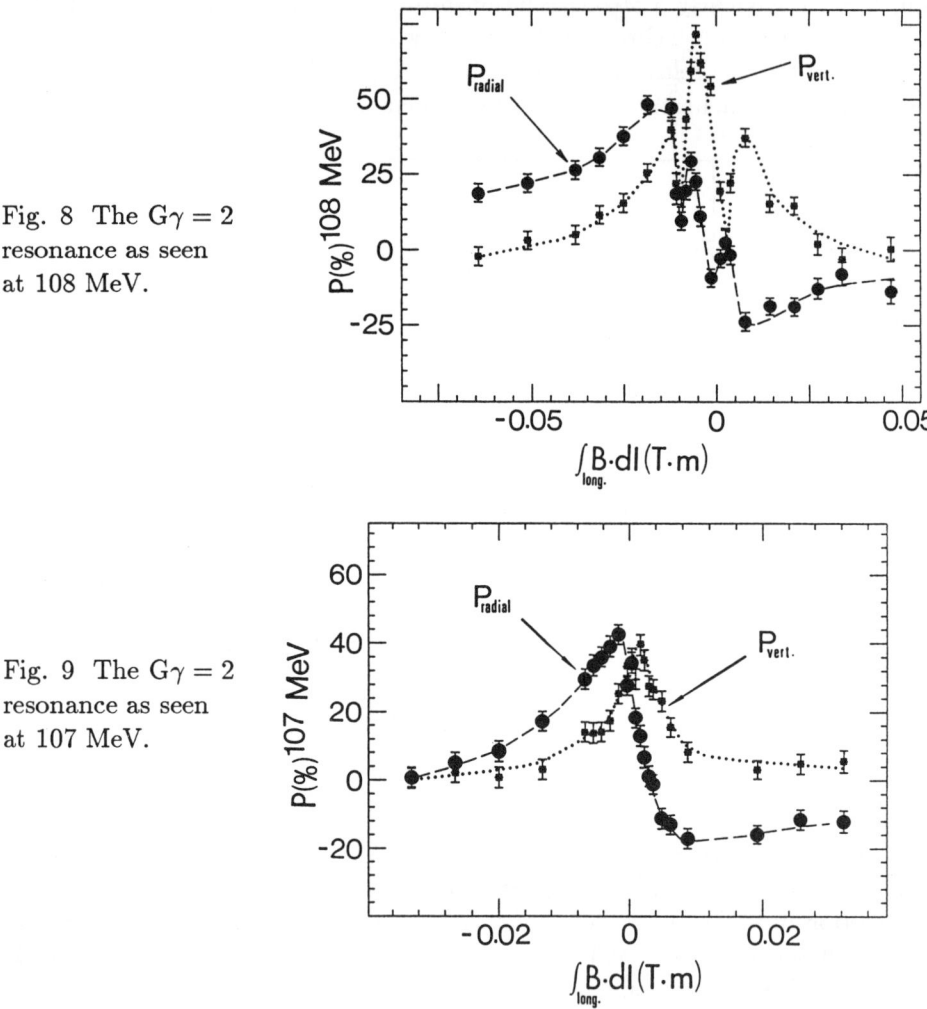

Fig. 8 The $G\gamma = 2$ resonance as seen at 108 MeV.

Fig. 9 The $G\gamma = 2$ resonance as seen at 107 MeV.

The 104 and 108 MeV curves suggest that the resonance might be near 106.5 MeV. Therefore, we studied the polarization at 107 MeV which is shown in Fig. 9. This curve was very narrow; it was about a factor of 2 sharper than the 108 MeV curve. Also note that one never reached full polarization in the vertical direction, apparently because 107 MeV was so close to the resonance. Moreover, the synchrotron depolarizing resonances now apparently either disappeared or became very narrow. Thomas Roser and S.Y. Lee believed that as one approached the resonant energy, the two synchrotron resonances should move together, eventually merge, and then cease to exist. [The 107 MeV data was certainly consistent with no synchrotron resonances; but the existence of very narrow dips can never be completely ruled out. However, we had about reached the stability limit of the Cooler Ring; probably we cannot reliably get points much closer together than in Fig. 9.]

We were eager to determine if a Siberian Snake could overcome an intrinsic depolarizing resonance; Fig. 10 shows the polarization at 177 MeV plotted against the vertical betatron tune, ν_y. The correction quads were not always used; however for this study of the $G\gamma = -3 + \nu_y$ intrinsic resonance we used the correction quads to insure that the vertical betatron tune, ν_y, was exactly right. It was difficult to ramp the Snake because the superconducting solenoid and the quads must then exactly match the Ring ramp. Therefore, S.Y. Lee proposed ramping the tune instead. Normally, during acceleration one holds the tune fixed and ramps the energy; here we held the energy fixed and changed ν_y in discrete steps. We did a measurement at one tune and then we changed the tune for the next measurement. Notice in Fig. 10 that with no Snake, the polarization was zero near the resonance. However, with the Snake on, full polarization was maintained over the entire ν_y range. This clearly demonstrated that the Snake was capable of overcoming a quite strong intrinsic depolarizing resonance.

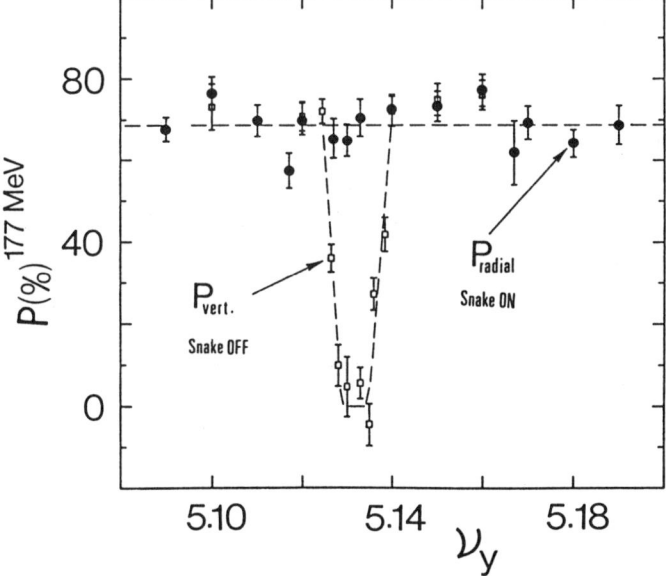

Fig. 10 The Siberian Snake overcoming the $G\gamma = -3 + \nu_y$ intrinsic resonance.

22 Siberian Snake Experiments

Now I will discuss partial Siberian Snakes. Many people have considered these partial Snakes; probably Derbenev and Kondratenko were first. Courant has been quite involved with them and Roser and Ratner have recently been actively studying partial Snakes. We made a study of partial Snakes by measuring the polarization at 104 MeV, while lowering the spin rotation angle in the Snake. As shown in Fig. 11, the Snake was able to maintain full polarization from a 100% Snake, down to about a 5% Snake. When the Snake was less than a few percent, then it was no longer strong enough and the polarization dropped rapidly. The broad flat region of full polarization seems to be clear evidence that an appropriate partial Snake can overcome imperfection resonances.

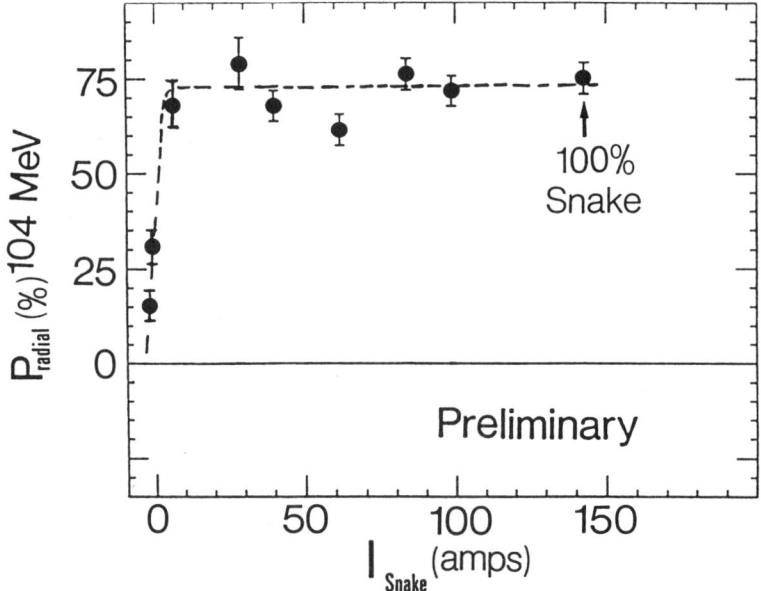

Fig. 11 A partial Snake overcoming the $G\gamma = 2$ intrinsic depolarizing resonance.

We now have rather extensive studies of the interactions of Siberian Snakes with various depolarizing resonances. We have about 50 data curves of one sort or another, but many of them are not yet organized. Our thesis student, Michiko Minty, is organizing all of this data. She has compiled a book about one inch thick with many cross references. It is important to keep track of all this data; several times we were almost ready to start some new measurement when we discovered that we had already done it some months earlier.

By Spring 1990 we had shown that a Siberian Snake was capable of overcoming an imperfection depolarizing resonance, an intrinsic depolarizing resonance, and a synchrotron depolarizing resonance. We had also shown that a partial Snake could overcome an intrinsic depolarizing resonance.

It was predicted that partial Snakes could only overcome imperfection depolarizing resonances; a full Snake is needed for an intrinsic resonance. I will not give a detailed theoretical discussion; however a partial Snake could apparently shift the spin tune and thus drive an intrinsic resonance to a different energy. We searched for this shift by using a 25% Snake while ramping the Cooler Ring energy from 104 to 120 MeV; we measured the radial polarization while injecting horizontally polarized protons. We changed the total percentage of Snake in the Ring by varying the correction solenoids; they were weak partial Snakes which could vary the percentage of total Snake by a few percent. As shown in Fig. 12 there was a strong depolarization in the predicted region. This seemed to be a clear indication that the spin tune shift in the partial Snake shifted the energy of the $G\gamma = -3 + \nu_y$ intrinsic resonance from 177 MeV into the 104 to 120 MeV ramp. There was another intrinsic resonance, $G\gamma = 7 - \nu_y$, which normally occurred well below injection. However the partial Snake apparently made both these intrinsic resonances occur at the same energy. In any case, we were apparently moving some intrinsic depolarizing resonance into the ramp region of 104–120 MeV; there was certainly a strong depolarization.

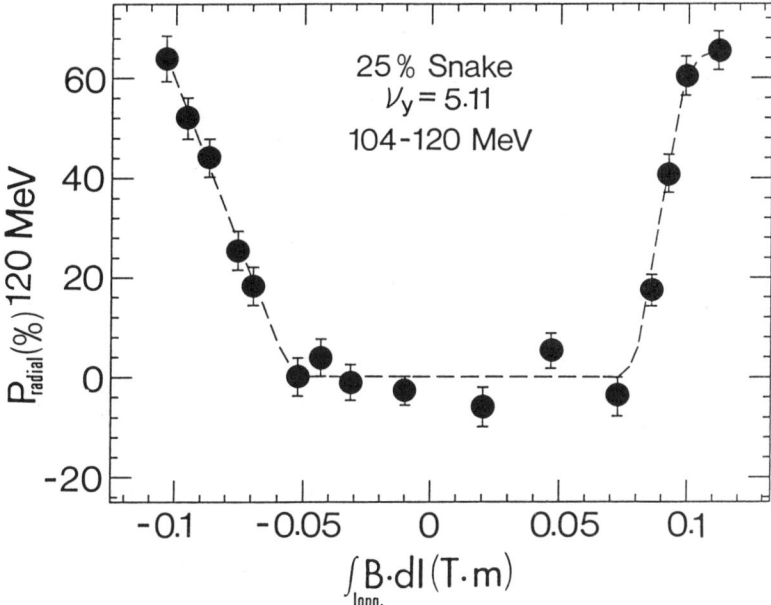

Fig. 12 The 25% partial Snake moving the intrinsic depolarizing resonances into the 104 to 120 MeV range.

During our August 1990 run, we found for the first time a depolarizing resonance that the Snake did not overcome. This surprising result is shown in Fig. 13 where we varied the current in the correction solenoids while sitting at 106 MeV near the $G\gamma = 2$ intrinsic resonance. There was a very clear dip in this curve.* We had a 100% Snake turned on, but this was the first time that we ran with the Snake quadrupoles turned on near the $G\gamma = 2$ imperfection resonance.† We had been looking for Snake resonances for some time, but never before saw any depolarization when the Snake was on. Derbenev and I now believe that the depolarization in Fig. 13 was probably due to a Snake resonance which was driven by the Snake quadrupoles.

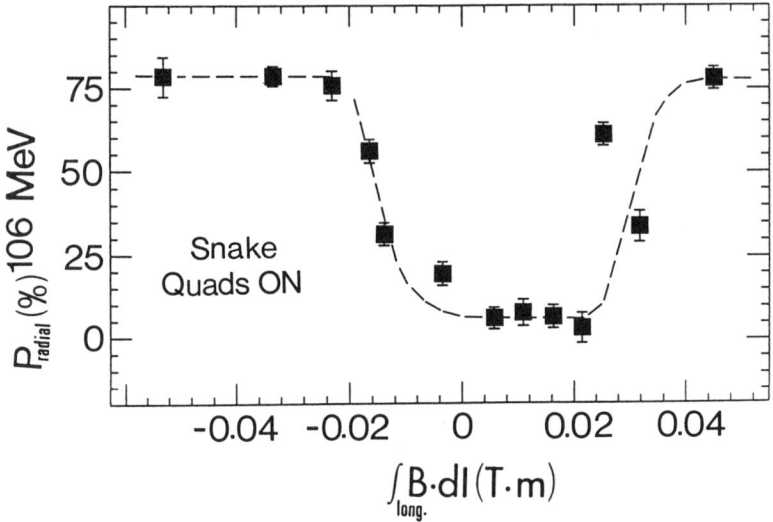

Fig. 13 A possible Snake resonance.

To understand something about a Snake resonance, let us first recall how a Snake works. Consider a vertically polarized proton circling a ring containing a Snake. Normally the spin gets slightly rotated during each turn around the ring by the horizontal imperfection fields. On one turn the vertical spin may get rotated from 0° to perhaps 3°; then it passes through the Snake which rotates it by 180° to 183°. On the next turn around the Ring, the imperfections act in the same direction as on the first turn; therefore, they now rotate the spin by 3° back to 180°. Finally the Snake again rotates the spin by 180° bringing it back to 0° after two full turns around the ring. Thus, the Snake makes each imperfection field cancel itself. The only exceptions to this cancellation are apparently the magnetic fields inside the Snake itself.

---

* Do not yet take too seriously the two points on the right which may not agree; one may just be due to an error in recording the data.

† Recall that at 104 MeV the Snake easily overcame the $G\gamma = 2$ resonance, as was shown in Fig. 5. We later made a similar Snake-on study at 108 MeV; there was again no depolarization. However these studies were done with the Snake quadrupoles turned off. Often we did not use the quadrupoles because they were so strong that they had to be finely adjusted or they would kill the beam. The quadrupoles shifted the horizontal tune, ν_x, by a full unit of one.

Recall that our Snake was not a normal high energy Snake constructed of 8 dipoles; it instead contained a solenoid and eight quadrupoles. To see how the Snake quadrupoles might drive a Snake resonance, first note that there was a sharp focus at the center of the solenoid. Therefore if a proton in the upstream quadrupole was above the beam axis, then in the downstream quadrupole it would be below the axis. Moreover, the proton's spin was rotated through 180° by the Snake. Finally, the quadrupoles were mirror symmetric. These three factors combined to force the two quadrupoles to have a constructive depolarizing effect on the spin. If in the upstream quadrupole the field pointed to the left and the spin was up, then in the downstream quadrupole the field pointed to the right and the spin was down. The two quadrupoles would therefore interact coherently with the proton's spin and cause a coherent depolarization inside the Snake. We think that the depolarization in Fig. 13 was probably a Snake resonance which was being driven by the Snake quadrupoles; these were much stronger than the normal Cooler Ring quadrupoles. The Snake apparently could not overcome the depolarizing quadrupole fields because they existed inside the Snake itself. The Snake could cure other problems, but it could not cure its own problems. This type of Snake resonance should not occur with a high energy Snake made of eight dipoles.

We studied this Snake resonance by slightly varying the strength of the Snake solenoid. As shown in Fig. 14, there was clearly a dip in the polarization as the solenoid current was varied near 144.2 amps which gives a 100% Snake at 106 MeV. Notice that the width of the dip was about 2%; this width may provide information about the nature of the Snake resonance.

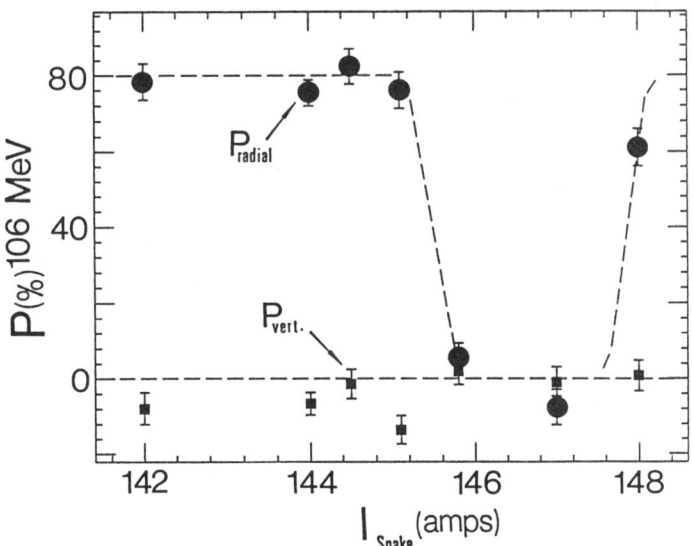

Fig. 14 Studying a Snake resonance by varying the Snake Solenoid.

We then studied the Snake resonance by varying the horizontal and vertical betatron tunes. As shown in Fig. 15, we again saw clear dips in the polarization. During the first data run, where we were changing the vertical tune, we were not careful enough about keeping fixed the horizontal tune. We did obtain good data where we held fixed the vertical betatron tune and varied the horizontal tune. Fig. 15 clearly shows that the resonance depended strongly upon the horizontal betatron tune, ν_x, and probably upon the vertical betatron tune, ν_y. Notice that the dip occurred near $\nu_x + \nu_y \approx 1.50$ for both curves. Since the spin tune, ν_s, was close to one-half, this data showed a strong depolarization near $\nu_x + \nu_y + \nu_s \approx 2$ which is an integer. The shift from an exact integer was probably due to the Type-3 Snake which we will soon discuss. This $\nu_x + \nu_y + \nu_s =$ integer behavior supports the Snake resonance hypothesis. Notice that the width of the dip in the horizontal curve, $\Delta\nu_x$, was considerably larger than the vertical width, $\Delta\nu_y$. These widths may provide important information about the nature of the Snake resonance. In other similar curves with the Snake solenoid on, there was no evidence of depolarization. Thus the unusually strong driving term due to the Snake quadrupoles may be essential for this Snake resonance.

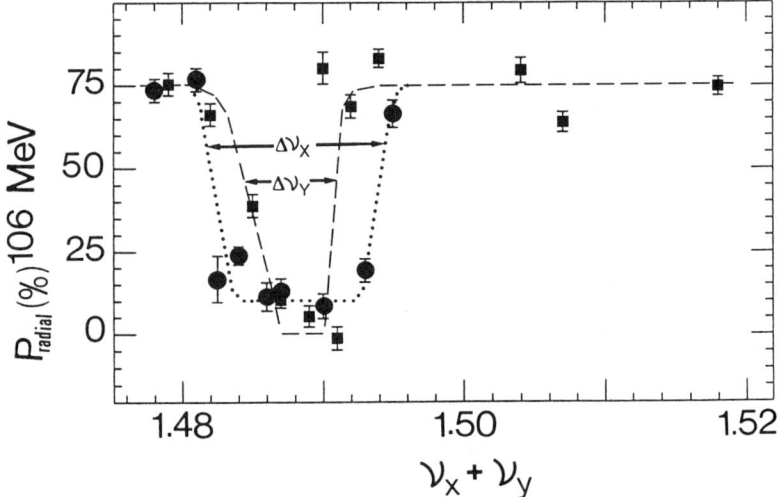

Fig. 15 Studying a Snake resonance by varying ν_x and ν_y.

Recall the concern about the possible miscalibration of the Cooler Ring energy. Our data indicated that the $G\gamma = 2$ resonance was occurring near 106.5 MeV, while much data was based upon a well established 108.4 MeV energy calibration for both the Cooler Ring and the Cyclotron. Pollock believed that this 108.4 MeV calibration was correct. Pollock, with some advice by Roser, then proposed that there might be a third type of Snake in the Cooler Ring. Type-1 Snakes rotate the spin about the longitudinal direction, while Type-2 Snakes rotate the spin about the radial direction. At the June meeting in Brookhaven on partial Snakes, Roser apparently suggested that there could also be Type-3 Snakes which could rotate the spin about the vertical direction. Just before our August run, Pollock submitted a paper[7] proposing that a Type-3 Snake would explain our apparent shift in the energy calibration. He suggested that the Electron Cooler magnet system was a Type-3 Snake, because it contained several solenoids and toroids. Spin operators do not commute, so these magnets could produce a weak Type-3 Snake which could shift the spin tune and thus the resonant energy. I afterwards found a paper by Shatunov and Skrinsky[8], which never refers to Type-3 Snakes, but stresses that a depolarizing resonance energy calibration is only reliable provided there are no longitudinal fields.

We tested the Type-3 Snake proposal by first measuring the vertical polarization at 106 MeV as a function of $\int B \cdot dl$ with the cooling magnets on; this data is shown in Fig. 16. As expected, we saw a sharp peak. This peak at 106 MeV had a width almost identical to the peak at 107 MeV shown in Fig. 9; this equality gave a calibration of the resonance energy at about 106.5 MeV. Then we turned the cooling system completely off. This made the measurements somewhat more difficult because the beam intensity and life-time both deteriorated; however, we were able to make some measurements. Notice that with the cooling magnets turned off and thus any Type-3 Snake turned off, the peak was much broader; in fact it was quite consistent with the resonance having shifted back up to near 108 MeV. This shift seems to be rather direct evidence for the existence of a weak Type-3 Snake which was accidentally built into the Ring's cooler section. This was a clever bit of detective work by Pollock and Roser.

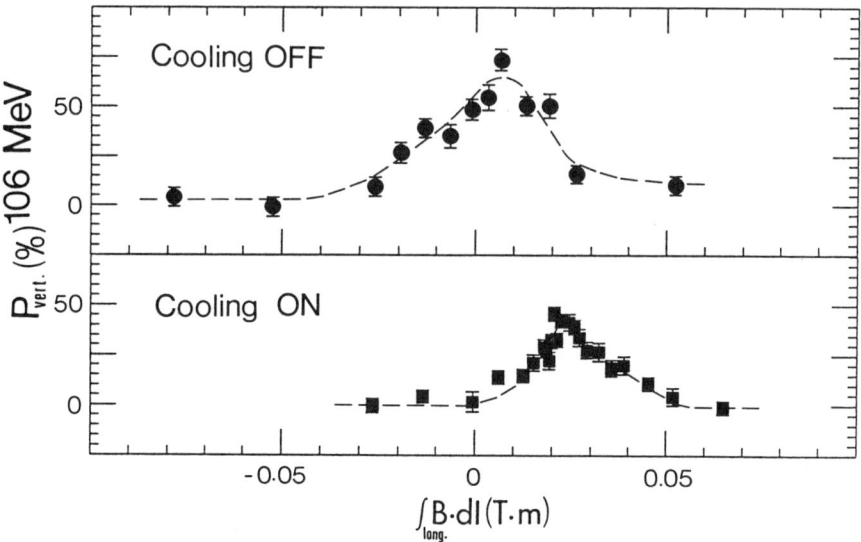

Fig. 16 Measuring the resonance width at 106 MeV with a Type-3 Snake on and off.

At multi-TeV facilities there will be a new problem called overlapping depolarizing resonances; these will occur as the various resonances get stronger and thus wider, while the 523 MeV spacing between the imperfection resonances remains fixed. This overlapping was no problem up to the maximum AGS energy of 22 GeV. However, at multi-TeV energies, overlapping depolarizing resonances should be common. Experts, such as Courant and Derbenev, believe that Siberian Snakes should be able to overcome the overlapping resonances at the SSC. We plan to experimentally study these overlapping resonances by creating an induced depolarizing resonance at the Cooler Ring as was done at Novosibirsk[8]. We will then move this induced resonance near either the $G\gamma = 2$ or the $G\gamma = -3 + \nu_y$ resonance. We are now building a high power RF solenoid to create a strong induced depolarizing resonance that can interfere with each of these two resonances.

During our August run we made a simple first study of an induced depolarizing resonance. We disconnected the RF knock-out system which was normally used for measuring the Cooler Ring betatron tunes; we then connected its power supply to one of our two kicker magnets. This produced a rather weak RF vertical field. With 106 MeV radially polarized protons injected, we then turned on the Snake and thus forced the spin tune to be about 1/2. The spin tune was not exactly 1/2 because the cooling was on, which presumably shifted ν_s slightly. Then we varied the RF frequency in an attempt to produce a weak induced depolarizing resonance. As shown in Fig. 17, this study was quite successful. We saw a clear depolarizing resonance when we varied the frequency around the calculated resonant frequency. We estimate that the center of the dip was about 771.0 ± 0.2 kHz; thus the relative error was about $3 \cdot 10^{-4}$. This seems a very precise way to directly measure the spin tune with a precision of about $3 \cdot 10^{-4}$.

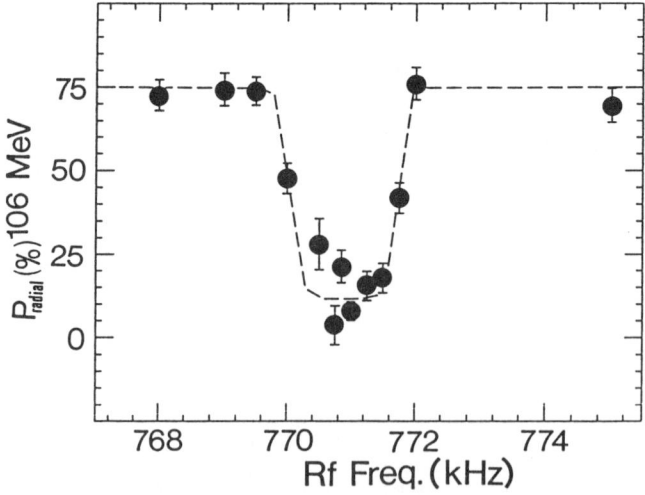

Fig. 17 A depolarizing resonance induced by an rf dipole field.

A major goal of our Siberian Snake studies was the acceleration of polarized protons to TeV energies at some facilities such as the SSC. Our SPIN collaboration recently submitted a preliminary proposal to install Siberian Snakes in the SSC and then accelerate polarized protons to 2 TeV and then to 20 TeV. At the June PAC meeting at SSC our collaboration contained 38 people from Michigan, Protvino and Dubna; recently 28 additional accelerator physicists and experimenters from Indiana, Moscow, KEK, and Kyoto joined our SPIN collaboration. We have had considerable interaction with the SSC people and they now seem rather interested in polarized proton beams but certainly not yet committed. This SSC project could be the first significant application of Siberian Snakes to a very high energy facility. Other possibilities include using Siberian Snakes to accelerate polarized protons at the Fermilab Main Ring, at RHIC, or at the 400 GeV and 3 TeV UNK rings.

REFERENCES

‡ Supported by research grants from the U.S. Department of Energy and the U.S. National Science Foundation.

1. Proc. of the 1985 Ann Arbor Workshop on Polarized Protons at the SSC, eds. A. D. Krisch, A.M.T. Lin, and O. Chamberlain, AIP Conf. Proc. **145** (New York, 1986).

2. Ya. S. Derbenev and A. M. Kondratenko, JETP **35**, 230 (1972); Part. Accel. **8**, 115 (1978).

3. A. D. Krisch, S. R. Mane, R. S. Raymond, T. Roser, J. A. Stewart, K. M. Terwilliger, J. E. Goodwin, H-O. Meyer, M. G. Minty, P. V. Pancella, R. E. Pollock, T. Rinckel, M. A. Ross, F. Sperisen, E. J. Stephenson, E. D. Courant, S. Y. Lee, and L. G. Ratner, Phys. Rev. Lett. **63**, 1137 (1989).

4. J. E. Goodwin et al., Phys. Rev. Lett. **64**, 2779 (1990).

5. T. Khoe et al., Part. Accel. **6**, 213 (1975).

6. F. Z. Khiari et al., Phys. Rev. **D39**, 45 (1989).

7. R. E. Pollock, submitted to Nucl. Instrum. and Meth.

8. Yu. M. Shatunov and A. N. Skrinsky, Part. World **1**, 35 (1989).

Prospects for Polarization at RHIC and SSC*

S.Y. Lee[†]
Department of Physics
Indiana University, Bloomington, IN 47405

ABSTRACT

In low to medium energy accelerators, betatron tune jumps and vertical orbit harmonic correction methods have been used to overcome the intrinsic and imperfection resonances. At high energy accelerators, snakes are needed to preserve polarization. We analyze the effects of snake resonances, snake imperfections, overlapping resonances on the spin depolarization. We discuss also results of recent snake experiments at the IUCF Cooler Ring. The snake can overcome various kinds of spin depolarization resonances. These experiments pointed out further that partial snake can be used to cure the imperfection resonances in low to medium energy accelerators. We also examine various snake designs. A new generalized snake concept allows for two possible configurations. The compact configuration offers the advantages of shorter total snake length and smaller horizontal orbit displacement. The split snake configuration allows for dual functions of a snake and a 90^0 spin rotator at the mid-section of the snake, which provides helicity state collisions. The requirements for obtaining high luminosity polarized protons at high energy colliders, such as RHIC and SSC, are reviewed.

1. Introduction

Spin is a fundamental property of the basic constituents of elementary particles. Studies of spin physics leads to important understanding of fundamental interactions between elementary particles. When machine energies is increased in order to probe into the inner structure of fundamental particles, spin degree of freedom remains as an important window for a detailed understanding of fundamental physics[1,2].

To achieve high energy polarized proton collision, an intense polarized proton source should be prepared. Many laboratories[3] have developed high polarization ($\geq$ 65%) and high current ($\geq 50\mu A$) polarized ion sources. These polarized protons should then be accelerated to high energy without losing their polarization[4]. During the acceleration process, these polarized protons encounter thousands of spin depolarization resonances. These resonances arise mainly from the the horizontal fields in focusing quadrupoles. The horizontal field rotates the spin away from the vertical axis, and becomes important if the kicks are correlated each turn, i.e. the resonance condition. There are two types of spin depolarization resonances[5], intrinsic resonance at $K = kP \pm \nu_y$ and imperfection resonance at $K = k$, where k is an integer and P is the superperiodicity of the accelerator. K is the resonance tune obtained from Fourier analyzing quadrupole kicks around the accelerator. The resonance width(strength), ϵ, is defined as the corresponding Fourier amplitude.

In circular accelerators, the spin vector precesses around the vertical axis with a frequency of $G\gamma$ per turn[6], where G=(g-2)/2=1.792846 is the anomalous g-factor of the proton. The spin tune, ν_s, of the polarized proton is therefore $\nu_s = G\gamma$. When spin tune equals resonance tune, successive kicks add up coherently to give rise to depolarization. Fig. 1 shows resonance strength as a function of the energy for various accelerators[7]. Observe that the intrinsic resonance strength is of the order of $|\epsilon| \leq 0.5$ for RHIC (Relativistic Heavy Ion Collider at Brookhaven National Laboratory) and $|\epsilon| \leq 5$ for SSC (Superconducting Super Collider in Texas, U.S.A).

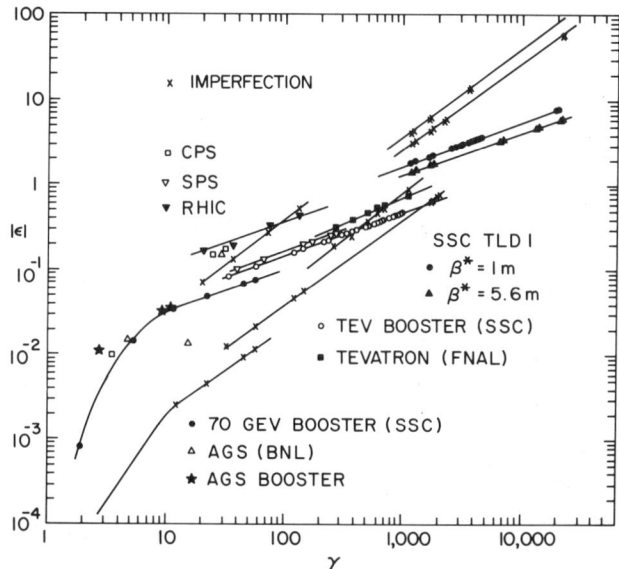

Fig.1 Compilation of intrinsic and imperfection resonance strengths for accelerators in the world. Note the scaling law of $\epsilon_{int} \sim \sqrt{\gamma}$ and $\epsilon_{imp} \sim \gamma$. The rms quadrupole misalignment assumed is ± 0.1 mm. The normalized emittance is $10\pi\mu$m-rad.

When spin particles are accelerated through depolarization resonances, the final polarization, P_f, is given by the Froissart-Stora Formula[8],

$$P_f = P_i \cdot (2e^{-\pi|\epsilon|^2/2\alpha} - 1) \quad (1.1)$$

where P_i is the initial polarization, $\alpha = dG\gamma/d\theta$ is the angular acceleration rate with θ as the azimuthal angle of the particle in accelerator and ϵ the resonance strength. Polarization is not affected for $\pi|\epsilon|^2/2\alpha \ll 1$ and spin flip occurs for $\pi|\epsilon|^2/2\alpha \gg 1$. Based on the Froissart-Stora formula, high polarization can be obtained with a high acceleration rate or small resonance strengths. At small or medium energy accelerators, such as ZGS, Saturn, KEK-PS, and AGS, tune jump to achieve high acceleration rate through intrinsic resonance and harmonic orbit correction to obtain small imperfection resonance strengths are successfully used. The highest energy polarized proton beam have been accelerated up to 22 GeV/c at AGS with 45% polarization.

The conventional tune jump correction is limited to a resonance strength, $|\epsilon| \leq 0.05$, by betatron half integer stopbands (see Section 2). The depolarization process arises from the resonance condition of spin tune equal the quadrupole kick tune. If particle spin tune can be maintained at a constant value away from the spin resonance tune, there will be no depolarization. Derbenev and Kondratenko[9] discovered that when a local spin rotator, called snake, is introduced in the accelerator, the spin tune becomes 1/2 and is independent of the energy. Therefore the spin tune does not cross any resonances. The depolarization process can be avoided during the acceleration. However subsequent studies indicate that if resonance strength is large, then depolarization will occur at a new resonance condition, called snake resonance[10].

Snake resonances occur at condition $m\nu_s \pm nK = integer$ with odd-integer m and n, where K is the resonance tune. A snake resonance arises from the coherency between the tune of the spin depolarizing kicks and the spin tune, which characterizes spin direction on each successive turn around the accelerator. For example, at the spin tune $\nu_s = 1/2$, spin vector returns to the same direction every two turns around the accelerator. Thus the spin depolarization kicks will add up coherently if the tune of kicks, K, is also 1/2. This means that no matter how far $G\gamma$ is away from the spin resonance tune ($K = 1/2$), the spin will be depolarized due to spin tune of 1/2 in the presence of a snake. Higher order snake resonances can be obtained similarly from the higher power kicker strength parameter. By avoiding low order snake resonances, we found[11,12] that the number of snakes needed to maintain polarization during acceleration is about $N_s \approx 5|\epsilon|$ for intrinsic resonances, whilst the vertical closed orbit is corrected to an rms value of less than 0.3 mm. This implies that RHIC and SSC need respectively 2 and 26 snakes. Similar analysis has also been worked out by Steffen[13].

This paper reviews prospects for polarized proton collision. We shall review (1) problems in polarized proton acceleration, (2) the effect of snake resonances, (3) the effect of overlapping resonances and snake imperfections, and (4) the achievable luminosity. The paper is organized as follows: Section 2 discusses the basic equation of motion, section 3 deals with the spin motion in the presence of snakes, section 4 reviews the snake design, section 5 evaluates the luminosity. The conclusion is given in Section 6.

2. Spin Equation of Motion

The spin equation of motion for a moving particle in a static magnetic field is given by[8]

$$\frac{d\vec{S}}{dt} = \frac{e}{\gamma m}\vec{S} \times [(1+G\gamma)\vec{B}_\perp + (1+G)\vec{B}_\parallel]. \quad (2.1)$$

where $\vec{B}_\perp$ and $\vec{B}_\parallel$ are the transverse and longitudinal components of the magnetic fields respectively. G is the anomalous gyromagnetic g-factor and γmc^2 is the energy of the moving particle. Particles in circular accelerator are guided by magnetic dipole fields in a enclosed circular path, which obeys the Lorentz force equation,

$$\frac{d\vec{v}}{dt} = \frac{e}{\gamma m}\vec{v} \times \vec{B}. \quad (2.2)$$

Using Eq.(2.2), Courant and Ruth[5] rewrite the spin equation of motion in terms of particle coordinates in the accelerator.

Let us use the coordinate system of reference orbit, where $\hat{x}, \hat{s}, \hat{z}$ are unit vectors corresponding to radial outward, longitudinal, and transverse vertical respectively. Whence we have $d\hat{x}/ds = \hat{s}/\rho$, $d\hat{s}/ds = -\hat{x}/\rho$, and $d\hat{z}/ds = 0$ for the planar orbital motion, where ρ is the local radius of curvature of the reference orbit. Particle motion in the accelerator is then characterized by the coordinate (x, s, z) as $\vec{r} = \vec{r}_0(s) + x\hat{x} + z\hat{z}$, where $\vec{r}_0(s)$ is the reference orbit. The velocity of a particle is then $\vec{v} = \frac{d\vec{r}}{dt} = \frac{ds}{dt}[x'\hat{x} + (1+x/\rho)\hat{s} + z'\hat{z}] \approx v(x'\hat{x} + \hat{s} + z'\hat{z})$, where the prime denotes differentiation with respect to the coordinate s and the magnitude of the particle velocity, $v \approx \frac{ds}{dt}(1 + x/\rho)$, is constant. One can obtain also $\vec{v}\,' = v[(x'' - 1/\rho)\hat{x} + \frac{x'}{\rho}\hat{s} + z''\hat{z}]$.

Thus the transverse magnetic field is given by $\vec{B}_\perp \equiv \frac{1}{v^2}(\vec{v} \times \vec{B}) \times \vec{v} = B\rho(1 - x/\rho)[(x'' - 1/\rho)\hat{z} + \frac{z'}{\rho}\hat{s} - z''\hat{x}]$, where $B\rho = \gamma mv/e$ is the magnetic rigidity of the particle.

Since the dipole guide field is given by $B_z = -\frac{B\rho}{\rho}$, the corresponding longitudinal field is then obtained from the Maxwell equation as $B_s = -B\rho z(1/\rho)'$. Thus the magnetic field parallel to the particle orbital direction is given by $B_\parallel = B_s + z'B_z$, i.e. $\vec{B}_\parallel = -B\rho(z/\rho)'\hat{s}$. Eq.(2.1) can then be transformed to[5,6],

$$d\vec{S}/d\theta = \vec{S} \times \vec{F}, \qquad (2.3)$$

with $d\theta = ds/\rho$, where s is longitudinal path length and ρ is the radius of curvature. The vector $\vec{F} = F_1\hat{x} + F_2\hat{s} + F_3\hat{z}$, can be expressed in term of particle coordinate as,

$$F_1 = -\rho z''(1 + G\gamma),$$

$$F_2 = (1 + G\gamma)z' - \rho(1 + G)(\frac{z}{\rho})',$$

$$F_3 = -(1 + G\gamma),$$

Defining a 2-component spinor, Ψ, with $S_i \equiv <\Psi|\sigma_i|\Psi>$, Eq.(2.3) becomes,

$$d\Psi/d\theta = -\frac{i}{2}(G\gamma\sigma_3 - F_1\sigma_1 - F_2\sigma_2)\Psi = -\frac{i}{2}H\Psi. \qquad (2.4)$$

The off-doagonal matrix element of H, $\xi(\theta) \equiv F_1 - iF_2$, characterizes the spin depolarization kick by coupling the up and down components of the spinor wave function. Given the repetitive nature of circular accelerators, $\xi(\theta)$ can be Fourier analyzed as

$$\xi(\theta) = \sum \epsilon_j e^{iK_j\theta}, \qquad (2.5)$$

where the Fourier amplitude, ϵ_j, is resonance strength and resonance tune, K_j, is given by $K_j = k \cdot P \pm m\nu_y$ for intrinsic resonances and $K_j = k$ for imperfection resonances. Intrinsic resonances arise from the vertical betatron motion of particle, and imperfection resonances arise from the vertical closed orbit distortion. In a real accelerator, synchrotron and transverse betatron motions may also be coupled. resonance tune becomes more generally as,

$$K_j = k \cdot P \pm m\nu_y \pm n\nu_x \pm l\nu_{syn}, \qquad (2.5a)$$

where k, l, m, and n are integers and ν_{syn} stands for synchrotron tune.

For a single resonance, i.e. $\xi(\theta) = \epsilon \cdot e^{iK\theta}$, Eq.(2.4) can then be solved to obtain Froissart-Stora formula (Eq.(1.1)) for a constant acceleration rate through a resonance. Another interesting analytic solution corresponds to slow or zero acceleration rate. At a constant spin tune, $G\gamma$, one obtains spin polarization as

$$<S_z(\infty)> = -\frac{G\gamma - K}{\sqrt{(G\gamma - K)^2 + |\epsilon|^2}} <S_z(-\infty)> \qquad (2.6)$$

where $<S_z(-\infty)>$ is initial polarization far away from a resonance. Eq.(2.6) manifests clearly $|\epsilon|$ as the resonance width. At a distance $G\gamma - K = \pm|\epsilon|$ away from resonance, polarization is reduced by a factor of $1/\sqrt{2}$. A minimum of $\pm 3|\epsilon|$ from a resonance is needed to maintain 95% polarization (see Eq.(2.6)). Thus to obtain a proper tune jump through a resonance, $\Delta\nu_y = 6|\epsilon|$ is needed. Since the maximum tune jump is limited by half integer stopbands to about $\Delta\nu_y \leq 0.3$, the maximum tolerable resonance strength in tune jump correction method is limited to about $|\epsilon| \leq 0.05$.

The spinor wave function for constant $G\gamma$ can be found easily (see Section (2.3a) of Ref.11).

$$\Psi(\theta_f) = e^{-\frac{i}{2}K\theta_f \sigma_3} e^{\frac{i}{2}[\delta\sigma_3 + \epsilon_R\sigma_1 - \epsilon_I\sigma_2](\theta_f - \theta_i)} e^{\frac{i}{2}K\theta_i \sigma_3} \Psi(\theta_i)$$
$$\equiv t(\theta_f, \theta_i)\Psi(\theta_i) \qquad (2.7)$$

Here $t(\theta_f, \theta_i)$ is the spin transfer matrix, whose components are given by

$$t_{11}(\theta_f, \theta_i) = ae^{i[c - K(\theta_f - \theta_i)/2]}, \qquad (2.7a)$$

$$t_{12}(\theta_f, \theta_i) = ibe^{-i[d + K(\theta_f + \theta_i)/2]}, \qquad (2.7b)$$

$$t_{21}(\theta_f, \theta_i) = -t_{12}^*(\theta_f, \theta_i) \quad ; \quad t_{22}(\theta_f, \theta_i) = t_{11}^*(\theta_f, \theta_i),$$

with

$$b = \frac{|\epsilon|}{\lambda}\sin[\lambda(\theta_f - \theta_i)/2] = (1 - a^2)^{1/2} \qquad (2.8a)$$

$$\lambda = (\delta^2 + |\epsilon|^2)^{1/2} \quad ; \quad \delta = K - G\gamma \qquad (2.8b)$$

$$c = \arctan[\frac{\delta}{\lambda}\tan(\lambda(\theta_f - \theta_i)/2)] \quad ; \quad d = \arg(\epsilon^*)$$

The off-diagonal matrix elements t_{12}, t_{21} are the depolarization driving terms, where parameter b oscillates with an amplitude $|\epsilon|/\lambda$. The effect of the depolarization kicks gives rise to an average polarization given in Eq.(2.6). When snakes are inserted into accelerator, the sine factor in Eq.(2.8a) for parameter b remains small until the next snake, which rotates spin by 180^0 around a horizontal axis. The depolarization driving terms can thus be arranged to cancel each other.

3. Spin Motion in accelerator with Snakes

A snake is a local spin rotator, which rotates particle spin by π radians about a horizontal axis locally without perturbing particle orbits outside snake region. A partial snake differs only in the amount of spin rotation angle, e.g. a 10% snake rotates spin by 0.1π radians. Thus a snake is characterized by the amount of spin rotation angle, ϕ, and the **snake axis angle,** ϕ_s, with respect to $\hat{x}$ (radial outward direction). Section 4 shall discuss snake design.

The spinor wave function at a snake will be transformed locally according to

$$\Psi(\theta^+) = e^{i\frac{\phi}{2}\hat{n}_s\cdot\vec{\sigma}}\Psi(\theta^-) = e^{i\frac{\pi}{2}\hat{n}_s\cdot\vec{\sigma}}\Psi(\theta^-) = S(\phi_s)\Psi(\theta^-) \qquad (3.1)$$

where $\phi = \pi$ is spin rotation angle and $\hat{n}_s = (\cos\phi_s, \sin\phi_s, 0)$ denotes the snake axis with respect to horizontal outward direction, $\hat{x}$. $\theta^\pm$ depict azimuthal orbit rotation angles just before and after snake.

Let us assume that there are N_s snakes with snake axes, $(\phi_1, \phi_2, \cdots, \phi_N)$ distributed in accelerator. Let $\theta_{i,i+1}$ be the azimuthal orbit rotation angle between the i-th, and (i+1)-th snakes. The distribution of snakes should satisfy following conditions

$$\sum_{k=odd}^{N_s} \theta_{k,k+1} = \sum_{k=even}^{N_s} \theta_{k,k+1} = \pi , \qquad (3.2a)$$

$$\nu_s = \frac{1}{\pi}\sum_{k=1}^{N_s}(-1)^k\phi_k = j + \frac{1}{2} , \qquad j = integer . \qquad (3.2b)$$

Eq.(3.2a) ensures that spin tune, ν_s, is independent of particle energy. Eq.(3.2b) can be used to set spin tune to a most favorable number in avoiding snake resonance, which will be discussed in the following section. As an example, an accelerator with two snakes, $N_s = 2$, these two snakes should be distributed at location of π orbital angle apart and the snake axes of these two snakes should be orthogonal to each other to obtain a spin tune of 1/2. For accelerators with a large number of snakes, there are many ways to organize snakes to obtain proper snake superperiodicity and proper spin tune.

The spin transfer matrix after passing through a pair of (ϕ_2, ϕ_1) snakes is given by

$$\tau(\theta_0 + \frac{4\pi}{N_s}, \theta_0) = S(\phi_2)t(\theta_0 + \frac{4\pi}{N_s}, \theta_0 + \frac{2\pi}{N_s})S(\phi_1)t(\theta_0 + \frac{2\pi}{N_s}, \theta_0). \qquad (3.3)$$

where $t(\theta_f, \theta_i)$ and $S(\phi_i)$ are given by Eqs.(2.7) and (3.1) respectively. The components of spin transfer matrix are given by

$$\tau_{11}(\theta_0 + \frac{4\pi}{N_s}, \theta_0) = -e^{-i(\phi_2-\phi_1)}(1 - 2b^2 e^{i\Phi}\cos\Phi), \qquad (3.4a)$$

$$\tau_{12}(\theta_0 + \frac{4\pi}{N_s}, \theta_0) = -2iabe^{-i(c-2K\pi/N_s+\phi_2)}\cos\Phi. \qquad (3.4b)$$

$$\tau_{21}(\theta_0 + \frac{4\pi}{N_s}, \theta_0) = -\tau_{12}^*(\theta_0 + \frac{4\pi}{N_s}, \theta_0) \; ; \; \tau_{22}(\theta_0 + \frac{4\pi}{N_s}, \theta_0) = \tau_{11}^*(\theta_0 + \frac{4\pi}{N_s}, \theta_0)$$

where $\Phi = K\theta_0 + 2K\pi/N_s + d - \phi_1$ and parameters, a, b, c, and d are defined in Eq.(2.8) with $\theta_f - \theta_i = \pi$ as

$$b = \frac{|\epsilon|}{\lambda} \sin \frac{\pi\lambda}{N_s} = (1 - a^2)^{1/2} \tag{3.5}$$

$$\lambda = (\delta^2 + |\epsilon|^2)^{1/2} \; ; \; \delta = K - G\gamma \; ; \; c = \arctan[\frac{\delta}{\lambda} \tan \frac{\pi\lambda}{N_s}] \; ; \; d = \arg(\epsilon^*)$$

The spin motion in accelerator can then be obtained iteratively by using the spin tracking equation through pairs of snakes:

$$T(\theta_{n+1}) = \tau(\theta_{n+1}, \theta_n) T(\theta_n) , \tag{3.6}$$

where $\theta_{n+1} = \theta_n + 4\pi/N_s$. Eq.(3.6) can be solved iteratively using a power series expansion in strength parameter b^2, i.e.

$$T_{11} = T_{11}^{(0)} + T_{11}^{(1)} + T_{11}^{(2)} + \cdots, \tag{3.6a}$$

$$T_{12} = T_{12}^{(1)} + T_{12}^{(2)} + T_{12}^{(3)} + \cdots, \tag{3.6b}$$

where $T_{11}^{(i)} = O(b^{2i})$ and $T_{12}^{(i)} = O(ab^{2i-1})$. A set of hierarchy equations to solve Eq.(3.6) iteratively can be found in References 11-12. Solving spin tracking equation, one can find spin tune and snake resonance. The final polarization is obtained from the expectation value of σ_3 in spinor wave function, i.e.

$$<S> = |T_{11}|^2 - |T_{12}|^2 = 1 - 2|T_{12}|^2 \tag{3.7}$$

where unitarity condition, $|T_{11}|^2 + |T_{12}|^2 = 1$, has been used in Eq.(3.7).

3.1 Snake resonances

Without losing generality, we shall first discuss an accelerator with two snakes ϕ_1, ϕ_2, located at an orbital angle of π from each other. The spin transfer matrix for passing two snakes (or equivalently a **one turn map(OTM)**) is given by

$$\tau_{11}(\theta_0 + 2\pi, \theta_0) = -e^{-i\pi\nu_s}(1 - 2b^2 e^{i\Phi} \cos \Phi) , \tag{3.8a}$$

$$\tau_{12}(\theta_0 + 2\pi, \theta_0) = -2iabe^{-i(c - K\pi + \phi_2)} \cos \Phi , \tag{3.8b}$$

where $\pi\nu_s = \phi_2 - \phi_1$ and $\Phi = K\theta_0 + K\pi + d - \phi_1$ is the characteristic phase of the orbital motion. The perturbed spin tune, Q_s, is given by the trace of OTM as (using $\nu_s = 1/2$)

$$\begin{aligned} \cos \pi Q_s &= -\cos \pi\nu_s + 2b^2 \cos \Phi \cos(\Phi - \pi\nu_s) \\ &= b^2 \sin(2\Phi), \end{aligned} \tag{3.9}$$

Fig. 2 show the perturbed spin tune obtained from spin tracking calculation, which agrees well with Eq.(3.9). The parameter b in Eq.(3.5) becomes 1 when

$|\epsilon| = N_s/2$. During acceleration through a resonance with strength $|\epsilon| \approx N_s/2$, perturbed spin tune, Q_s, will cover whole integer unit and cross intrinsic resonance many times. Naturally polarization can not be preserved. When the maximum of $|Q_s - \frac{1}{2}|$ equals the deviation of vertical betatron tune from half integer, the resonance condition recurs. Thus the maximum tolerable resonance strength will be given by

$$<\epsilon_c> = \frac{\arcsin(|\cos \pi K|^{1/2})}{\pi} N_s , \qquad (3.10)$$

where $<\epsilon_c>$ is the critical resonance strength, which depends on resonance tune K or equivalently betatron and/or synchrotron tunes of an accelerator. Eq.(3.10) indicates that the tolerable critical resonance strength will be larger when betatron tune is nearer to integer. Fig. 3 compares the tolerable resonance strength obtained from numerical simulation with the critical resonance strength of Eq.(3.10).

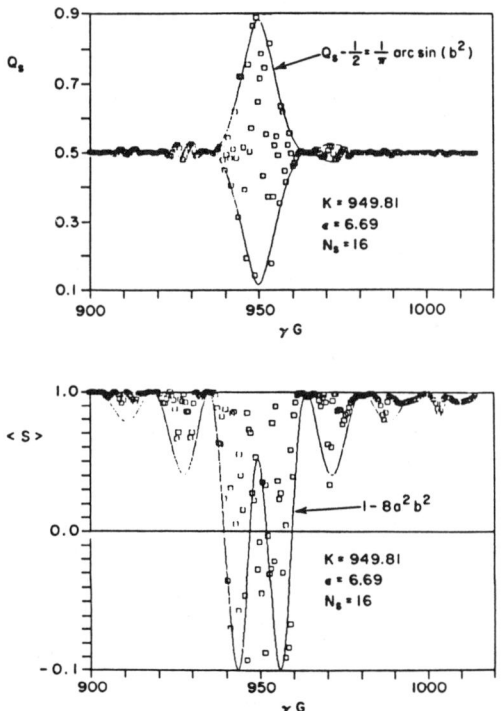

Fig.2 The perturbed spin tune Q_s obtained from a numerical tracking calculation at intrinsic resonance region is compared with that of analytical calculation of Eq.(3.9).

On the other hand, when $K = integer$, phase Φ is constant (modulo 2π). The perturbed spin tune of Eq.(3.9) varies smoothly across the reosnance. The resonance condition will not occur.

The spin motion in accelerator can then be obtained iteratively from spin tracking equation, Eq.(3.6). To first order, we obtain easily

$$T^{(1)}_{12}(\theta_{n+1}) = 2iab(-1)^n e^{-i(c-K\pi+\phi_2)} \times$$
$$\times \frac{1}{2}\{e^{i(\Phi+nK\pi)}\zeta_{n+1}(K+\nu_s) + e^{-i(\Phi+nK\pi)}\zeta_{n+1}(K-\nu_s)\} \quad (3.11)$$

where $\zeta_n(q)$, enhancement function, is given by,

$$\zeta_n(q) = \frac{\sin nq\pi}{\sin q\pi}. \quad (3.12)$$

At $q = integer$, we find that $\zeta_n(q) \to n$. This means that the off-diagonal kicks add up coherently each turn through snake pairs. This condition is indeed a nominal resonance condition, since the spin tune equals a spin resonance position $\pm K$. However betatron tunes of accelerator are not half integers, the condition that $K \pm \nu_s = integer$ will never occur. Avoiding snake resonances, polarization will fall within the envelope of

$$\ll S \gg = 1 - 8a^2 b^2 \quad (3.13)$$

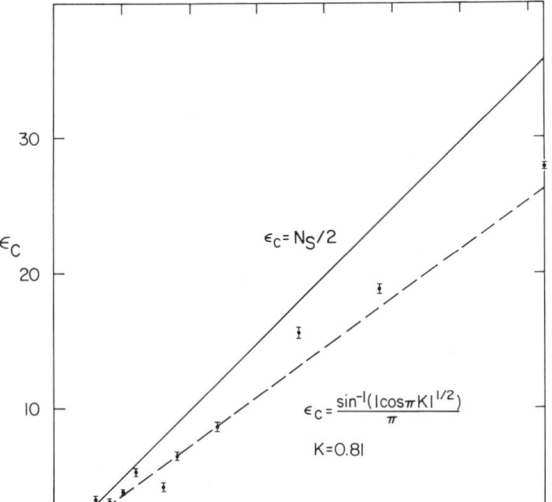

Fig.3 The critical (tolerable) resonance strength obtained from tracking calculation is compared with that derived from the perturbed tune spread.

The lower part of Fig.3 compares the polarization obtained from a spin tracking simulation with envelope function (Eq.3.13). We observe that agreement is good. A few important observations can be drawn from Eqs(3.11-13):

1. At an imperfection resonance, $K = integer$, $T_{12}^{(1)}(\theta_{n=even}) \equiv 0$. This means that imperfection kicks cancel each other every two turns around accelerator. Thus snakes cure most effectively imperfection resonances.
2. At $2nK = integer, K \neq 1/2$, similar cancellation of $T_{12}^{(1)}(\theta_m)$ occurs at $m = 2n\ turns$ around accelerator.
3. The envelope function $\ll S \gg$ has many nodal points, where the depolarization driving term vanishes, i.e. $b = 0\ or\ 1$. These nodal points corresponds to the spin matching condition[13], where $G\gamma = K \pm \sqrt{(integer \cdot N_s)^2 - |\epsilon|^2}$. Thus these nodal locations are separated approximately by N_s units of $G\gamma$. These nodal points play an essential role in spin restoration during the passage through a depolarization resonance.
4. The width of envelope function is about $12|\epsilon|$ for 95% polarization. However, one can choose a nodal point of spin matching condition to obtain 100% polarization (see Fig. 3).

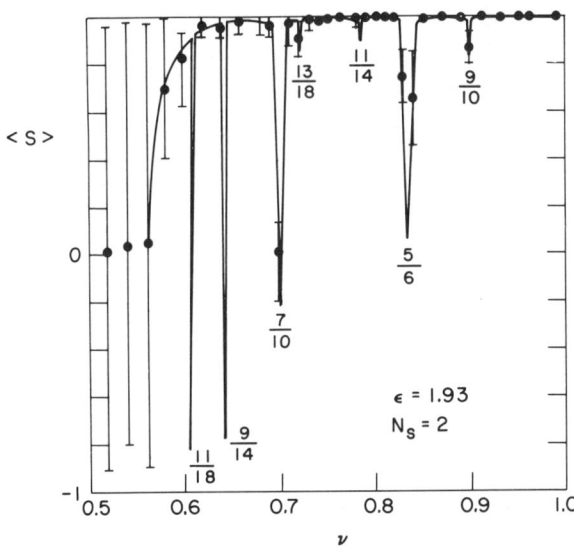

Fig.4 Beam polarization after passage through a single spin resonance is shown as a function of the fractional part of spin resonance tune. Higher order snake resonances are seen clearly.

Based on the linear response theory of Eq.(3.11), we expect that snakes will not work at a betatron tune equal to a half integer. Fig. 4 shows polarization vs the fractional part of the vertical betatron tune.

A surprisingly many depolarization resonances appear at a betatron tune of rational numbers, e.g. 1/6, 5/6, 1/10, 3/10, etc.. To understand these resonances, we have to study the spin tracking equation beyond linear order in b. References (10) and (11) studied higher order resonances. These higher order snake resonances can also be studied through solving spin hierarchy equations (Eq.3.6). In general, the snake resonance condition is given by

$$m\nu_s + nK = integer \quad ; \quad m, n = odd\ integers \qquad (3.14)$$

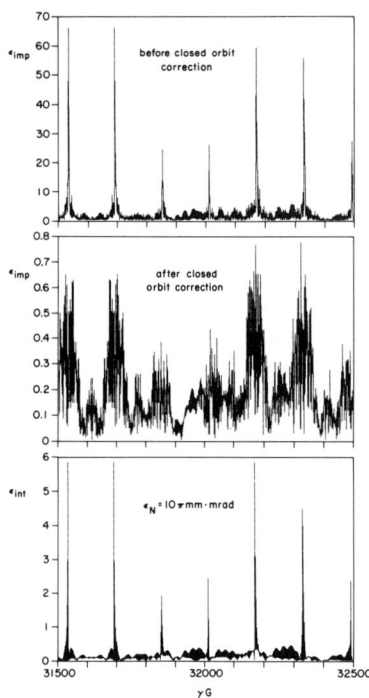

Since betatron tunes of colliders, such as RHIC, SPS, Tevatron, and SSC, have to avoid similar low order betatron resonances for a long term orbital stability, snake resonances does not impose further constraints to the operational condition of the colliders. The resulting tolerable resonance strength agrees well with the critical resonance strength of Eq.(3.10)[11]. One can generalize the discussion to multi-snake accelerators, where the resonance condition of Eq.(3.14) will be modified by snake superperiodicity P_s. At higher snake superperiodicity, there are fewer snake resonances, yet resonance width is also increased. Basic physics remains unchanged.

Fig.5 Intrinsic and imperfection resonance strengths calculated for SSC lattice are shown. The top figure displays imperfection resonance strength before closed orbit correction with quadrupole misalignment rms error of 0.1 mm. The middle figure shows that after a closed orbit correction. The corresponding intrinsic resonance for a normalized emittance of $10\pi\mu$m-rad is shown at the bottom for comparison.

3.2 Overlapping Resonances

Basic accelerator theory[14] indicates that a closed orbit distortion has largest amplitude at a harmonic number nearest to betatron tunes. We thus expect a large imperfection resonance next to an important imperfection resonance (see references 7,11,12 and 15). Fig. 5 shows the resonance strength as function of energy for SSC lattice. We observe a clear correlation between intrinsic and imperfection resonances. The correlation remains important even after closed orbit corrections. Since the width of a resonance is about $6|\epsilon|$, we thus have to study the overlapping resonances.

The spin depolarization resonance tune and strength are intrinsic properties of lattice design as well as beam emittance. Important intrinsic resonances are normally well separated and can be treated as isolated resonances. However intrinsic and imperfection resonances overlap with each other. Section 3.1 shows that imperfection kicks cancel each other every two turns around accelerator. When an intrinsic resonance is present, the self cancellation mechanism of depolarization kicks disappears. Spin becomes susceptible to depolarization kicks.

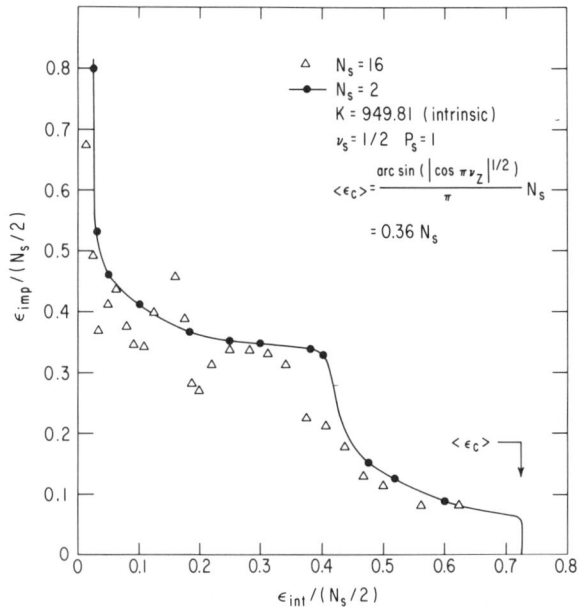

Fig.6 We show the correlation between tolerable intrinsic and imperfection resonance strengths for $N_s = 2$ and $N_s = 16$. See Section 3.2 for further discussion.

Fig. 6 shows tolerable resonance strengths of overlapping imperfection and intrinsic resonances. When the strength of intrinsic resonance, $|\epsilon_{int}|$, is very small, the tolerable imperfection resonance strength, $|\epsilon_{imp}|$, becomes very large due to the self cancellation mechanism of Section 3.1. However, when $|\epsilon_{int}|$ is slightly increased, tolerable $|\epsilon_{imp}|$ decreases drastically until about $|\epsilon_{imp}|/\frac{N_s}{2} \leq 0.3$, where the imperfection resonance strength plays minor role in depolarization process. Tolerable intrinsic resonance can then be increased until about $|\epsilon_{int}|/\frac{N_s}{2} \leq 0.4$. Beyond this intrinsic resonance strength, the perturbed spin tune plays a decisive role in determining the tolerable intrinsic and imperfection resonance strengths until the critical resonance strength $<\epsilon_c>$ is reached. The relationship between tolerable intrinsic and imperfection resonances is also valid for an accelerator with multi-snakes. A pleateau for a limiting imperfection resonance is clearly seen on Fig. 6 indicating the sensitivity of spin to imperfection errors when an intrinsic resonance is present nearby. The sensitivity is clearly due to the disappearence of the self cancellation mechanism. To achieve a higher tolerance to imperfection resonances, we can set a limit to tolerable intrinsic resonance strength as $|\epsilon_{int}|/\frac{N_s}{2} \le 0.4$, or

$$N_s \ge 5|\epsilon_{int}| \qquad (3.15)$$

Fig. 6 indicates that the tolerable imperfection strength will be $|\epsilon_{imp}|/\frac{N_s}{2} \le 0.3$. For SSC and RHIC, we expect $|\epsilon_{imp}|/\frac{N_s}{2} \le 0.05$ with 0.3 mm rms closed orbit distortion[12].

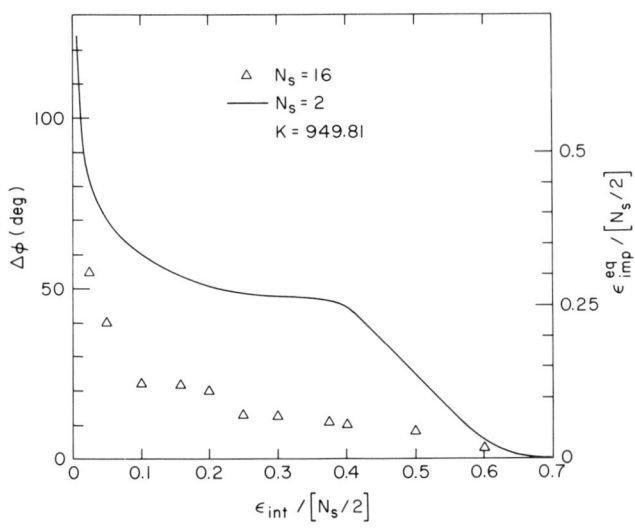

Fig.7 Tolerable error in the snake spin rotation angle is shown as a function of intrinsic resonance strength. Note that the tolerable error $\Delta\phi$ is greatly reduced for a large number of snakes due to spin tune modulation of Eq.(3.18).

3.3 Snake Imperfections

When the spin rotation angle ϕ of Eq.(3.1) deviates from π by $\Delta\phi = \pi - \phi$, the spin transfer matrix of snake becomes,

$$S(\phi_s) = e^{-i\frac{\Delta\phi}{4}\hat{n}_s \cdot \vec{\sigma}} e^{i\frac{\phi}{2}\hat{n}_s \cdot \vec{\sigma}} e^{-i\frac{\Delta\phi}{4}\hat{n}_s \cdot \vec{\sigma}}. \qquad (3.16)$$

Comparing Eqs.(2.7) with (3.16), one obtains an equivalent imperfection spin resonance, ϵ_{imp}^{eq}, as,

$$\epsilon_{imp}^{eq} = \frac{\Delta\phi}{\pi} \qquad (3.17)$$

Therefore errors in snake spin rotation angle are equivalent to integer imperfection resonances. The corresponding spin tune becomes energy dependent. To the leading order, we obtain

$$\cos\pi\nu_s = \frac{N_s}{2}\cos(2G\gamma\pi/N_s)\sin^2(\frac{\Delta\phi}{2}), \qquad (3.18)$$

where the linear dependence of Eq.(3.18) on N_s is due to the assumption that each snake has identical systematic error in the spin rotation angle. In reality, the snake rotation angle may deviate from π randomly. The resulting spin tune modulation will not increase linearly with the number of snakes. Fig. 7 shows the tolerable $\Delta\phi$ vs intrinsic resonance strength. The corresponding equivalent resonance strength is also shown for comparison. For $N_s = 2$, the characteristics of Fig. 7 are the same as that of Fig. 6. For $N_s = 16$, the tolerable error is greatly reduced due to the spin tune modulation of Eq.(3.18).

Besides the error in ϕ, the snake axis angle, ϕ_s may also deviate from the ideal situation. The resulting spin tune is energy independent (Eq.(3.2b)). The snake resonance condition of Eq.(3.14) determines the tolerable snake axis angle[12].

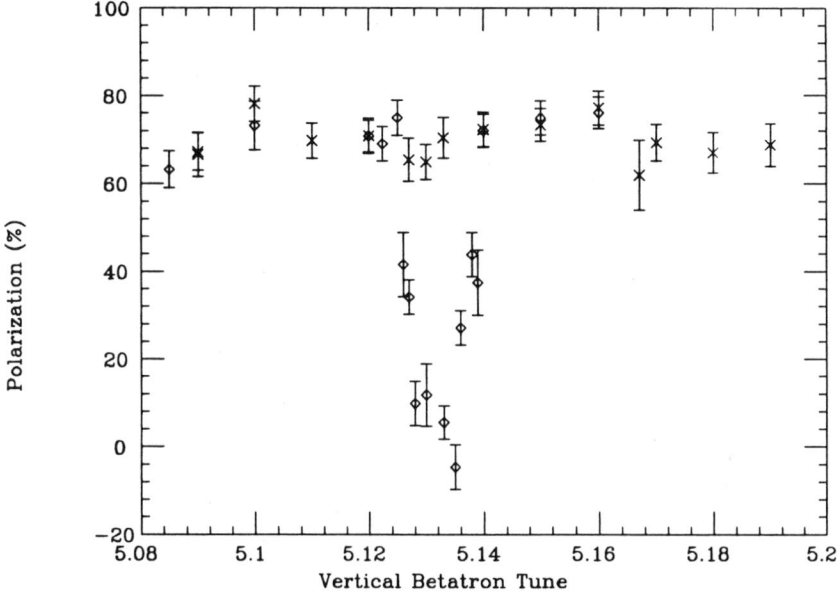

Fig.8 The beam polarization in each stable spin direction at 177 MeV is plotted against the vertical betatron tune ν_y. The data marked x's are the radial polarization with snake on and horizontal polarization injection. The open diamonds are polarization with snake off and vertical polarization injection.

3.4 Snake Experiments

Recently, Krisch et al.[16] have carried out successfully a series of experiments in the IUCF Cooler Ring to test snake concept. Using a single solenoid snake, polarized protons have been accelerated through imperfection and intrinsic resonances without losing polarization. The experiments also discovered synchrotron spin resonance in a proton storage ring. Synchrotron spin resonances have played important roles in electron storage rings, but have never been found in the proton storage ring before. Snake cures synchrotron spin resonance as well. Fig. 8 shows polarization vs. vertical betatron tune at proton energy of 177 MeV, where $G\gamma = 2.13$. Note that polarization becomes 0 at $G\gamma = \nu_y - 3$ resonance. However there is no depolarization, when snake is turned on.

Besides experimental tests of full snake, partial snake idea of Roser[17] has also been studied extensively. Indeed partial snake can be used in low to medium energy machines for correcting imperfection resonances. AGS has provisions for using partial snake to ease polarized proton operation[18]. An interesting question involves the evolution of spin tune when snake is adiabetically turned on. Numerical tracking calculations have been performed to study the problem. A new series of experiments have been approved at the Cooler Ring to study more complex problems, such as overlapping resonances, spin tune, etc. using a solenoid rf kicker[19].

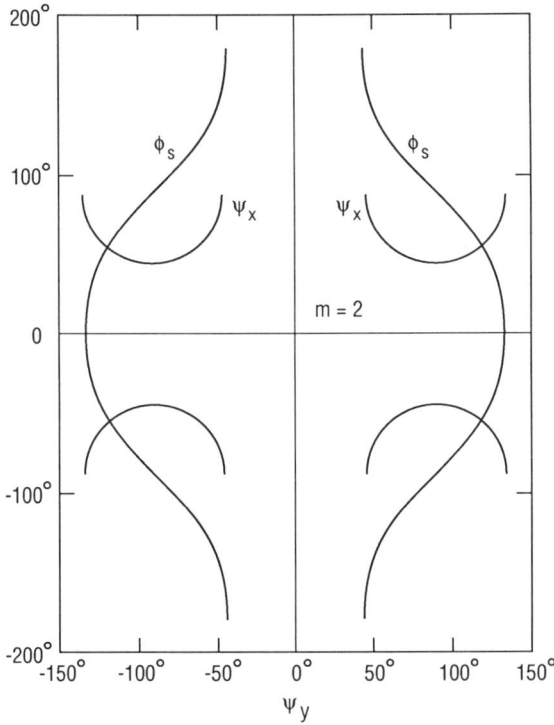

Fig.9 Relation between ψ_x and ψ_y is shown for the Steffen snake configuration. The corresponding snake axis angle is also given.

4. Snake Design Issues

Since the invention of the snake idea by Derbenev and Kondratenko, the design of snake and/or spin rotator has become an interesting task. There are many varieties of snake designs[13,20-22]. For low to medium energy accelerators, Helical type snakes[21] seems to offer advantages in obtaining smaller transverse orbit displacements. At high energy, flexibility in snake design is available. Steffen[13] made a major advancement in snake design concept to be discussed in Section 4.1. Further flexibility of snake design has been discovered by Lee[22] to be discussed in Section 4.2.

4.1 Steffen Snake

Steffen[13] has discovered a family of snakes with the following magnet sequence: $S = (-H, -V, 2H, 2V, -2H, -V, H)$, where H and V are respectively the horizontal and vertical bending magnets. To satisfy the snake criteria, the sequence of magnets must not alter the particle orbit outside the snake and the spinor of the particle is transformed according to $e^{i\frac{\phi}{2}\hat{n}_s \cdot \vec{\sigma}}$, where ϕ is the spin rotation angle and $\hat{n}_s = (\cos\phi_s, \sin\phi_s, 0)$ is the snake axis.

An interesting feature of the Steffen's snake is that the snake axis, ϕ_s, and the spin rotating angle, ϕ, can be varied by varying the excitation of H and V magnets, i.e. for $\phi = \pi$, we obtain

$$\cos^2\psi_y + \cos 2\psi_x \sin^2\psi_y = 0 , \qquad (4.1)$$

$$\sin\phi_s = \sqrt{2}\cos\psi_x , \qquad (4.2)$$

where ψ_x and ψ_y are respectively the spin rotation angle of H and V magnets. The relation between ψ_x and ψ_y in Eq. (4.1) ensures the snake condition of the magnet sequence and Eq. (4.2) determines the snake axis. Fig. 9 shows ψ_x, ψ_y relationship of Eq. (4.1) and ϕ_s vs. ψ_y. When $\phi_s = 0$, or π, the snake axis is along the radial $\hat{x}$ axis.

The total integrated magnet strength is given by $\int Bd\ell = 1.746(6\psi_x + 4\psi_y)$ [Tesla-meter] and the corresponding orbit displacements are given by

$$D_x = (\ell_x + \ell_y + 2\,\ell_g)\frac{\psi_x}{G\gamma} \; ; \; D_y = (2\,\ell_x + \ell_y + 2\,\ell_g)\frac{\psi_y}{G\gamma} ,$$

where ℓ_x, ℓ_y and ℓ_g are respectively the length of H and V magnets and the distance between the adjacent magnets. Since the lengths of the magnets ℓ_x, ℓ_y are given by $\ell_{x,y} = 1.746\psi_{x,y}/B$, where the magnetic flux density B is in Tesla, the orbit displacement is inversely proportional to the magnetic field strength of the dipole. The advantage of the Steffen snake is that the snake axis $\hat{n}_s$ can be changed continuously by proper ψ_x, ψ_y excitations. However, the snake configuration suffers from the rigid structure of magnet position, i.e. the total length of the snake is given by $L = 6\,\ell_x + 4\,\ell_y + 6\,\ell_g + [\ell_x + 2\,\ell_g]$, where space in the bracket is wasted.

On the other hand, if a snake can be divided into two parts, then the snake can be fitted into two adjacent straight sections. In this case, helicity state of particles at mid-section of snake can be achieved. Such a modified snake configuration has been worked out recently[22].

4.2 the Modified Snake Configuration[22]

The essential feature of Steffen snake is the symmetric arrangement of vertical bending magnets and anti-symmetric horizontal bending magnets. These features can be preserved in following modified snake configuration

$$S_m = (-H, -V, mH, 2V, -mH, -V, H)$$

where the number m is determined by geometry.

$$(m-1)\left(d+\frac{1}{2}(m-1)\ell_x+\ell_y+\ell_g\right)=\ell_x+\ell_y+2\,\ell_g\,. \qquad (4.3)$$

The spin rotation angle, ϕ, and snake axis angle, ϕ_s, are given by

$$\cos\frac{\phi}{2}=\cos^2\psi_y+\cos m\psi_x\sin^2\psi_y \qquad (4.4)$$

$$\cos\phi_s=\frac{-\sin\frac{m\psi_x}{2}\cos\psi_y}{\sqrt{\cos^2\frac{m\psi_x}{2}+\sin^2\frac{m\psi_x}{2}\cos^2\psi_y}} \qquad (4.5)$$

Note that $m\psi_x$ and ψ_y are the relevant variables in determining ϕ and ϕ_s. For a partial snake, we have $\phi<\pi$. When ψ_x,ψ_y are small, one obtain then $\phi\approx 2m\psi_x\psi_y$ and $\phi_s\approx\frac{\pi}{2}+\frac{m\psi_x}{2}$.

4.2a The Compact Snake Configuration

The compact snake configuration corresponds to parameter $d=0$. Depending on the values of ℓ_x, ℓ_y and ℓ_g, m can be determined from geometric considerations. For example, assuming $B=2$ Tesla, $\ell_g=0.15$ m and $\phi_s=180^0$, we obtain m=2.334 with a total length of 11.48 meters for the compact snake configuration. The Steffen snake with m=2 requires a total length of 13.54 meters. The compact snake configuration saves about 15% of total length. Besides, total $\int Bd\ell$ and orbit displacement, D_x, are also reduced.

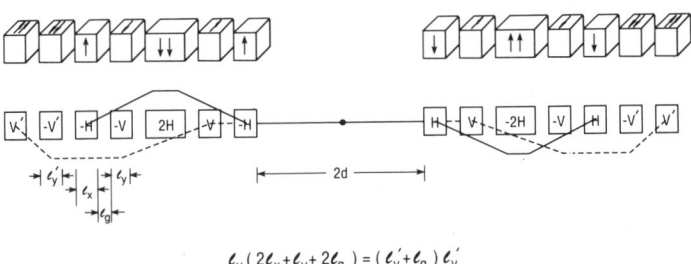

$\ell_y(\,2\ell_x+\ell_y+2\ell_g\,)=(\,\ell_y'+\ell_g\,)\,\ell_y'$

Fig.10 Schematic drawing of the split snake configuration.

4.2b Split Snake Configuration

By adjusting the parameter m, we can obtain a proper distance $2d$ between two halves of a snake (see Eq.4.3). The orbit displacements at middle of two halves of a snake can be corrected by a orbit shifter of $(-V',\,V')$ at both ends of a snake. However when the distance $2d$ becomes large, we will have m$\simeq$1. Obviously, the total length and the radial orbital displacement of a snake increases as well. Such a split snake configuration is not practical for an insertion detector area. It can be used in accelerators with two adjacent straight sections separated by a quadrupole. Such a snake configuration eases design criteria of low energy (≤ 30 GeV) accelerators.

To fit a collider interaction region (IR) onto space between the split snake, the snake configuration shown in Fig. 10 is most appropriate. The advantage of the split snake configuration is that the spin in the mid section of a snake will be in the horizontal plane. Such a snake therefore serves a dual purpose of being a snake and a spin rotator for helicity state experiments. For a spin up particle passing through a half snake, the spin orientation becomes

$$S_x = -\sin m\psi_x \sin \psi_y \; ; \; S_y = \sin^2 \frac{m\psi_x}{2} \sin 2\psi_y \; ; \; S_z = 0 \qquad (4.6)$$

where $m = 2$ for snake configuration shown in Fig. 10. Because of the horizontal orbit compensation magnet, spin is further rotated by ψ_x angle around the vertical axis. Thus by changing ψ_x, ψ_y excitation, one can study helicity experiments or transverse spin experiments in same IR. Such a scheme can save the need of four spin rotators in polarized proton experiment.

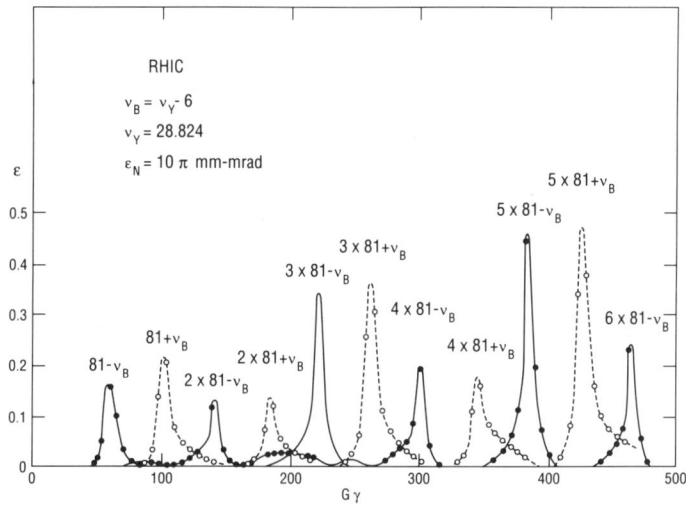

Fig.11 Intrinsic resonance strength for RHIC for $\epsilon_N = 10\pi mm - mrad$

5. Luminosity

To achieve a good luminosity in polarized proton collision in RHIC, we must try to achieve the following parameters:

$$N_B \geq 10^{11}; \epsilon_N \leq 20 \; \pi \text{mm} - \text{mrad}$$

where N_B is the number of particles per bunch and ϵ_N is the 95% normalized emittance. To achieve these goals, 20 linac pulses accumulation in the Booster are needed. These polarized protons are then accelerated in AGS to reach $G\gamma \simeq 48$. Partial snake in AGS was studied recently[18] to correct the imperfection resonances. We foresee no difficulty in reaching $G\gamma = 48$ with about 50-70% polarization.

Once the polarized proton reach $G\gamma \simeq 48$, the beam can be transferred to RHIC without depolarization[23]. The depolarization resonance strength in RHIC is shown on Fig. 11, i.e. the important resonance location will be $K = 81n \pm \nu_B$, where $2\pi\nu_B$ is the phase advance accumulated across the dipole cells, $\nu_B = \nu_y - 6$. The maximum resonance strength will be of the order $\epsilon_K \leq 0.5$ for $\epsilon_N \leq 10\ \pi$ mm-mrad. The imperfection resonance will be of the order 0.06, when the closed orbit is corrected to within 0.3 mm rms. Numerical simulation and analytic study[12] indicates that two snakes will be sufficient to maintain spin polarization in RHIC.

The corresponding luminosity, at the top energy of 250 GeV, is given by

$$\mathcal{L} = 5.7 \times 10^{31} \frac{\left(\frac{N_B}{10^{11}}\right)^2 \left(\frac{B}{57}\right)^2}{\left(\frac{\epsilon_N}{20\pi}\right)^2} \text{ cm}^{-2}\text{sec}^{-1}$$

where B is the number of bunches in each ring. Note the importance of smaller emittance in obtaining higher luminosity. The luminosity will gain a factor of 4 when the emittance is smaller by a factor of 2.

For SSC, the chain of the SSC boosters are shown in Table 1. Assuming a similar source as that of AGS, one needs 70 pulses accumulation in the LEB to achieve 10^{10} particles per bunch. A conventional tune jump method and/or a partial snake can be used in LEB. For MEB and HEB, we need two and four snakes respectively. For the SSC, we need about 26 snakes per ring. The final luminosity is 10^{33} cm^{-2}sec^{-1} at the rms emittance of $1\pi\mu$m.

Table 1. Beam parameters for the SSC accelerator complex

Accelerators	SSC	HEB	MEB	LEB	Linac
Injection momentum(GeV/c)	2,000	200	12	1.69	
Final momentum(GeV/c)	20,000	2,000	200	12	1.69
Circumference(km)	87.12	10.89	3.96	0.54	
Harmonic number	17424	2178	792	108	
Normalized rms emittance($\pi\mu$m) @ 10^{10} particles per bunch	1.0	0.8	0.7	0.6	≤ 0.5
Cycle time(sec)		120	3	0.1	

6. Conclusion

We have made a feasibility study for the possible polarized proton operation in RHIC and SSC.

For RHIC, the important tasks are:

1. multi-pulses accumulation in the AGS–Booster with small emittance;
2. acceleration through AGS to reach $G\gamma \simeq 48$, where the transfer line between the AGS and RHIC is spin transparent; and

3. construction of four snakes in RHIC.

When these tasks are accomplished, the polarized proton luminosity will be larger than 5×10^{31} cm^{-2}sec^{-1} at $\epsilon_N = 20\pi$ mm-mrad. By carefully maintaining the emittance at 10π mm-mrad, the luminosity can achieve 2×10^{32} cm^{-2}sec^{-1}, which corresponds to 2 interactions per crossing.

For SSC, the important tasks can be listed as following.

1. Design SSC low energy booster (LEB) with enough straight section for solenoid type snake, which may need to be rampable. Design SSC medium energy booster with enough straight sections for 2 snakes. Design SSC high energy booster with 4 snakes. Finally, the present SSC lattice is compatible with 26 snakes[24].
2. SSC demands more on the source development program.
3. Transfer lines between accelerators should also be spin transparent.

---

* Work performed under the auspices of the U.S. Department of Energy and Indiana University
† On leave of absence from Accelerator Development Department, Brookhaven National Laboratory, Upton, NY 11973.

References

1. K.J. Heller ed., High Energy Spin Physics, AIP Conf. Proc. No.187, (1988).
2. A.D. Krisch ed., Proc. of Workshop on Polarized Beams at SSC, AIP Conf. Proc. No.145, (1985).
3. A.N. Zelenskii, et al., p.1208 in Ref. 1 (1988).
   Y. Mori, p. 1200 in Ref. 1 (1988).
   C.D.P. Levy, et al., p.1210 in Ref. 1 (1988).
   J.G. Alessi, et al., p.1221 in Ref. 1 (1988).
4. T. Khoe et al., Part. Accel. 6, 213 (1975).
   F.Z. Khiari et al., Phys. Rev. D39, 45 (1989).
   L. Ahren, p. 1068 in Ref. 1 (1988).
5. E.D. Courant and R. Ruth, BNL report, BNL-51270 (1980).
6. L.H. Thomas, Phil. Mag. 3, 1 (1927).
   V.Bargmann, L. Michel, and V.L. Telegdi, Phys. Rev. Lett. 2, 435 (1959).
7. E.D. Courant, S.Y. Lee and S. Tepikian, p. 174 in Ref. 2 (1985).
8. M. Froissart, and R. Stora, Nucl. Inst. Meth. 7, 297 (1960).
9. Ya.S. Derbenev, and A.M. Kondratenko, Sov. Phys. Doklady, 20, 562 (1976).
   Ya.S. Derbenev et al., Particle Accelerators, 8, 115 (1978).
10. S.Y. Lee and S. Tepikian, Phys. Rev. Lett. 56,1635(1986).
    S.Tepikian, Ph. D. Thesis, S.U.N.Y. Stony Brook, (1986)
11. S.Y. Lee, p. 1105 of Ref. 1 (1988).
12. S.Y. Lee and E.D. Courant, Phys. Rev. D41,292(1990).
13. K. Steffen, Particle Accelerator 24, 45 (1989).
14. E.D. Courant and H.S. Snyder, Ann. Phys. 3,1(1958).
15. S.Y. Lee, p. 189 in Ref.2, (1985).
16. A.D. Krisch et al., Phys. Rev. Lett. 63, 1137(1989).
    A.D. Krisch, in the Proc. of Spin Workshop, Protvino, USSR(1989)

J.E. Goodwin et al., Phys. Rev. Lett. 64,2779(1990).
J.E. Goodwin, Ph.D. Thesis, Indiana University, (1990).
17. T. Roser, "Properties of Partially Excited Siberian Snakes", AIP Conf. Proc. No. 187, p. 1442 (1988).
18. L. Ratner, ed., Workshop on Partial Snake in AGS. June(1990).
19. A.D. Krisch, et al. IUCF Cooler Snake Experiments.
20. K. Steffen, DESY Report 83-124(1983).
D.G. underwood, Nucl. Inst. and Methods,173,351(1980).
21. E.D. Courant, p. 1085 in Ref. 1(1988).
U. Wienands, Proc. of 1st European Part. Acc. Conf., Rome, p. 905, (1988).
22. S.Y. Lee, "Snakes and Spin Rotators", BNL 52248/UC–414 (1990).
23. S.Y. Lee and E.D. Courant, "Effect of the Vertical Bends in ATR Transfer Line on the Polarized Proton Operation", AD/RHIC–63 (1990).
24. E.D. Courant, private communications.

THE SPIN DEPENDENT STRUCTURE FUNCTIONS OF THE NUCLEON

Vernon W. Hughes
Yale University, Physics Department
J.W. Gibbs Laboratory
New Haven, CT 06520

ABSTRACT

A brief review is given of our knowledge of the spin dependent structure functions of the nucleon and of some plans for future experiments.

INTRODUCTION

The interpretation of experiments with a polarized collider having a polarized proton or deuteron beam requires understanding of the spin dependent structure functions of the nucleon. In this article I will review briefly our present knowledge of this topic and indicate some plans and possibilities for future experiments.

The proton and neutron each have two spin dependent structure functions which are often designated by the asymmetries A_1 and A_2. Direct information on these spin dependent structure functions of the nucleon requires the measurement of spin dependent asymmetries in the scattering of high energy polarized electrons or muons by polarized nucleons. This far two such experiments have been done with polarized protons - the first at SLAC with polarized electrons and the second at CERN with polarized muons. In both longitudinally polarized leptons were scattered from longitudinally polarized protons and the asymmetry A_1 was determined.

1. THE SLAC EXPERIMENT ON THE PROTON SPIN DEPENDENT STRUCTURE FUNCTION.[1]

It is well known that we can study the structure function of the proton by deep inelastic inclusive e-p scattering in which only the outgoing electron is measured. The kinematic variables are Q^2 (4-momentum transfer squared), ν (energy loss) and the dimensionless variable $x = Q^2/2M\nu$ in which M = proton mass. The proton tensor $W_{\mu\nu}$ involves four independent structure functions - F_1 and F_2 which are spin independent structure functions and g_1 and g_2 which are spin dependent. In the deep inelastic or scaling regime where Q^2 and ν are large compared to the proton mass, the photon can be considered to act on an individual quark and these functions depend only on the variable x which can be interpreted as the fraction of the momentum of the proton in the infinite momentum frame that is carried by the struck quark. If both electron and proton are polarized, we can determine the spin dependent structure functions from the spin dependent asymmetries in the differential scattering cross sections. In particular the asymmetry A between the antiparallel and parallel ep spin cases determines the virtual photon-proton asymmetry A^1 which leads to the spin dependent structure function g_1. The definitions and relationships are given in Eq. 1, where D and η are kinematic factors.

$$A \equiv \frac{\frac{d^2\sigma}{d\Omega dE'}(\uparrow\downarrow) - \frac{d^2\sigma}{d\Omega dE'}(\uparrow\uparrow)}{\frac{d^2\sigma}{d\Omega dE'}(\uparrow\downarrow) + \frac{d^2\sigma}{d\Omega dE'}(\uparrow\uparrow)} \quad (1)$$

$$A_1 = D(A_2 + \eta A) \simeq DA_1 \qquad g_1(x) = \frac{A_1(x)F_2(x)}{2x(1+R(x))}$$

$$R = \sigma_L/\sigma_T$$

Further aspects of polarized deep inelastic scattering can be discussed in terms of the absorption of the polarized virtual photon by the polarized proton. For a z-axis collision between a polarized photon and a polarized proton, the total absorption cross section is $\sigma_{3/2}$ for the case of total angular momentum component + 3/2 and $\sigma_{1/2}$ for total angular momentum component 1/2. The asymmetry $A_1 = \frac{\sigma_{1/2} - \sigma_{3/2}}{\sigma_{1/2} + \sigma_{3/2}}$ is the virtual photon-proton asymmetry corresponding to the cases of components with total photon and proton angular momentum along the collision axis equal to 3/2 and 1/2. At the quark level there is incoherent absorption of the photons by the quarks, and in the absence of orbital angular momentum the virtual polarized photon can only be absorbed when the quark and photon spins are oppositely directed, in order to conserve the component of total angular momentum. A_1 is proportional to the probability that a quark of type i has its spin along the proton spin direction multiplied by the square of its charge and summed over all types of quarks, or equivalently to the spin dependent structure function g_1, as given in Eq. 2.

$$A_1(x) = \frac{\sum_i e_i^2 (q_i^\uparrow(x) - q_i^\downarrow(x))}{\sum_i e_i^2 (q_i^\uparrow(x) + q_i^\downarrow(x))} \quad (2a)$$

$$g_1(x) = \frac{1}{2} \sum_i e_i^2 (q_i^\uparrow(x) - q_i^\downarrow(x)) \quad (2b)$$

For quark models of the proton $A_1 > 0$.

In 1971 SLAC approved experiment E80 to do the first measurement of polarized deep inelastic e-p scattering. The polarized electron source for this experiment was built at Yale and was based on our atomic physics research at Yale on the production of polarized electrons with an atomic beam. (Figure 1). By magnetic deflection in an inhomogeneous magnetic field atoms with a particular component of electronic polarization were selected and photoionization of these atoms then led to a source of polarized electrons. The direction of H in the photoionization region determines the electron spin direction, which can therefore be modulated by changing the direction of the current in the large coil. Lithium-6 atoms were chosen for technical reasons and an intense UV flash lamp was used for photoionization. The source was installed at SLAC in 1973.

The electron polarization was measured at the injection energy of 70 keV by Mott scattering and at the high energy output (10-20 GeV)

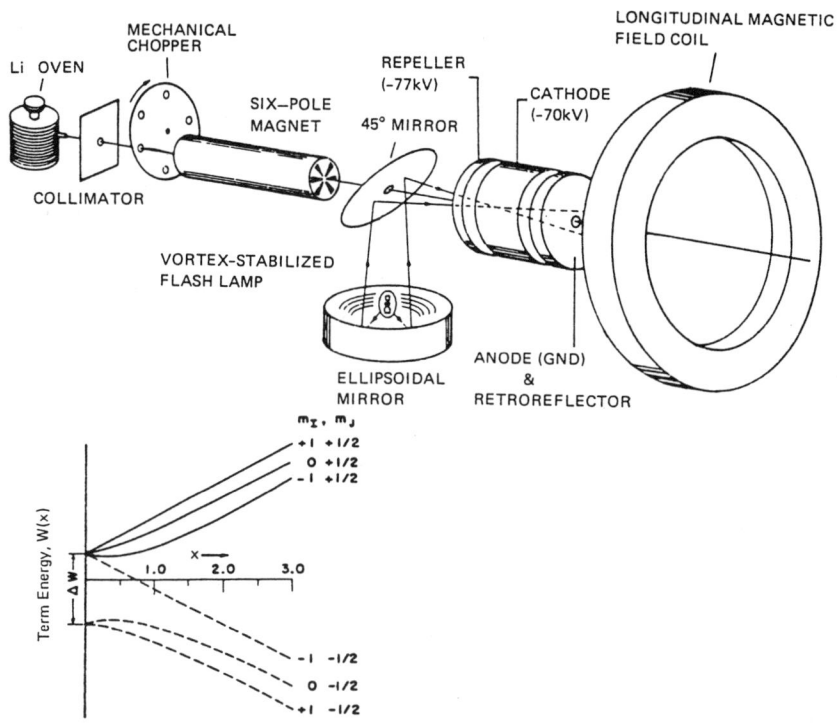

Figure 1. Schematic diagram of PEGGY showing the principal components of the lithium atomic beam, the uv optics and the ionization region electron optics; Energy level and magnetic moments of ^{6}Li (nuclear spin I=1) in the ground $^2S_{1/2}$ atomic state as a function of magnetic field H.

of the linac by Möller scattering by polarized electrons in an iron foil (Figure 2). The analyzing power is large, indeed about 0.8 for 90° scattering in the CM system, and the cross section is also large. Because the electron beam from the linac must be deflected about 24° into the experimental area (end station A), the polarization of the beam on the target for a given polarization of the linac output depends on the energy because of the electron g-2 precession. The data agree well with expectation. The operating characteristics of the polarized electron source (designated PEGGY) are given Table I.

TABLE I: OPERATING CHARACTERISTICS OF POLARIZED ELECTRON BEAM

Characteristic	Value
Pulse length	1.5 μs
Repetition rate	180 pps
Average intensity at GeV energies	5×10^8 e$^-$ per pulse
Pulse-to-pulse intensity variation	< 5%
Polarization	0.8 ± 0.03
Polarization reversal time	3s
Intensity difference upon reversal	< 5%

54 Spin Dependent Structure Functions of the Nucleon

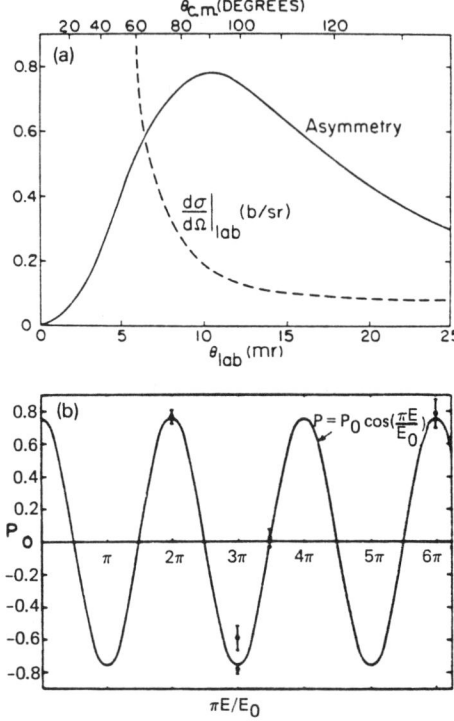

Figure 2. Measurement of electron beam polarization. The Möller asymmetry and laboratory cross section plotted versus scattering angle for the representative incident energy of 9.712 GeV. The longitudinal component, P, of the beam polarization plotted versus $\pi E/E_0$ with $E_0 = 3.237$ GeV.

The polarized proton target utilized the method of dynamic nuclear polarization with the hydrocarbon butanol and the paramagnetic impurity porphyrexide (Figure 3). At the high field ($B \approx 5T$) and low temperature ($T \approx 1$ K) the unpaired electron spins in the porphyrexide are highly polarized according to the Boltzman factor. Due to spin-spin coupling of the electrons and protons, microwave transitions can be driven which transfer the electron polarization to the protons. The target volume was 25 cm^3. The operating characteristics of the target are given in the Table II.

The data for SLAC E80 were taken with the high resolution 8 GeV/c SLAC spectrometer. Most of the data on polarized electron-proton scattering were obtained in a subsequent experiment, SLAC E130, for which the large acceptance spectrometer shown in Figure 4 was used. This spectrometer consisted of two dipoles, PWC chambers, a gas Cerenkov threshold counter and a Pb glass shower counter. The spectrometer acceptance $\delta p/p$ was 50%, its resolution $\frac{\Delta p}{p}$ was 1%, and the π^-/e^- rejection factor was 10^{-3}.

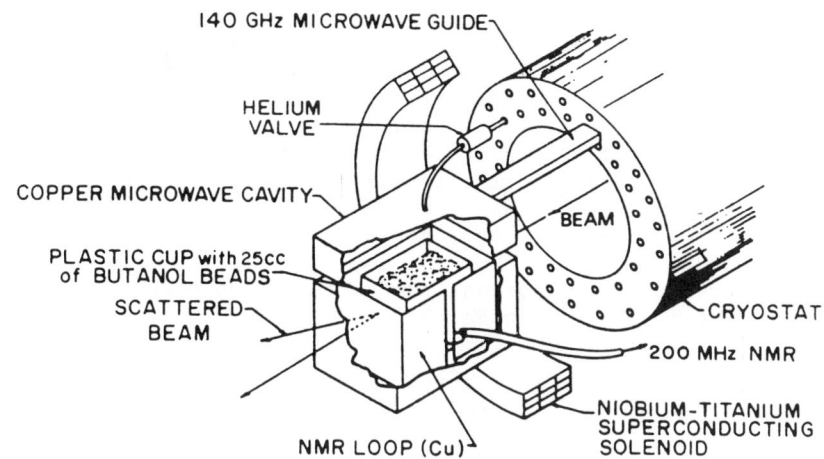

Figure 3. Schematic diagram of the Yale-SLAC polarized proton target.

TABLE II. OPERATING CHARACTERISTICS OF POLARIZED PROTON TARGET

Characteristic	Value
Magnetic field (superconducting)	50 kG
Temperature	1 K
Target material	25 cm^3 of butanol-porphyrexide
Maximum polarization, P_p	0.75
Depolarizing dose (1/e)	3×10^{14} e$^-$ cm^{-2}
Polarizing time (1/e)	~4 min
Anneal or target change time	~45 min

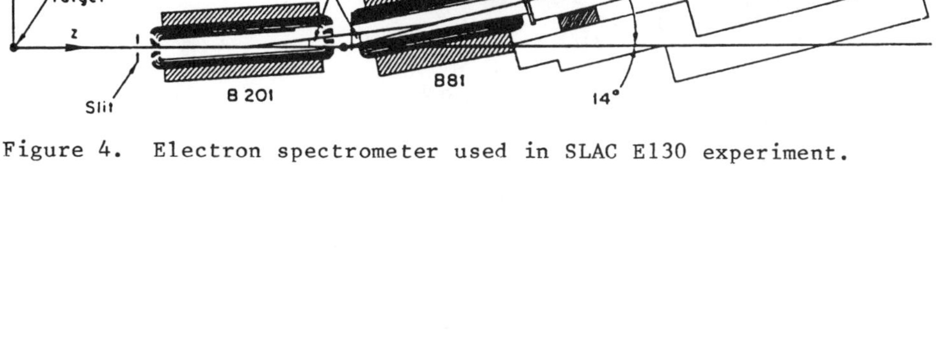

Figure 4. Electron spectrometer used in SLAC E130 experiment.

The counting rate asymmetry Δ is given by $\Delta = P_e P_p f A$ in which P_e = polarization of electron beam, P_p = polarization of the H protons in butanol, f = fraction of nucleons in butanol associated with H and A = intrinsic e-p asymmetry. Since $P_e \approx 0.8$; $P_p \approx 0.6$; $f \approx 0.1$, the asymmetries Δ are in the range of 0.001 to 0.01 and statistical errors are dominant.

Figure 5 shows the kinematic range covered which included elastic, resonance region, and deep inelastic scattering. Figure 6 indicates that A_1^p obeys the scaling relation i.e. it depends only on x and not on Q^2.

There are several important sum rules for spin dependent structure functions given in Eq. 3.

$$S_{Bj} = \int_0^1 dx(g_1^p - g_1^n) = \int_0^1 \frac{dx}{2x}\left(\frac{A_1^p F_2^p}{1+R^p} - \frac{A_1^n F_2^n}{1+R^n}\right) = \frac{1}{6}\left|\frac{g_A}{g_V}\right| = 0.209(1)$$

With QCD correction

$$S_{Bj} = \int_0^1 dx(g_1^p - g_1^n) = \frac{1}{6}\left|\frac{g_A}{g_V}\right|\left(1 - \frac{\alpha_s(Q^2)}{\pi}\right) = 0.191(2) \quad (3)$$

$$S_{EJ}^p = \Gamma_1^p = \int_0^1 dx\, g_1^p = \frac{1}{12}\left|\frac{g_A}{g_V}\right|\left[1 + \frac{5}{3}\frac{3F/D - 1}{F/D + 1}\right] + O(\alpha_s) = 0.189 \pm 0.005$$

$$S_{EJ}^n = \Gamma_1^n = \int_0^1 dx\, g_1^n = \frac{1}{12}\left|\frac{g_A}{g_V}\right|\left[-1 + \frac{5}{3}\frac{3F/D - 1}{F/D + 1}\right] + O(\alpha_s) = -0.002 \pm 0.005$$

where

$$g_1 = \frac{A_1 F_2}{2x(1+R)}$$

$$R = \frac{\sigma_L}{\sigma_T}$$

$g_A = 1.254(6)$

$\alpha_s = 0.27(2)$ at $Q^2 = 11(GeV/c)^2$

$F/D = 0.631 \pm 0.018$

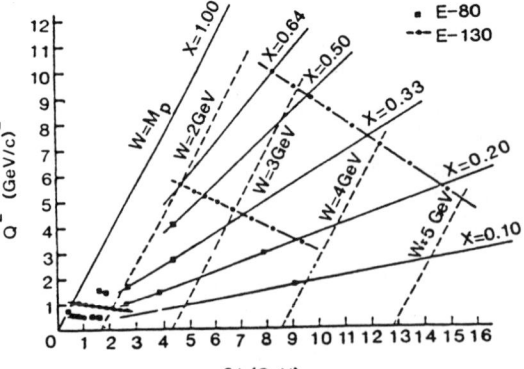

Figure 5. Kinematic points measured.

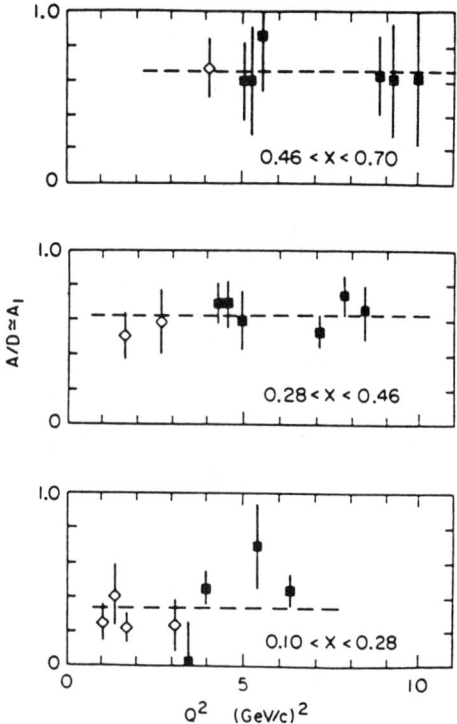

Figure 6. Radiatively corrected values of $A/D \approx A_1$ obtained in SLAC E80 (open diamonds) and in SLAC E130 (closed squares) plotted vs Q^2. The fits to horizontal lines demonstrate scaling of A_1.

The Bjorken polarization sum rule is a basic relation originally derived from current algebra for a quark model of the nucleon and with incorporation of the view, now well established, that the weak interactions of quarks and leptons are the same. This remarkable relation between structure functions and constants characterizing nuclear beta decay was derived in 1966 and played a seminal role in stimulating our experimental program. This sum rule can be derived in the quark-parton model by evaluating the expectation value of the axial vector beta decay operator for n→p decay. It is now recognized as a rigorous consequence of QCD in the scaling limit and the $O(\alpha_s)$ correction has been computed by perturbative QCD. First moments of the spin dependent structure functions for proton and neutron separately were given by Ellis and Jaffe. These sum rules are model dependent, with the principal assumption being that the strange quark sea is unpolarized.

The measured values of A_1^p are shown in Figure 7. At low x where the scattering is expected to be predominantly by unpolarized sea quarks, A_1^p is small. As x increases, scattering by the valence quarks which should carry the spin of the proton becomes more important and A_1^p increases. Indeed as x→1, perturbative QCD predicts that $A_1^p \to 1$ since the struck quark should carry the entire spin of the proton. A test of the Ellis-Jaffe sum rule can be made with these data. The value obtained for the first moment of the proton spin dependent structure function is

$$\Gamma_1^p = \int_0^1 g_1^p(x)\, dx = 0.17 \pm 0.05 \qquad (4)$$

The error in the experimental value is relatively large because the data only extend down to $x = 0.1$ and hence a large extrapolation is required to determine the first moment of g_1. Agreement between experiment and theory is satisfactory within the experimental error.

Comparison of the data with various theoretical models is shown in Figure 8. Best agreement is obtained with the quark model of Carlitz and Kaur. It is an unsymmetrical model which satisfied the Ellis-Jaffe sum rule, Regge theory at low x, perturbative QCD at high x and includes an adjustable parameter to account for the transfer of the spin of valence quarks to sea quarks.

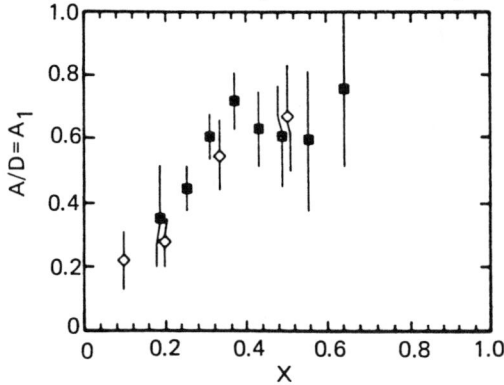

Figure 7. Measured values of $A/D \simeq A_1$ vs x.

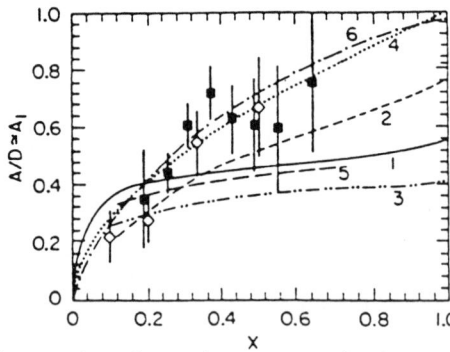

Figure 8. Experimental values A_1 compared with theories. 1. Symmetrical valence-quark model. 2. Current quarks. 3. Orbital angular momentum. 4. Unsymmetrical model. 5. MIT bag model. 6. Source theory.

Data was also taken in the resonance region at low Q^2 which indicated that scaling still applied except at the Δ resonance.

SLAC E130 (1976 proposal) originally included a plan to measure the neutron spin dependent structure function with a polarized deuteron target, but running time was not provided for this measurement. A more ambitious experiment designated Son of E130 (1980 proposal) to measure the spin dependent asymmetries A_1 for the proton and neutron with a longitudinally polarized target, and A_2 for the proton with a transversely polarized target, and (in its final proposal form) the SLAC GaAs polarized electron source PEGGY II was not approved by the Program Advisory Committee.

2. THE CERN POLARIZED MUON-PROTON SCATTERING EXPERIMENT[2]

The second experiment on the proton spin dependent structure function was done at CERN. The experimental setup is shown in Figure 9 with the polarized proton target, the EMC forward spectrometer and the incident polarized muon beam. The positive muon beam energy was between 100 and 200 GeV, its intensity was about 2×10^7 μ^+/pulse, and its polarization P_μ was 0.8. The EMC polarized target is shown in Figure 10. This enormous target used irradiated NH_3 as the target material. The magnetic field was 2.5T and the temperature was

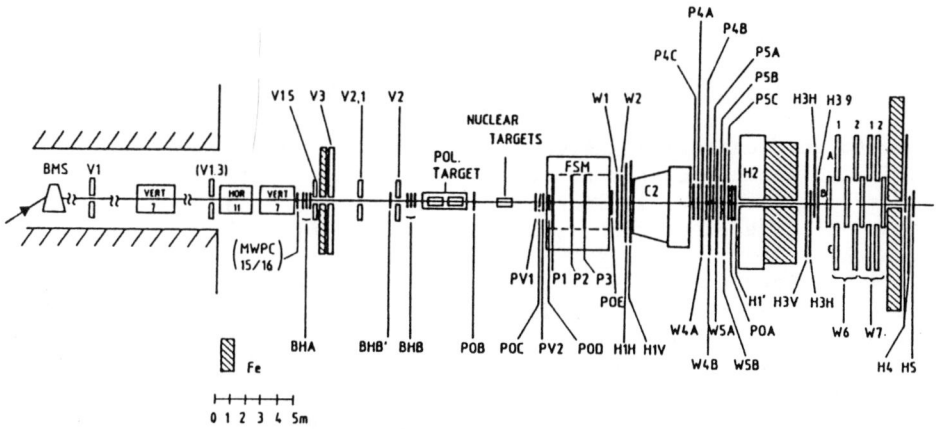

Figure 9. The EMC forward spectrometer for the CERN EMC polarized-target experiment. Key: H=scintillator trigger hodoscope, V-veto hodoscope, BH=beam hodoscope, P=multiwire proportional chamber, W=drift chamber, FSM=forward spectrometer magnet C_2=Cerenkov counter, H_2=calorimeter.

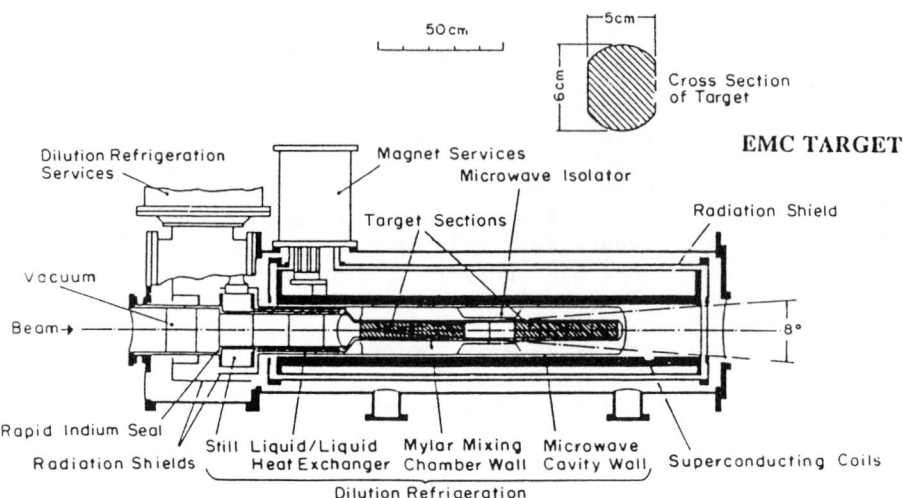

Figure 10. The EMC polarized target.

maintained at 0.5 K in the dynamic nuclear polarization mode and 0.05 K in the frozen spin mode by a dilution refrigerator with 1 W capacity. The target was divided into two halves each 40 cm in length and 5cm in diameter which were polarized in opposite directions using slightly different microwave freqencies. Hence data could be obtained simultaneously for the cases of parallel and antiparallel beam and target polarization directions.

The spectrometer detected the scattered muons and measured their momenta and scattering angles. The principal data obtained were for inclusive scattering and about 10^6 scattered μ^+ were observed. The spectrometer could also identify hadrons in the final state and some semi-inclusive scattering asymmetries mostly with π^+ and π^- were reported.

The virtual photon-proton asymmetry A_1^p determined from the measured asymmetries are shown in Figure 11 together with the SLAC data points. In the region of overlap the EMC and the SLAC data agree well. The principal contribution of the CERN data is to determine A_1 to lower values of x, indeed down to x = 0.01. This is a very important contribution because it allows a much more sensitive test of the Ellis-Jaffe sum rule as indicated in Figure 12. The conclusion is that the data disagree with the Ellis-Jaffe sum rule.

The implication based on the naive quark-parton model is that very little of the proton spin is contributed by the quark spins. This surprising conclusion has attracted great interest and has led to many theoretical papers. Possible reasons for the discrepancy and mechanisms by which gluons or orbital angular momentum could contribute to the proton spin have been discussed extensively.
Also many suggestion have been made for further measurements in this field in addition to the obvious one of improving and extending the determination of A_1 for the proton and neutron. The most fundamental theoretical relation in this field is the Bjorken polarization sum

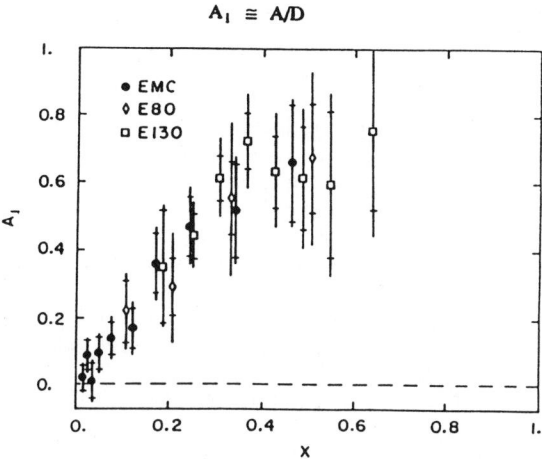

Figure 11. Compilation of all the data of A_1^p as a function of x. The EMC points are shown as full circles while the SLAC points are shown as open diamonds (experiment E-80), and open squares (E130). Inner error bars are the statistical errors and the outer error bars are the total errors (statistical plus systematic added in quadrature). The systematic errors include uncertainties in the values of R and A_2.

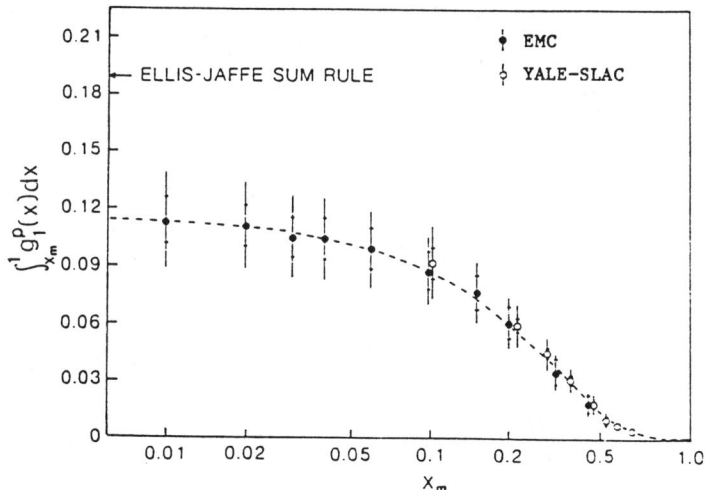

Figure 12. The value of the integral $\int_{x_m}^{1} g_1^p dx$ as a function of x_m, the value of x at the low edge of each bin.

rule, and its accurate verification will provide an important test of
quantum chromodynamics. The asymmetry A_2 is expected to be zero in
the scaling limit; it is similar to R and must presumably be
understood in terms of nonperturbative QCD. Studies of asymmetries
in semi-inclusive scattering such as the EMC measurements of π^+ or π^-
but particularly with J/ψ or jets in the final hadron state would be
useful for understanding the gluon contribution to the spin structure
functions. The many theoretical interpretations and suggestions for
experiments are reviewed and referenced in this Volume, in the
Proceedings of the High Energy Spin Conference in Bonn[3] and in
several review articles.[4,5]

The present high interest in the nucleon spin dependent
structure functions has led to plans for new experiments. In
particular, a new collaboration at CERN (SMC) will do CERN experiment
NA47[6] to measure the neutron spin dependent structure function and
also to improve knowledge of A_1^p. A μ^+ beam intensity of 4×10^7
μ^+/pulse is planned and the projected operating conditions for a
modified target with butanol (hydrogen and deuteron) as the target
material is shown in Table III. A major new solenoid-dipole magnet
with a homogeneity of 10^{-5} in the longitudinal solenoid field and a
target length of 120 cm (two 60 cm halves) is being constructed. The
dipole field will be used to achieve rapid (< 30 min) target
polarization reversal and to provide a holding field for the
transverse polarization case. The spectrometer has been upgraded
from the EMC and NMC one in both major and minor ways including a
revised muon beam line, several new drift chamber planes and a
complete replacement of the W6/7 drift tube chamber by streamer tube
and drift tube arrays. A major polarimeter to measure the muon beam
polarization is being installed. Extensive modifications in on-line
data acquisition system and plans for rapid data analysis are being
made. The experiment should obtain 20×10^6 scattered muons and
cover the x range from 0.005 to 0.7. The overall systematic and
statistical error should be less than 10%. A test of the Bjorken
polarization sum rule at the 10% level should be achieved and
accurate information on the neutron and proton spin dependent
structure function will be obtained. In addition, exploratory
measurements are planned of A_2 for the proton using a transversely
proton taret.

In addition, an experiment is planned at HERA[7] to use the
electron ring at about 30 GeV with an internal polarized H or D gas
target to measure A_1^p and A_1^n. An experiment has been approved at

TABLE III: PROJECTED OPERATING CONDITIONS FOR SMC POLARIZED TARGET

Target Material	Butanol and deuterated butanol (Doped with EHBA-Cr (V))
Magnet Field	2.5T
Temperature	0.5 K for DNP mode
	0.05 K for frozen spin mode
Polarization	0.8 for proton; 3% rel. error
	0.4 for deuteron; 3% rel. error

SLAC[8] to use a polarized ^{3}He target to measure A_1^n. A Letter of Intent at CERN[9] "Polarized Electroproduction at LEP, was submitted to use a polarized electron beam in LEP with an internal polarized jet H or D target for such studies.

3. SUMMARY, CONCLUSIONS AND FUTURE

A summary of the present situation and a listing of some important experimental and theoretical problems is given below.

EMC(CERN) data on polarized μ-p DIS extended SLAC polarized e-p data to $x \tilde{=} 0.01$; CERN and SLAC data agree from $x=0.1$ to $x=0.7$
(1) Violation of Ellis-Jaffe sum rule for Γ_1^p.
(2) If Bjorken sum rule is valid, implication is that A_1^n and Γ_1^n are much larger than previously predicted.
(3) According to the interpretation with naive quark-parton model, quark spins seem to contribute very little of the proton spin.

Important Experimental Problems:
(1) Measure A_1^n, Γ_1^n for the first time and improve data on A_1^p and Γ_1^p Particularly to test Bjorken polarization sum rule.
(2) Measure A_2^p and A_2^n
(3) Measure spin-depdendent effects in exclusive channels, e.g. (J/Ψ).

Theoretical Problems:
(1) Ellis-Jaffe sum rule.
(2) Carriers of proton spin.
(3) Nucleon spin dependent structure functions.

Implications:
(1) Polarized hadron-hadron scattering.
(2) Parity violation in atoms.
(3) Dark matter in universe.

It is quite clear that the subject of the nucleon spin dependent structure functions is a rich and active one.

Research supported in part by the Department of Energy under Contract No. DE-AC02-76-ER03075.

4. REFERENCES

1. V.W. Hughes and J. Kuti, Ann. Rev. Nucl. Part. Sci 33, 611 (1983). This review article includes a quite complete list of references on the SLAC experiment.
2. J. Ashman et al., Nucl. Phys. B328, 1 (1989). This article includes a quite complete list of references on the CERN experiment.
3. Proceedings of the 9th International Symposium on High Energy Spin Physics, Bonn, September, 1990.
4. G. Altarelli and W.J. Sterling, Partical World 1, 40 (1989).
5. PANIC VII Conference, MIT, June, 1990.
6. "Measurement of the Spin-Dependent Structure Functions of the Neutron and Proton." SMC Collaboration. 12/22/88, CERN Experiment NA47.

7. "A Proposal to Measure the Spin-Dependent Structure Functions of the Neutron and the Proton at HERA", The Hermes Collaboration, January, 1990.
8. "A Proposal to Measure the Neutron Spin Dependent Structure Function" SLAC Proposal E-142, October, 1989.
9. "Polarized Electroproduction at LEP" CERN/LEPC 88-16, 1988.

SPIN EFFECTS AT COLLIDER AND SUPERCOLLIDER ENERGIES

Jacques SOFFER
Centre de Physique Théorique, CNRS Luminy
Case 907, 13288 Marseille Cedex 9, France

ABSTRACT

Assuming that future hadron machines will have polarized proton beams we show the relevance of spin effects for the next generation of *pp* colliders and supercolliders. We calculate various helicity asymmetries in the multi-TeV energy range and also below 1 TeV for a selection of processes in the appropriate kinematic regions. We will see that some of our results are spectacular and can help to pin down fundamental issues in the framework of the Standard Model.

MOTIVATION

The main reason for particle physics to build the next generation of high-energy hadron colliders SSC, LHC and RHIC is to improve our present understanding of the dynamical theory which describes quarks, leptons, gauge bosons and their interactions. In spite of the present success of the Standard Model many fundamental questions have not been answered yet and in particular the existence of a large number of parameters (masses, coupling constants, etc.) which are inputs to the theory and cannot be calculated in the framework of the Standard Model. Another key question is whether or not parity violation is a low-energy feature of the theory which might disappear at higher energies. Clearly new hadron facilities in the multi-TeV energy range could also allow the production of new heavy objects, which might be related to new additional symmetry in nature or to quark and lepton substructure, both supersymmetry and compositeness being interesting theoretical speculations. At this stage it is natural to ask if spin effects will be important in the multi-TeV energy range and whether or not physics will require polarized proton beams in a supercollider. These questions have been studied in great detail in a dedicated work[1] but here we will only cover some selected topics in the framework of the Standard Model. We will leave aside all the key issues related to the physics beyond the Standard Model which will be fully discussed this afternoon by P. Taxil[2]. We will also see that for a *pp* collider below 1 TeV, e.g. a machine like RHIC, one expects many interesting effects with polarized proton beams[3]. In particular, it will give us the opportunity to check the sign and the magnitude of the gluon helicity asymmetry $\Delta G(x)$.

In the next section we recall some basic definitions of the observables and the main features of polarized parton distributions and parton-parton luminosities. Then we will discuss dynamical tests based on perturbative QCD in particular hadronic jet production and direct photon production. Finally in the last section we will study specific issues related to standard electroweak interactions.

OBSERVABLES, PARTON DISTRIBUTIONS AND LUMINOSITIES

A high energy proton beam is an unseparated beam of constituent partons and all fundamental hadronic interactions, which are probed in pp collisions by testing the Standard Model or by producing new particles, involve the collisions of quarks and gluons at short distances. As an example, let us consider the hard scattering hadronic process

$$a + b \to c(\text{or jet}) + X \qquad (1)$$

which is described in terms of two to two parton subprocess in the QCD parton model, and whose corresponding inclusive cross section, provided factorization holds, is given by

$$d\sigma(a+b \to c+X) = \sum_{ij} \frac{1}{1+\delta_{ij}} \int dx_a dx_b \left[f_i^{(a)}(x_a, Q^2) f_j^{(b)}(x_b, Q^2) d\hat{\sigma}_{ij} \right.$$

$$\left. + (i \leftrightarrow j) \right]. \qquad (2)$$

The summation runs over all contributing parton configurations; the parton distribution $f_i^{(a)}(x_a, Q^2)$ is the probability that hadron "a" contains a parton "i" carrying a fraction x_a of the hadron's momentum. It represents the parton flux available in the colliding hadron which is universal, that is process independent. Clearly the parton distributions play a crucial role because they allow the connection between hadron-hadron collisions and elementary subprocesses. $d\hat{\sigma}_{ij}$ is the cross section for the interaction of two partons i and j which can be calculated perturbatively and whose expressions at the Born lowest order are given in BRST. The total energy of the partons in the subprocess center of mass frame is $\sqrt{\hat{s}} = \sqrt{x_a x_b s}$ where $\sqrt{s}$ denotes the total center of mass energy of the initial hadrons. Finally Q^2 which is defined in terms of the invariants of the subprocess, characterizes the physical momentum scale. The distributions $f_i(x, Q^2)$ are extracted from deep inelastic data at low Q^2 and their Q^2 dependence, which is logarithmic, is predicted in perturbative QCD by the Altarelli-Parisi equations[4] based on the renormalization group. For supercollider energies the relevant Q^2 range is $10^2 < Q^2 < 10^8 \text{GeV}^2$. The distributions fall rapidly with x at fixed Q^2 and for increasing Q^2, $f_i(x, Q^2)$ becomes larger and larger at small x where more and more partons can be produced. If we want to detect a small invariant mass, produced in the parton subprocess, say 100GeV, one sees from the above expression of $\sqrt{\hat{s}}$ that we need small x values, say $x \leq 10^{-3}$ for $\sqrt{s} = 40\text{TeV}$; so at supercollider energies the hadron energy is less efficiently used and one should keep in mind that the kinematic region of interest is $10^{-5} < x < 10^{-1}$.

These small values of x, even at low Q^2, are not probed by current experiments which only extend to $x > 10^{-2}$. However this present limited knowledge is not so critical because uncertainties are reduced for higher Q^2. We will take for granted that, modulo some reasonable approximations, the small x behavior of the parton distributions at large Q^2 can be safely obtained analytically from perturbative QCD. Rather than referring us to any of the various numerical solutions available in the literature, we will use our own set of parton distributions in terms of a simple analytic parametrization of their x and Q^2 dependences, already given in BRST and which agrees to a good approximation, up to $x = 0.1$, with the solutions provided in ref.5.

So far we have ignored the spin of the initial protons, but if we consider the case of polarized beams, one can define correspondingly polarized parton distributions. For a given parton (quark, antiquark or gluon) we denote $f_{\pm}(x, Q^2)$ the parton distributions in a polarized nucleon either with helicity parallel (+) or antiparallel (−) to the parent nucleon helicity. As usual, we define the *unpolarized distribution* $f = f_+ + f_-$ and the *parton helicity asymmetry* $\Delta f = f_+ - f_-$. Let us recall that for the inclusive reaction (1) with both initial hadrons longitudinally polarized, the double helicity hadron asymmetry A_{LL} defined as

$$A_{LL} = \frac{d\sigma_{a(+)b(+)} - d\sigma_{a(+)b(-)}}{d\sigma_{a(+)b(+)} + d\sigma_{a(+)b(-)}} \quad (3)$$

is given by

$$A_{LL}d\sigma = \sum_{ij} \frac{1}{1+\delta_{ij}} \int dx_a dx_b \left[\Delta f_i^{(a)}(x_a, Q^2) \Delta f_j^{(b)}(x_b, Q^2) \hat{a}_{LL}^{ij} d\hat{\sigma}_{ij} \right.$$

$$\left. + (i \leftrightarrow j) \right] \quad (4)$$

assuming the factorization property, where $d\sigma$ is given by (2) and $\hat{a}_{LL}^{ij}$ denotes the subprocess double helicity asymmetry for initial partons i and j. The explicit expressions of these quantities at the Born lowest order for various subprocesses are also given in BRST. To get a rough estimate of A_{LL} one can use the following approximation

$$A_{LL} \sim \sum_{ij} \langle \frac{\Delta f_i}{f_i} \rangle \langle \frac{\Delta f_j}{f_j} \rangle \hat{a}_{LL}^{ij} \quad (5)$$

in terms of the average of the *parton polarizations* defined as $\frac{\Delta f_i}{f_i}$. It shows that, even if at the parton level $\hat{a}_{LL}^{ij}$ is as large as ±100%, it is expected to be diluted twice at the hadron level since, as we will see below, the parton polarizations are much less than one in most relevant kinematic regions. We also refer to BRST for the simple expressions of the different $\Delta f_i'$s and for the corresponding parton polarizations.

Concerning the gluon helicity asymmetry $\Delta G(x, Q^2)$ one possible interpretation of the EMC result[6] is to assume that gluons carry a large fraction of the proton spin[7]. So in addition to the standard choice (see eq.(3.12) in BRST), we will also use following ref.8, a simple parametrization

$$\Delta G(x, Q_0^2) = \begin{cases} G(x, Q_0^2) & x_c \geq x \geq 1 \\ (x/x_c)G(x, Q_0^2) & 0 \geq x \geq x_c \end{cases} \quad (6)$$

which leads to a much larger value of the integral $\Delta G = \int_0^1 \Delta G(x)dx$, typically $2 - 3$ for $x_c \sim 0.2$, compared to $0.2 - 0.3$ for the standard choice.

To evaluate hadronic cross sections let us consider a convenient quantity introduced by EHLQ, the differential parton-parton luminosity defined as

$$\tau \frac{d\mathcal{L}_{ij}}{d\tau} = \frac{\tau}{1+\delta_{ij}} \int_\tau^1 \frac{dx}{x} \left[f_i^{(a)}(x, \hat{s}) f_j^{(b)}(\tau/x, \hat{s}) + (i \leftrightarrow j) \right]. \quad (7)$$

It represents the number of parton i-parton j collisions per unit τ with subprocess energy square $\hat{s} = \tau s$. Thus the differential cross section with respect to the scaling variable τ for the reaction (1), using eqs.(2) and (7) is given by

$$\frac{d\sigma}{d\tau} = \sum_{ij} \tau \frac{d\mathcal{L}_{ij}}{d\tau} \hat{\sigma}_{ij}. \quad (8)$$

By dimensional analysis one has $\hat{\sigma}_{ij} = k/\hat{s}$, with coupling strength k, so the quantity $(\tau/\hat{s})d\mathcal{L}_{ij}/d\tau$ which has the dimension of a cross section, can be used to estimate the total number of events/year N_{ij} for a given subprocess

$$N_{ij} = L_h k \left(\frac{\tau}{\hat{s}} \frac{d\mathcal{L}_{ij}}{d\tau} \right), \quad (9)$$

knowing k and the hadron-hadron luminosity L_h, which is typically $10^{40}\,\text{cm}^{-2}$ at SSC. This notion of parton luminosity can be generalized to the case of hadronic reaction with *one or two* polarized beams by replacing in eq.(7) one or two unpolarized distributions f by the corresponding parton helicity asymmetries Δf. Therefore we define the two following useful quantities, namely the single-polarized luminosity

$$\tau \frac{d\mathcal{L}_{\Delta ij}}{d\tau} = \tau \int_\tau^1 \frac{dx}{x} \Delta f_i^{(a)}(x, \hat{s}) f_j^{(b)}(\tau/x, \hat{s}) \quad (10)$$

when hadron a only is polarized and the double-polarized luminosity

$$\tau \frac{d\mathcal{L}_{\Delta i \Delta j}}{d\tau} = \frac{\tau}{1+\delta_{ij}} \int_\tau^1 \frac{dx}{x} \left[\Delta f_i^{(a)}(x, \hat{s}) \Delta f_j^{(b)}(\tau/x, \hat{s}) + (i \leftrightarrow j) \right] \quad (11)$$

when both initial hadrons a and b are polarized.

We show these parton luminosities for gluon-gluon interactions at $\sqrt{s} = 40$TeV in Fig.1 where we should notice a strong reduction of the (single or double) polarized luminosity at small $\sqrt{\hat{s}}$. We also show in Fig.2 the case of quark-quark interactions and in Fig.3 some cases with two unlike partons giving very small double-polarized luminosities. These calculations were done with the standard choice for $\Delta G(x, Q^2)$, i.e. eq.(3.12) of BRST, but by using eq.(6) we would not get such a strong reduction of the polarized luminosities.

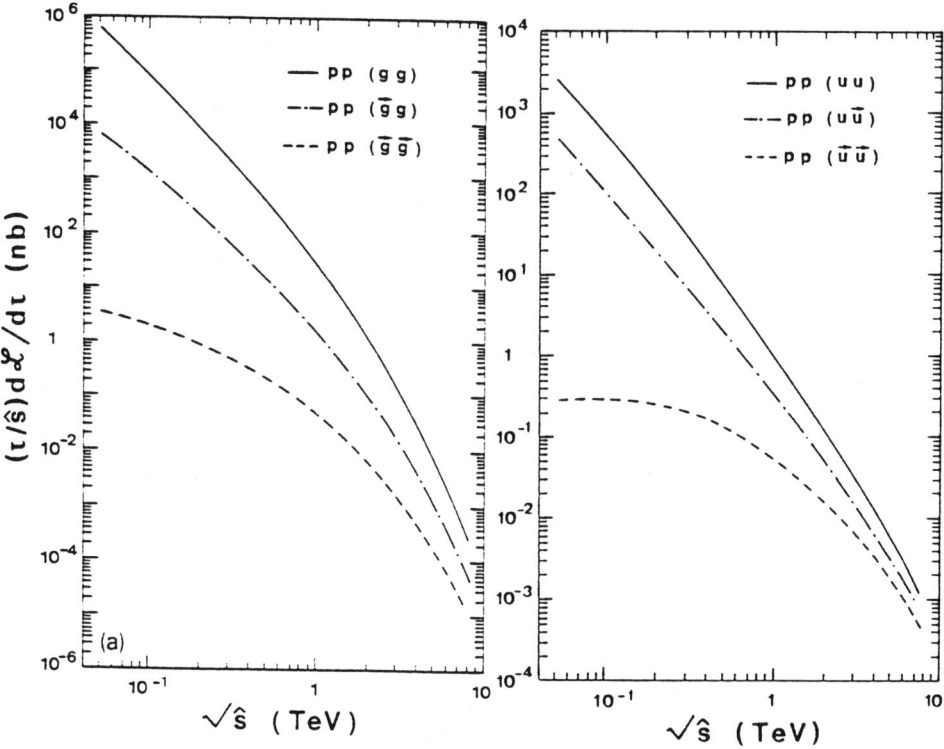

Fig.1 : The quantity $(\tau/\hat{s})d\mathcal{L}/d\tau$ (in nanobarns) versus $\sqrt{\hat{s}}$ for gg interactions in pp collisions at $\sqrt{s} = 40$TeV. Solid curve, unpolarized gluon distributions; dashed-dotted curve, one polarized gluon distribution; dashed curve, two polarized gluon distributions.

Fig.2 : Same as fig.1 for uu interactions.

By making ratios of the (single or double) polarized luminosity to the unpolarized luminosity, one can estimate how much a (single or double) subprocess asymmetry will contribute to the corresponding helicity asymmetry in hadron-hadron collisions due to the dilution effect of the parton distributions. This is more accurate than using the parton polarizations mentioned above (see eq.(5)). As a simple example let us consider a reaction with one beam longitudinally polarized whose cross sections are $d\sigma^{(\pm)}/d\tau$ (see eq.(8)). The single helicity hadron asymmetry defined as

$$A_L = \frac{d\sigma^{(-)}/d\tau - d\sigma^{(+)}/d\tau}{d\sigma^{(-)}/d\tau + d\sigma^{(+)}/d\tau} \qquad (12)$$

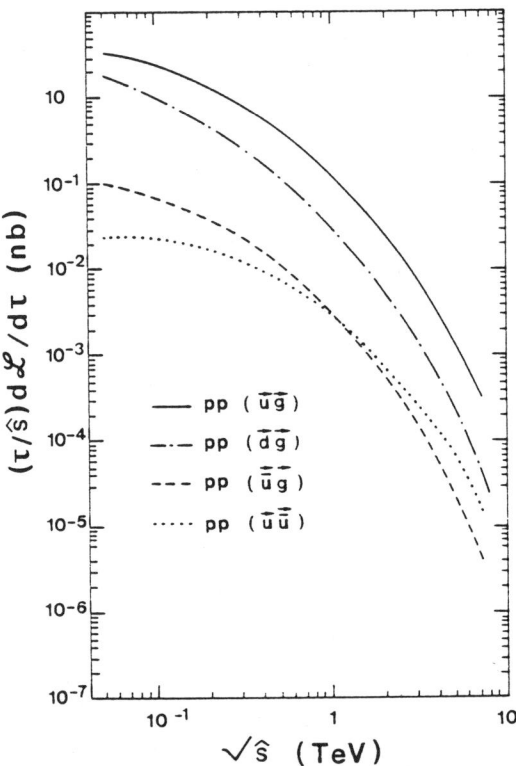

Fig.3 : Doubly polarized luminosities for various parton-parton interactions in pp collisions at $\sqrt{s} = 40$ TeV. Note that the dashed-dotted curve has been changed in sign because the d-quark helicity asymmetry is negative.

can be evaluated by means of these ratios. Let us denote by $\hat{a}_L^{ij}$ (resp. $\hat{a}_L^{ji}$) the subprocess helicity asymmetry where parton i (resp. j) is polarized. If the cross section is dominated by a subprocess induced by the collision of parton i and parton j and if $\hat{a}_L^{ij}$ and $\hat{a}_L^{ji}$ are constant one has

$$A_L \simeq a_i(j)\hat{a}_L^{ij} + a_j(i)\hat{a}_L^{ji} \tag{13}$$

where

$$a_i(j) = \frac{\tau d\mathcal{L}_{\Delta ij}/d\tau}{\tau d\mathcal{L}_{ij}/d\tau} \tag{14}$$

represents the polarization of parton "i" in interaction with the unpolarized parton "j". This quantity is not necessarily related to single helicity hadron asymmetry and its usefulness will become obvious in the specific cases we will study in the following. Some examples of $a_i(j)$ are displayed in Fig.4 for two different energies and we remark that the gluon polarization $a_g(u)$ is rather small since we have used the standard choice for $\Delta G(x, Q^2)$. It would be considerably larger if we had used instead eq.(6).

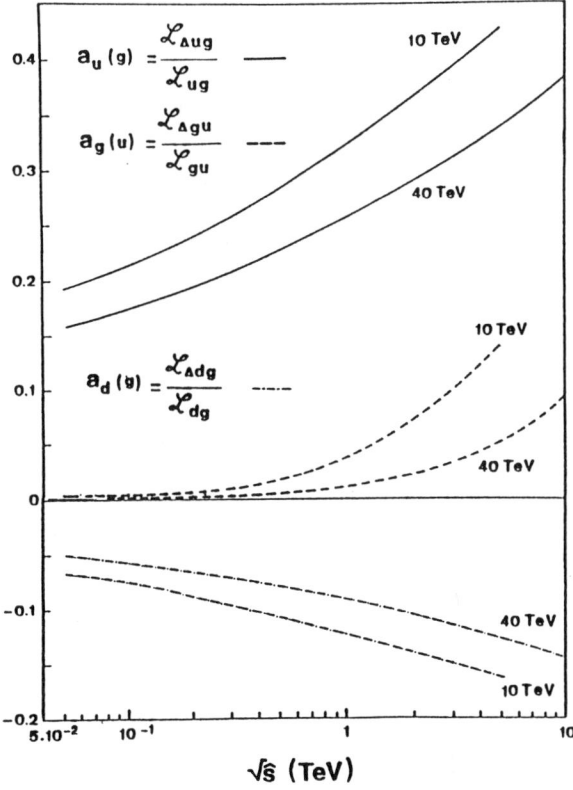

Fig.4 : $a_u(g), a_g(u)$ and $a_d(g)$ versus $\sqrt{\hat{s}}$ for $\sqrt{s} = 10$ and 40TeV.

HADRONIC JET AND DIRECT PHOTON PRODUCTION

One of the major difficulties for testing QCD is the fact that experiments cannot be performed directly with the basic constituents quarks and gluons. Hadron-hadron processes are rather complicated at the level of the constituents because many subprocesses have to be summed over in the evaluation of the cross sections and the double helicity asymmetries as shown by eqs.(2) and (4). These subprocesses lead to different cross sections and double helicity asymmetries and it might be possible, in a given restricted range of the large transverse momentum kinematic region, to know which subprocesses are dominant. This is one way to test QCD and of course the correct knowledge of the parton distributions is also part of this test. At high energies most of the information at short distances is contained in hadronic jet physics which has already exhibited some simple aspects at the CERN $Sp\bar{p}S$ collider[9]. Jet production will be very copious at the SSC and with its expected luminosity, there will be about one event per second yielding at least one jet with transverse momentum larger than $1 \text{TeV}/c$. Therefore event rates will be very high and jets will be the main source of conventional background to new physics, so it is essential we understand as well as possible both their cross sections and their helicity asymmetries. By reaching enormous values of transverse momentum it will be possible to probe very short distances and, for example, to detect some evidence for quark compositeness which leads to a definite signature in single helicity asymmetries resulting from the presence of an axial coupling which violates parity[2]. Here, since we assume that parity is preserved, we will compute only double helicity asymmetries which are expected to be small as explained previously. Although direct photon production has a much lower cross section, it will be also considered because it is simpler and allows a direct probe of a smaller number of specific hard scattering processes.

The cross section for the production of a single jet of transverse momentum p_T is given in BRST and, except for large p_T, it is dominated by gluon-gluon and quark-gluon interactions whose corresponding asymmetries $\hat{a}_{LL}$ are positive with a similar magnitude, 0.6 to 0.7, near $\theta^* = 90°$, where θ^* is the c.m. scattering angle of the subprocess. So we expect A_{LL} to be positive and the result of our calculation for a jet produced at zero rapidity is shown in Fig.5 versus p_T for three different energies. We have used the standard choice for $\Delta G(x, Q^2)$ and, as expected, the effect which grows with p_T, is of the order of a few percent in agreement with a rough estimate based on eq.(5). A_{LL} decreases with increasing energy due to the rapid growth of the single jet cross section with energy, driven by the small-x behavior of the gluon distributions. At RHIC energies A_{LL} will be larger and, moreover if one does the calculation assuming the large ΔG, i.e. eq. (6), one anticipates values around 20% for small values of p_T which are very easy to reach experimentally, as reported in ref.9.

In the physics of hadronic jets one can also consider the production of two jets in the final state i.e. $p + p \to \text{jet}_1 + \text{jet}_2 + X$. If M denotes the jet pair mass, the invariant mass distribution $d\sigma/dM$ will correspond to a rather large cross section. Since the underlying dynamics is the same as for the case of single jet production, the general pattern of A_{LL} corresponding to $d\sigma/dM$ is expected to be very similar as shown in Fig.6.

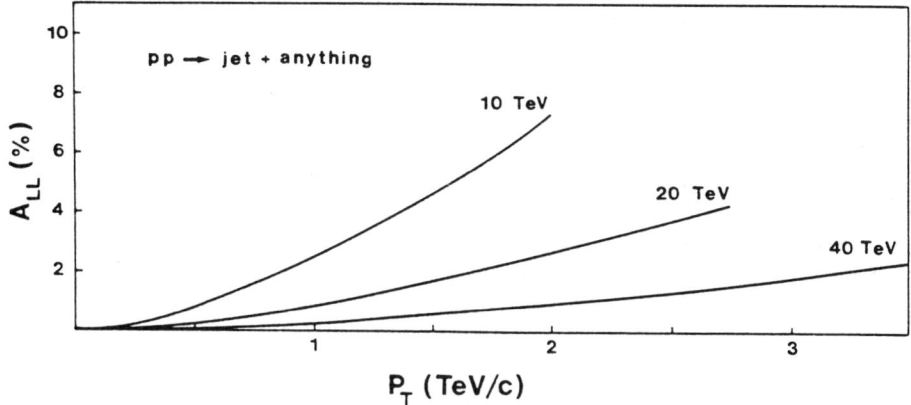

Fig.5 : Predictions for the double helicity hadron asymmetry A_{LL} in $pp \to \text{jet}+X$ at $y = 0$ versus p_T for three different energies.

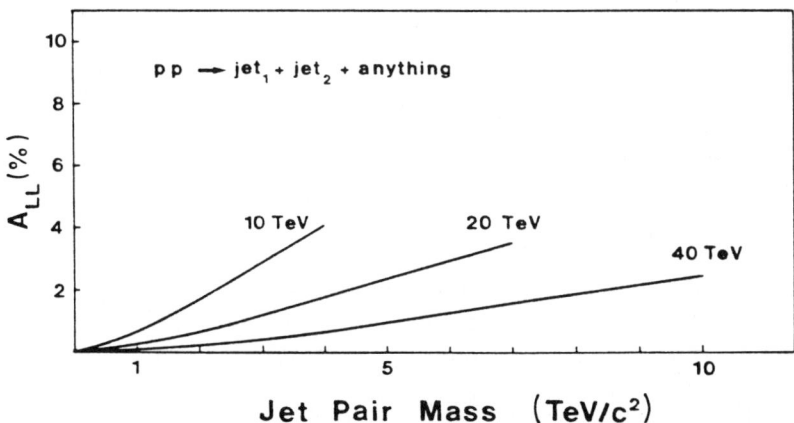

Fig.6 : Predictions for the double helicity hadron asymmetry A_{LL} for dijet events in pp collisions versus the jet-pair mass for three different energies.

Direct photon production at high p_T is also a useful probe of the underlying parton-parton interactions. In the QCD parton model, in the absence of photon bremsstrahlung contributions, direct photons are produced via the quark-antiquark annihilation subprocess $q\bar{q} \to \gamma g$ and the quark-gluon Compton subprocess $qg \to q\gamma$. The Compton subprocess has a positive $\hat{a}_{LL}$ whereas for the annihilation one has $\hat{a}_{LL} = -1$. We have calculated the unpolarized cross section at 90° and for different values of the c.m. production angle, and we found consistency with the results of ref.10. Event rates are very small and for example one has $d^2\sigma/dy dp_T = 4.10^{-40} cm^2/\text{GeV}$ for $p_T = 1\text{TeV}/c$ and $\theta_{cm} = 90°$. As it is well known, away from 90° the Compton subprocess dominates largely over annihilation except at very large p_T. The results we obtained with the standard distribution ΔG for the double helicity asymmetry are shown in fig.7 for $\theta_{cm} = 45°$ and 90°. In both cases A_{LL} is small and positive which reflects the dominance of the Compton subprocess. However u quark-gluon and d quark-gluon lead to opposite sign effects which reduce the magnitude of A_{LL}. We also see that A_{LL} scales with $x_T = 2p_T/\sqrt{s}$ but since the cross section does not, the measurement of A_{LL} at fixed x_T will be easier to measure for $\sqrt{s} = 10\text{TeV}$ than for $\sqrt{s} = 40\text{TeV}$. At 45° A_{LL} rises with x_T, whereas it is flatter at 90°. This is due to the fact that at 90° the positive Compton asymmetry is reduced by the contribution of the negative annihilation asymmetry which becomes more and more important for increasing x_T.

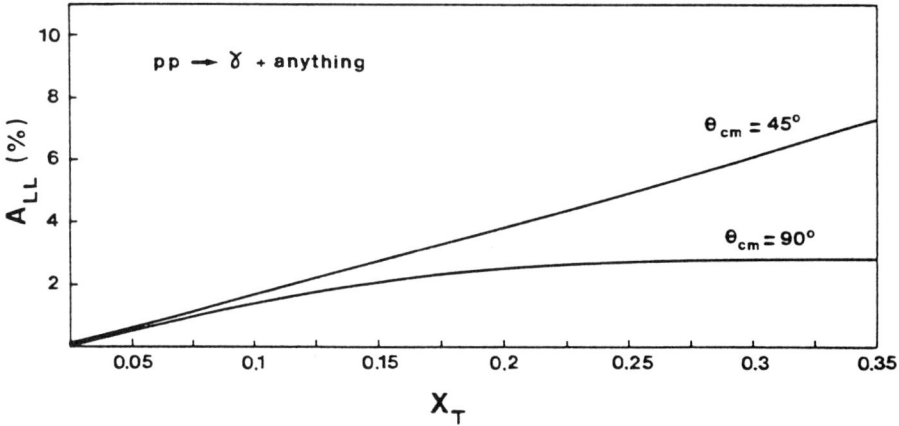

Fig.7 : Predictions for the double helicity hadron asymmetry A_{LL} in $pp \to \gamma + X$ versus x_T, at very high energies ($\sqrt{s} \geq 10\text{TeV}$), for two different values of the c.m. production angle $\theta_{cm} = 45°$ and 90°.

Given the high designed luminosity at RHIC[11], the cross sections at 90° and 45° will be measurable up to fairly high p_T values as shown in fig.8. We also displayed our predictions for A_{LL} in fig.9 and we see once more that the standard distribution ΔG leads to an effect of the order of a few percent. However by using the large ΔG one observes that A_{LL} rises with p_T, reaching values of the order of 20% or more for $\theta_{cm} = 90°$ and much larger for $\theta_{cm} = 45°$. So this shows that the p_T dependence of A_{LL} in direct photon production is very sensitive to the gluon polarization and that it should be best measured with photons of rapidity of the order of 1.

All these double helicity asymmetry A_{LL} in jet and direct photon production would be essentially zero if gluons were not polarized.

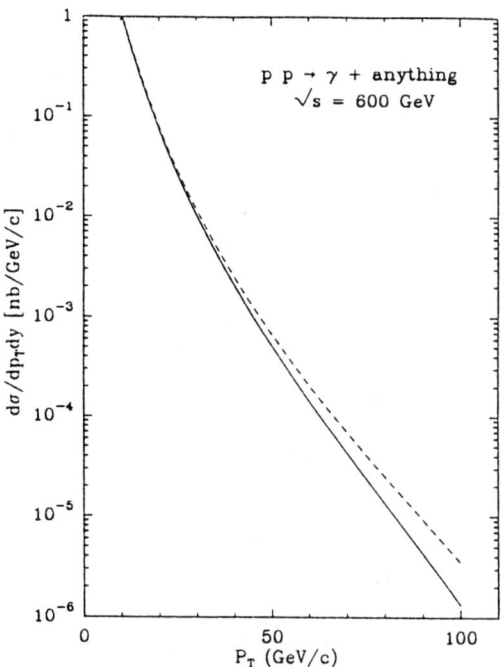

Fig.8 : Predicted unpolarized cross sections for $\sqrt{s} = 600$ GeV as a function of p_T, for two different values of the c.m. production angle : $\theta_{cm} = 90°$ (dashed curve), $\theta_{cm} = 45°$ (solid curve).

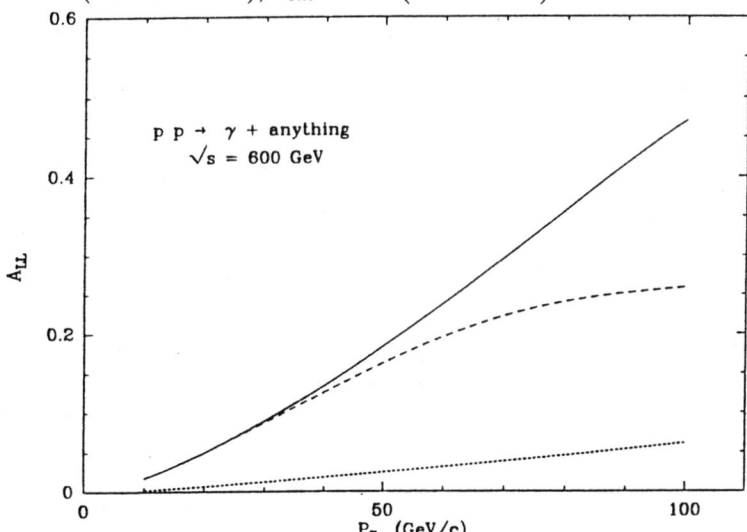

Fig.9 : Predicted A_{LL} with a large ΔG for $\sqrt{s} = 600$ GeV as a function of p_T, for two different values of the c.m. production angle : $\theta_{cm} = 90°$ (dashed curve), $\theta_{cm} = 45°$ (solid curve). The small-dashed curve corresponds to a standard ΔG.

STANDARD ELECTROWEAK INTERACTIONS

At SSC with $\mathcal{L} = 10^{33} \text{cm}^{-2}.\text{sec}^{-1}.$, one will be able to produce every second about $60 Z°$'s and $200 W^\pm$'s, so these impressive production rates will be also of great interest because the standard production mechanism being known, they will allow a *precise calibration* of the parton distributions. In addition, from gauge boson pair production it will be possible to study the influence of the three gauge boson couplings which is a characteristic of non abelian gauge theories. For example, approximately one thousand W^+W^- pairs per day will be produced at SSC and this rate, which is a direct consequence of the damping effect of the $Z°$ contribution, is an obvious test to perform.

The use of polarized beams to study reactions within the Standard Model will also allow a good calibration of the polarized parton distributions. It is expected to reveal large single helicity asymmetries, some of which are simply expressed in terms of parton polarizations $a_i(j)$ defined in eq.(14). For illustration let us consider single $W^\pm$ production which can be computed directly in the Drell-Yan picture in terms of the dominant quark-antiquark fusion reactions $u\bar{d} \to W^+$ and $\bar{u}d \to W^-$. After integration over the rapidity y of $W^\pm$, one can show (see BRST) that A_L reads

$$A_L \sim a_u(\bar{d}) - a_{\bar{d}}(u) \quad \text{at} \quad \sqrt{\hat{s}} = M_W \tag{15}$$

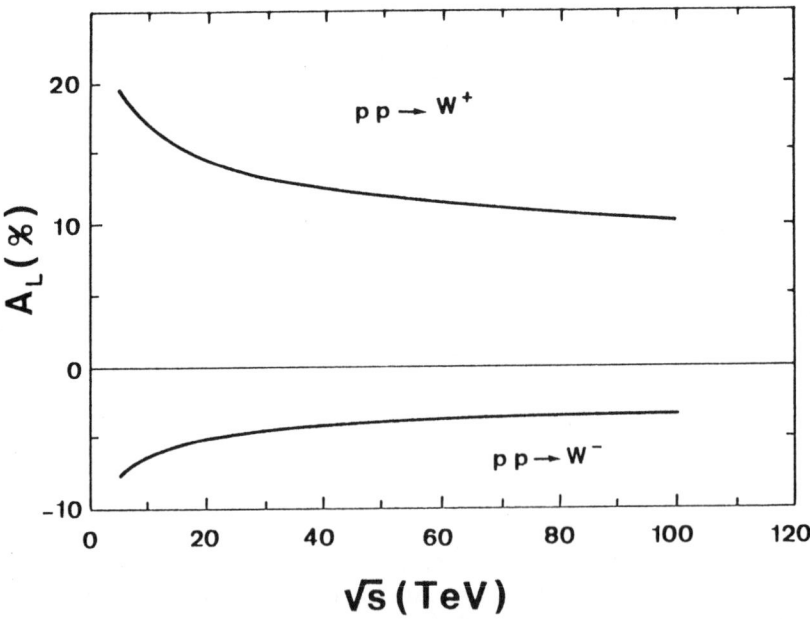

Fig.10 : Integrated A_L in $pp \to W^+$ and W^- versus $\sqrt{s}$.

for W^+ and a similar expression holds for W^- by permuting u and d. The resulting A_L are shown in Fig.10 and they decrease very slowly with increasing energy. The asymmetries in the production of a $W^\pm$ with a large p_T are even larger and grow with p_T. We show in Fig.11 some of our predictions at RHIC energies. For W^+ the trend is that of a_u whereas for W^- it is that of a_d. The production of W^+W^- pairs is also very interesting because it involves the WWZ coupling g_z. For a large invariant mass M of the pair compared to M_W, we have a simple expression for A_L

$$A_L \sim \frac{2}{3}a_u(\bar{u}) + \frac{1}{3}a_d(\bar{d}) - a_{\bar{q}}(q) \tag{16}$$

if g_z has the standard value $g_z = cotg\theta_W$. It is shown in Fig.12 together with the resulting A_L in the absence of the trilinear coupling i.e., $g_z = 0$. Various standard electroweak processes have been studied in BRST where one can find more details, in particular, A_L for other pair production of gauge bosons, which can be simply described in terms of the useful concept of the universal curves.

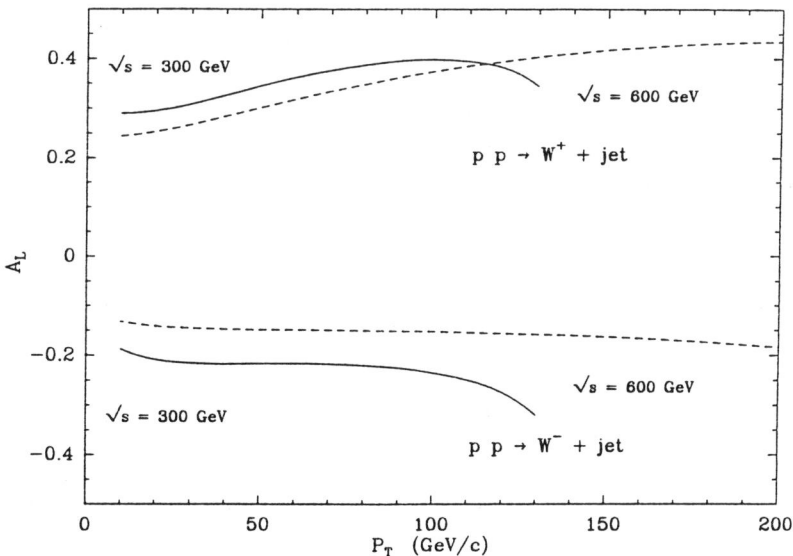

Fig.11 : Predicted A_L in $pp \to W^\pm+$ jet at $y = 0$ versus p_T for $\sqrt{s} = 300$GeV (solid curves) and $\sqrt{s} = 600$GeV (dashed curves).

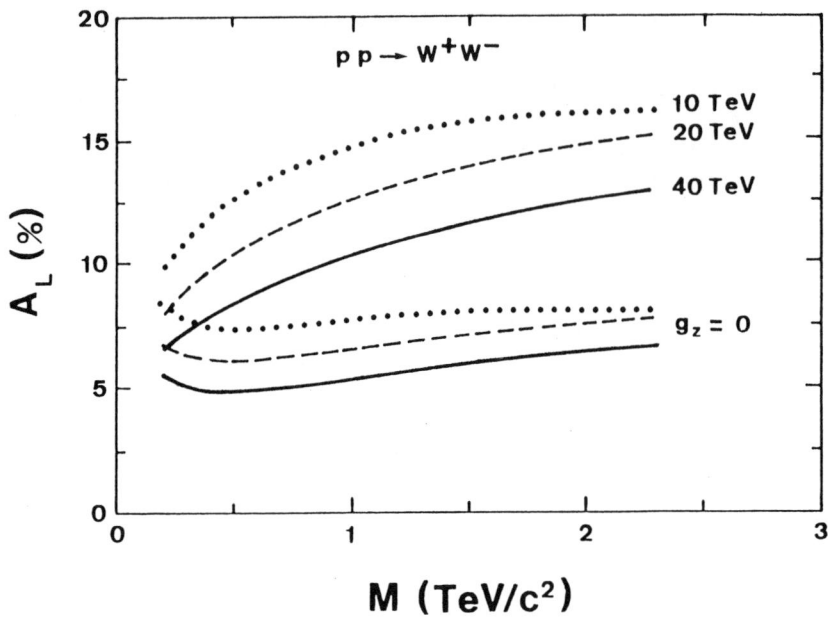

Fig.12 : A_L in $pp \to W^+W^-$ versus M at three different energies. Some with $g_z = 0$.

ACKNOWLEDGEMENTS

I would like to thank all the people who made this conference possible, in particular the local organizers and the Pennsylvania State University for its generous hospitality.

REFERENCES

1. C. Bourrely, J. Soffer, F.M. Renard and P. Taxil, Physics Reports, **117**, 319 (1989) and Erratum to preprint CPT-87/P.2056 (december 1990) (referred below as BRST).
2. P. Taxil, these proceedings, preprint CPT-91/2503 (january 1991).
3. C. Bourrely, J.Ph. Guillet and J. Soffer, preprint CPT-90/P.2474 (december 1990).
4. G. Altarelli and G. Parisi, Nucl. Phys. **126**, 298 (1977).
5. E. Eichten, I. Hinchliffe, K. Lane and C. Quigg, Reviews of Modern Physics **56**, 579 (1984) and Erratum **58**, 1065 (1986) (referred below as EHLQ).
6. EMC Collaboration, J. Ashman et al., Nucl. Phys. **B328**, 1 (1989).
7. A. Efremov, these proceedings.
8. E. Berger and J. Qiu, Phys. Rev. **D40**, 778 (1989).
9. J.Ph. Guillet, these proceedings.
10. J.F. Owens, T. Ferbel, M. Dine and I. Bars, Proceedings of the 1984 Summer Study on the Design and Utilization of SSC (Snowmass, Colorado), editors R. Donaldson and J. Morfin, p. 218.
11. M. Tannenbaum, these proceedings.

THEORETICAL INTERPRETATION OF POLARIZED LEPTOPRODUCTION DATA

Anatoli V. Efremov
Joint Institute of Nuclear Research, Dubna
Head Post Office, P.O. Box 79, 101000 Moscow, USSR

ABSTRACT

We examine the role of quarks and gluons in polarized leptoproduction in connection with the famous EMC experimental result. The gluon contribution generated by the axial anomaly is shown to have a well defined physical meaning and its gauge invariance properties are fully clarified. We also emphasize the role of the ghost contribution in the generalized Goldberger-Treiman relation.

INTRODUCTION

The naive parton interpretation of the famous EMC-result[1] that the total contribution of the quarks to the proton spin $\Delta\Sigma = \sum_f \int dx (q_f^+(x) - q_f^-(x) + \bar{q}_f^+(x) - \bar{q}_f^-(x))$ is compatible with zero is not only in contradiction with the naive quark model but also with whole of our understanding of baryon spectroscopy. As emphasized by Lipkin[2] using the Wigner-Ekhardt theorem, the total spin of the sea quarks and gluons plus orbital momentum has to be very large (> 6 !) in order to obtain this small value. It is then difficult to understand why the nucleon is the only stable state with $J = I = 1/2$.

The EMC-result has produced a stream of theoretical papers[3] with a broad spectrum of ideas from doubts in the experiment to questioning the applicability of perturbative QCD. In our opinion, however, this "Spin crisis" can be turned into a spectacular success of QCD. It has been demonstrated[4] that the factorization theorem in QCD leads to the expression, for the first moment of the flavour singlet part of the structure function $g_1^p(x)$,

$$\Gamma_1^{singl.} = \frac{<e^2>}{2}\left(\Delta\Sigma - \frac{\alpha_s}{2\pi} N_f \Delta g\right), \qquad (1)$$

where the second term is due to a short-range interaction of photons with polarized gluons via the quark box diagram (see Fig.1). For the first moment it reduces to the contribution of the triangle axial anomaly. It need not be a small correction in spite of α_s because $\Delta g \sim \alpha_s^{-1}$ due to evolution equations[5] : a zero mass quark emitting a polarized gluon conserves its helicity and again emits gluons of the same polarization. The more gluons are emitted (the smaller x), the higher the gluon polarization (of course, this increase is always compensated by an orbital momentum[6]). So, the EMC-effect could result from a compensation between $\Delta\Sigma$ and $\widetilde{\Delta g} = \frac{\alpha_s}{2\pi} N_f \Delta g$.

This approach has been criticized recently in the literature[7-10]

i) because of the nonlocality (k^2-dependence) of the box diagram together with a dependence of this contribution on the regularization procedure[8,9] and,

ii) for the absence of a local gauge invariant gluon spin operator.

These critisms were partly answered in refs.11-13 which we here want to elaborate upon and to present more clearly. This is done in the next section where we also show that the commonly used definitions of $\Delta\Sigma$ and Δg through the quark and gluon axial currents J_ν^5 and K_ν are in fact gauge invariant although they have to be more accurately defined, due to a nonperturbative ghost contribution. The generalized Goldberger-Treiman (GGT) relation is also discussed here. In the last section we give a general discussion on the interpretation of the EMC-result and our concluding remarks.

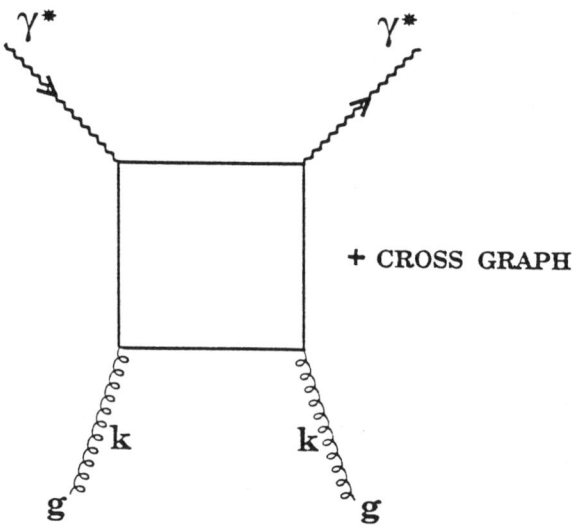

Fig.1 The photon-gluon scattering graphs.

THE AXIAL ANOMALY CONTRIBUTION
AND THE QUARK AND GLUON CONTENT OF THE PROTON SPIN

Consider first the ambiguity of the box contribution. In the QCD-modified parton picture there is no difficulty and no ambiguity in defining the first moment of the gluon spin distribution function (except the usual ambiguity in renormalization scheme in higher orders of α_s). This picture is based on the factorization theorem which has the same theoretical foundation as the OPE (i.e. proved in any order of perturbation theory) but many more applications. The object which generates the gluon distribution function there (polarized and unpolarized) is not a product of gauge invariant operators but rather an ultraviolet regularized matrix element of a product of gluon fields $<p|A_\rho^a(0)A_\sigma^a(\xi)|p>_{Reg}$. convoluted with a contribution from a short range, infrared regularized subprocess :

$$\int \sigma_{\rho\sigma}(\xi,q,\mu^2) <p|A_\rho^a(0)A_\sigma^a(\xi)|p>_{Reg.} d^4\xi .$$

The gluon distribution functions are defined through the Taylor series expansion of this matrix element in the limit $\xi \to 0$ and its gauge invariance is guaranteed by the gauge invariance of the subprocess which is on the mass shell, in the leading twist approximation. So the matrix element of the gluon axial current $K_\nu = \varepsilon_{\nu\mu\sigma\rho} A_\mu^a F_{\sigma\rho}^a$ (or more precisely its projection onto the gauge vector n in the axial gauge $A_\nu^a n_\nu = 0$) will appear as the first moment of the gluon spin distribution function[11]. It is gauge invariant in PQCD since it contains only the transversal component with respect to n of the gluon field A. However, as we will see, beyond PQCD, it requires further specification.

A real ambiguity exists, however, in defining the first moment of the quark distribution. Actually, the first moment of the box diagram contribution (see Fig.1) off the mass shell ($k^2 \neq 0$ and $m_q \neq 0$) has the form[9] :

$$\Gamma_1^{Box} = -\frac{\alpha_s}{2\pi} N_f \left[1 + \frac{2m_q^2/k^2}{\sqrt{1+4m_q^2/k^2}} \ln\left(\frac{\sqrt{1+4m_q^2/k^2}-1}{\sqrt{1+4m_q^2/k^2}+1}\right) \right] . \quad (2)$$

There are two gauge-invariant local (k^2-independent) limits of this quantity, which can be considered as contributions of parton subprocesses,

$$\Gamma_1^{Box} \to \begin{cases} -\frac{\alpha_s}{2\pi} N_f &, m_q^2/k^2 \to 0 \\ 0 &, k^2/m_q^2 \to 0 \end{cases} .$$

They correspond to the cross section of the $\gamma^* g \to q\bar{q}$ subprocess for either $m_q = 0$ or $m_q \neq 0$ on the mass shell $k^2 \to 0$, giving two different definitions of the quark contribution $\Delta\Sigma$ and different evolution equations. In the first case, the anomaly part (first term in eq.(2)) is subtracted from the quark distribution function and included into the subprocess. It gives eq.(1) for $\Gamma_1^{singl.}$ and the evolution equations

$$\Delta\dot{\Sigma} = 0 \quad \text{and} \quad \Delta\dot{\Sigma} - \widetilde{\Delta g} = \gamma(\Delta\Sigma - \widetilde{\Delta g}), \quad (\dot{} \equiv \frac{d}{d\ln Q^2}) \quad (3)$$

where γ is an anomalous dimension. In the second case, the whole contribution in eq.(2) is in the quark distribution function, i.e. $\Gamma_1^{singl.} \sim \Delta\Sigma' = \Delta\Sigma - \widetilde{\Delta g}$ and the evolution equations become : $\Delta\dot{\Sigma}' = \gamma\Delta\Sigma'$ and $\Delta\dot{\Sigma}' + \widetilde{\Delta g} = 0$.

Which of these definitions is better? We believe that for light quarks the first one is better because

i) When the mass m_q is small the cancellation of the anomaly by the mass term in eq.(2) occurs only in a small part $\left(|k^2|\ll m_q^2\right)$ of the whole integration region $|k^2|\le Q^2$. However in this small part, other effects (in particular nonperturbative) are expected to be more important than the cancellation.

ii) Equations (3) are just the analytic continuation of the evolution equation for higher moments $m\ge 2$. This is a necessary condition for the existence of soft x-dependent distribution functions $\Delta\Sigma(x)$ and $\Delta g(x)$ which are measured experimentally.

iii) Since $\Delta\Sigma$ is independent of Q^2 one has a closer connection between low- (naive quark model) and high-energy (parton) pictures of the proton

iv) Due to a possible cancellation of the two terms in eq.(1), the contradiction of the EMC-result and the quark model need not be so severe.

Now turn to the gauge invariance problem. There is no such problem for higher-moments $m\ge 2$ because there are two towers of local gauge-invariant operators which are built from quark and gluon fields, q and A_ν. So, being defined as analytic continuation of moments of x-dependent spin distribution functions to $m=1$, $\Delta\Sigma$ and $\widetilde{\Delta g}$ also have to be gauge invariant. The formal continuation of the local operators gives however, as was discussed earlier,

$$\Delta\Sigma = \frac{1}{2}<p\,|\,J_\nu^5 - \widetilde{K}_\nu\,|\,p>n_\nu \quad \text{and} \quad \widetilde{\Delta g} = -\frac{1}{2}<p\,|\,\widetilde{K}_\nu\,|\,p>n_\nu \quad (4)$$

where n_ν is a light-like vector satisfying $n^2=0, np=1$ and where $J_\nu^5 = \Sigma_f \bar{q}_f \gamma_\nu \gamma_5 q_f$, $\widetilde{K}_\nu = N_f \frac{\alpha_s}{2\pi} K_\nu$ are proportional to the quark and gluon spin operators (strictly speaking, K_ν is not the gluon spin, nevertheless, $K_\nu n_\nu$ is its projection). The subtraction of $\widetilde{K}_\nu$ in the definition of $\Delta\Sigma$ in eq.(4) is done in accordance with eq.(3) to make $\Delta\Sigma$ independent of Q^2 as a consequence of the Adler-Bardeen relation

$$\partial_\nu J_\nu^5 = \partial_\nu \widetilde{K}_\nu = N_f \frac{\alpha_s}{2\pi} F_{\rho\sigma}^a \tilde{F}_{\rho\sigma}^a \,. \quad (5)$$

The problem with gauge invariance arises since the operator K_ν in eq.(4) is apparently gauge variant at least with respect to large gauge transformations. Of course in the projection $K_\nu n_\nu$ remains only the variance on the large (homotopically nontrivial) gauge transformations which are absent in perturbation theory. However, beyond perturbation theory eq.(4) can be wrong due to nonperturbative contributions. So the resolution of the problem lies in a term nonanalytic in m in the matrix elements of eq.(4) due to a nonperturbative contribution[14]. The most important part of this contribution is tightly connected with the eminent U(1)-problem in QCD[15,16].

Actually, in any covariant gauge

$$< p' \mid J^5_\nu \mid p > = \bar{u}(p') \left[\gamma_\nu \gamma_5 G_1(q^2) + q_\nu \gamma_5 G_2(q^2)\right] u(p)$$
$$< p' \mid \widetilde{K}_\nu \mid p > = \bar{u}(p') \left[\gamma_\nu \gamma_5 \widetilde{G}_1(q^2) + q_\nu \gamma_5 \widetilde{G}_2(q^2)\right] u(p) \tag{6}$$

The same expressions can be written also in the form

$$< p' \mid J^5_\nu \mid p > = 2M_N s_\nu G_1(q^2) + q_\nu(sq) G_2(q^2)$$
$$< p' \mid \widetilde{K}_\nu \mid p > = 2M_N s_\nu \widetilde{G}_1(q^2) + q_\nu(sq) \widetilde{G}_2(q^2) \tag{6'}$$

where M_N is the proton mass, s_ν its spin 4-vector, and the contributions parallel to it are, in terms of quark and gluon distributions,

$$G_1(0) = \Delta\Sigma - \widetilde{\Delta g} \quad \text{and} \quad \widetilde{G}_1(0) = -\widetilde{\Delta g} \ .$$

Due to the gauge invariance of $\partial_\nu \widetilde{K}_\nu$ and the absence of a zero mass axial pole state in PQCD both $q^2 G_2$, $q^2 \widetilde{G}_2 \to 0$ when $q^2 \to 0$. So both $G_1(0)$ and $\widetilde{G}_1(0)$ are gauge invariant if one disregards nonperturbative effects. However, this is not correct in general. The current K_ν has to couple with a zero mass ghost pole[17] whose mixing with Nambu-Goldstone η'_0 supplies it with an additional mass $\Delta m^2_{\eta'} = \lambda^4/f^2_{\eta'}$, where λ^2 is the $\widetilde{K}_\nu$-ghost coupling and $f_{\eta'}$ is the η'-decay parameter $< 0|J^5_\nu|\eta' > = f_{\eta'} q_\nu$. This pole contributes to $< p' \mid \widetilde{K}_\nu \mid p >$ through the diagrams shown in Fig.2 and gives

$$q_\mu \widetilde{G}_2(q^2) = \frac{q_\mu \lambda^2}{q^2} \sqrt{N_f} \left(\frac{\Delta m_{\eta'} g_{\eta' NN}}{m^2_{\eta'} - q^2} - g_{QNN}\right) \tag{7}$$

where g_{QNN} is the ghost nucleon coupling constant $(\partial_\nu \widetilde{K}_\nu \equiv Q)$. Then

$$\lim_{q^2 \to 0} q^2 \widetilde{G}_2(q^2) = G = \sqrt{N_f} f_{\eta'} \left(\frac{\Delta m^2_{\eta'}}{m^2_{\eta'}} g_{\eta' NN} - \Delta m_{\eta'} g_{QNN}\right) . \tag{8}$$

This ghost contribution is nonanalytic in m (the moment number) and nonperturbative in nature. Its physical meaning is in a periodic dependence of the QCD potential $(\sim Tr(H^2))$ on a collective variable $X = \int d^3x K_0(x,t)$, which changes by an winding number under a homotopically nontrivial gauge transformation[18], $X \to X + n$. The reason of the pole is the same as that of a gapless excitation in an ideal conductor.

An important comment is now in order. Notice that the ghost pole can contribute only to the formfactor $\widetilde{G}_2$ and does not contribute to $\widetilde{G}_1$. Otherwise we would have an unphysical pole in the gauge invariant combination $2M_N \widetilde{G}_1 + q^2 \widetilde{G}_2$. Due to this, the effective ghost nucleon vertex (and η'-ghost vertex) has to contain a derivative, $\bar{N}\partial_\nu \gamma_5 N G_\nu$ where G_ν is the ghost field. Therefore the dependence on the gauge parameter drops out from the ghost propagator (see Fig.2) $(g_{\mu\nu} + aq_\mu q_\nu/q^2)(-q^2(a+1))^{-1}$, where a is a gauge parameter, and similarly from the $\widetilde{K}_\nu G_\nu$-vertex. That is the reason why the expression (7) is in fact gauge invariant. This means however that $\widetilde{G}_1$ has to be gauge invariant also. So, the whole matrix element of K_ν, for symmetric states where $q^2 = 0$, has to be gauge invariant.

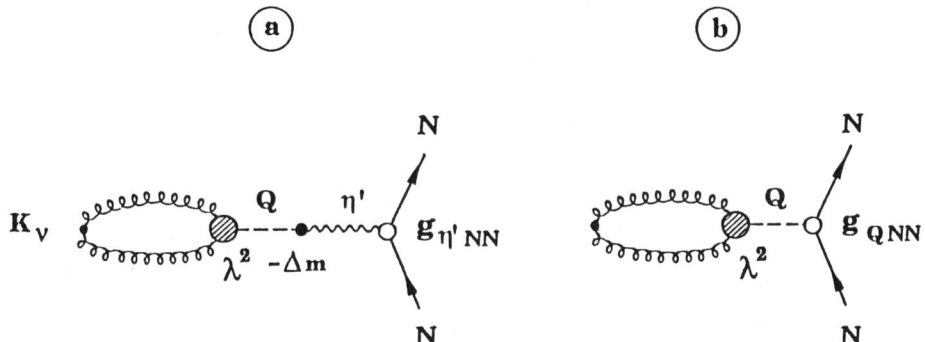

Fig.2 The two different ghost contributions to $\bar{N}N$
a) coupling through the physical η' b) direct coupling.

Turn now to the renormalization properties of the $G's$. The quark current J_ν^5 is known to be renormalized multiplicatively with an anomalous dimension γ, i.e. $\dot{J}_\nu^5 = \gamma J_\nu^5$ whereas $J_\nu^5 - \widetilde{K}_\nu$ is not renormalized, therefore $\dot{\widetilde{K}}_\nu = \gamma J_\nu^5$. Using eq.(6) one finds

$$\dot{G}_{1,2} = \gamma G_{1,2} \quad \text{and} \quad \dot{\widetilde{G}}_{1,2} = \gamma G_{1,2} .$$

The Adler-Bardeen relation eq.(5) leads to the equality

$$2M_N(G_1(q^2) - \widetilde{G}_1(q^2)) = q^2(\widetilde{G}_2(q^2) - G_2(q^2)) ,$$

both sides of which are now seen to be Q^2-independent. In the limit $q^2 \to 0$ it gives[16]

$$G_1(0) - \widetilde{G}_1(0) = \Delta\Sigma = \frac{G}{2M_N} = \frac{\sqrt{N_f} f_{\eta'} g_{\eta_0' NN}}{2M_N} , \qquad (9)$$

where $g_{\eta_0' NN} = g_{\eta' NN} - \Delta m_{\eta'} g_{QNN}$ is the η_0'-nucleon interaction coupling constant. This we believe to be the correct form of the GGT-relation. The difference compared to ref.15 lies in $\widetilde{G}_1$ in the l.h.s. of eq.(9). The source of this difference is the inclusion in the same one particle irreducible (1PI) vertex Γ_{QNN} in ref.15 of two terms. The first term is a truely 1PI multimeson contribution from the continuum spectrum, which we call $\widetilde{G}_1 = -\widetilde{\Delta g}$ and the second term, which is reducible with respect to the ghost, is the contribution containing the direct ghost nucleon interaction (see Fig.2b). They have different kinematic structures in $< p'|\widetilde{K}_\nu|p >$ but the same structure in $< p'|\partial_\nu \widetilde{K}_\nu|p >$. So the term $\widetilde{G}_1$, in ref.15 is in the r.h.s. of eq.(9) and, in fact, included in $g_{\eta'_0 NN}$ whose definition is different from ours. In our opinion these two terms are different in nature and this difference, can be easily seen in comparing QCD with QED where we have the axial anomaly but no ghost pole due to the Abelian character of the gauge group. Indeed, in QED $q^2 \widetilde{G}_2 \to 0$ in the symmetric limit and if one would forget about the first term $\widetilde{G}_1$, one would obtain $< e|\partial_\mu K_\mu|e >= 0$, which is certainly not correct. So, in our opinion the $\widetilde{G}_1$ term has to be taken into account separately from the ghost contribution.

In conclusion, we must stress once more that quark and gluon spin first moments defined through a hard subprocess in the QCD-improved parton model are not exactly represented by matrix elements of the currents $J_\nu^5 - \widetilde{K}_\nu$ and $\widetilde{K}_\nu$. The distinction is due to the nonperturbative ghost contribution. The same contribution determines the total quark spin fraction via the GGT-relation (9).

DISCUSSION AND CONCLUDING REMARKS

We have tried to elucidate that our choice of including the axial anomaly into the parton subprocess rather than into the quark spin distribution function is more of a physical than a mathematical problem. The main physical argument is that the anomaly is connected with the spin asymmetry for $q\bar{q}$ production in polarized photon-gluon scattering when the gluon is on the mass shell and the mass of the quarks is neglected. It is natural that the contribution of such a process is proportional to the difference $\Delta g(x)$ of the number of gluons polarized in opposite directions, with respect to the proton spin.

We give a clear understanding of how Δg is related to the matrix element of the gauge variant axial gluon current $\widetilde{K}_\nu$ in QCD. These quantities are proportional to each other in perturbative QCD. However beyond perturbative theory $< \widetilde{K}_\nu >$ contains an additional contribution from the zero mass ghost pole relevant to the resolution of the U(1)-problem in QCD. It is precisely this pole which helps to resolve the contradiction between the QCD modified parton model and the OPE. One important result is the proof that the forward matrix element of $\widetilde{K}_\nu$ is, in fact, gauge invariant. Thus, the main objection against the picture where gluons contribute to the EMC-result, i.e. that the separation of the quark and gluon contributions is gauge dependent, is no longer valid.

The same ghost contribution plays also the main role in obtaining the generalization of the Goldberger-Treiman relation to the flavour singlet channel. A detailed calculation of the corrections from finite quark masses and from $\eta' - \eta - \pi^\circ$ mixing has been discussed by N. Törnqvist[19]. Although both corrections are large, they cancel each other to first order in the expression for $\Delta\Sigma$ and only a modest correction (~ 0.2) remains. As a side product of this analysis, one also finds new simple algebraic expressions for the $\pi^\circ - \eta$ and $\pi^\circ - \eta'$ mixing angles which include the effect of the anomaly pole contribution to the singlet mass.

Concerning the evaluation of $\Delta\Sigma$ by means of the GGT relation (see eq.(9)), unfortunately neither $g_{\eta'NN}$ nor g_{QNN} are well enough known experimentally. There are two sources of information on the value of $g_{\eta'NN}$. The first comes from NN scattering phase shift analysis within the framework of the OBEP model[20]. It gives $g_{\eta'NN} \simeq 7.3$ with presumably a large error. The second comes from $\eta' \to 2\gamma$ decay calculated through the baryon triangle loop contribution[21] and it gives $g_{\eta'NN} = 6.3 \pm 0.4$. Both estimates are close to the $SU6$ value[22] $g_{\eta'NN} = 6.5$. However in the OBEP analysis of NN scattering, no ghost pole exchange was taken into account. This contribution would generate a contact NN interaction and for small squared momentum transfer $|t| \ll m_{\eta'}^2$, it would change the $g_{\eta'NN}$ obtained to $\sqrt{g_{\eta'NN}^2 - m_{\eta'}^2 g_{QNN}^2}$. If the second term is large the two methods would give two fairly different values. So it seems reasonable to assume that $m_{\eta'}g_{QNN} \ll g_{\eta'NN} \simeq 6.3$ (the sign is unknown and assumed to be positive). Using this, together with $f_{\eta'} = f_\pi = 132$ MeV, one obtains from eq.(9)

$$\Delta\Sigma = 0.77 \pm 0.06 \ . \tag{10}$$

Of course, we are aware of the fact that both sources are uncertain and therefore this value of $\Delta\Sigma$ is questionable. In order to obtain some new information on both couplings constants, new experiments are necessary. One of the experiments would be the energy behaviour of $\Delta\sigma_L = \sigma_{tot}(+-) - \sigma_{tot}(++)$, the total cross sections difference where the $+$ and $-$ refer to helicities of the incoming particles, in proton-deuteron (or deuteron-deuteron) collisions. The ghost exchange will correspond to a $j = 0$ fixed singularity which is not reggeized, in contrast to η', η exchanges. Assuming the Pomeron-nucleon vertex does not contain a spin-flip part and since ρ- and π-trajectories do not contribute for an isoscalar target (and/or beam), the main contribution comes from the ghost-η- and η'-exchanges. Due to the fact that the intercepts of the η and η' trajectories are below $j = 0$, the ghost exchange must dominate and we would expect a non-Regge behaviour of $\Delta\sigma_L$ if g_{QNN} is large enough. However a significant energy interval is necessary to check this.

The same ghost contribution to $< \widetilde{K}_\nu >$ has also helped us for understanding[23] the true meaning of the result of J. Mandula on a lattice simulation for this matrix element[24]. The smallness of the number he obtained implies, in fact, not necessarily that the gluon contribution is small but rather that it is cancelled by the ghost contribution. In fact two different matrix elements involving $\widetilde{K}_\nu$ were computed[25], namely $< s', \vec{o}|\widetilde{\vec{K}}\vec{s}|\vec{o}, s >$ and $\lim_{p \to 0} \frac{(s.p)}{p^2} < s', \vec{o}|\partial_\nu \widetilde{K}_\nu|\vec{p}, s >$ which lead to the following bounds

$$|\Delta\tilde{g} - \frac{1}{3}G/2M_N| < \frac{1}{20} \quad \text{and} \quad |\Delta\tilde{g} - G/2M_N| < \frac{5}{20} \tag{11}$$

respectively.

If one takes these results at face value, firstly it is clear that the ghost contribution G is not negligible compared to $\Delta \tilde{g}$ and secondly, due to the Adler-Bardeen relation, i.e. by using eq.(9), the second bound reads

$$|\Delta \Sigma - \Delta \tilde{g}| < 0.25$$

which is in agreement with the EMC experimental result. However by combining the two bounds of eq.(11) one gets

$$|\Delta \Sigma| < 0.45 \quad \text{and} \quad |\Delta \tilde{g}| < 0.20 . \tag{11'}$$

This means that both $\Delta \Sigma$ and $\Delta \tilde{g}$ are small, $\Delta \Sigma$ being less than our previous estimate eq.(10). Nevertheless it is not clear at all that the lattice size was large enough to take into account long range instantons which build the contribution to G. So more accurate computations are certainly required in order to get a more truthful conclusion.

Finally a limitation on the value of $\Delta \tilde{g}$ also follows from the x-behaviour of the structure function $g_1^p(x)$. If one tries to use a reliable parametrization of $\Delta g(x)$ (constrained by the positivity condition $|\Delta g(x)| < g(x)$) convoluted with $\sigma(\gamma^* g \to q\bar{q})$ to fit the x-dependence of the EMC data, one finds that either the gluon contribution lies mainly in a region below the existing data ($x < 0.01$) [9,26] or one needs some contribution from the strange quark[13]. However this last possibility of a substantial Δs seems to be excluded from a very recent analysis[27] of the CCFR data which has measured accurately the strange quark content of the nucleon.

ACKNOWLEDGMENTS

I would like to thank my collaborators J. Soffer and N.A. Törnqvist for useful discussions. I also thank the organizing committee for providing generous financial support.

REFERENCES

1. J. Ashman et. al. (EMC) Nucl.Phys. **B328**, 1 (1989).
2. H. Lipkin, Phys. Lett. **B237**, 130 (1990).
3. H. Rollnik, Ideas and Models for the Proton Spin, invited talk presented at the 9th Int. Symp. on High Energy Spin Physics, Bonn (10-15 Sept. 90), to be published in the proceedings.
4. A. V. Efremov, O. V. Teryaev, Preprint JINR, E2-88-287 (1988);
   G. Altarelli, G. G. Ross, Phys. Lett. **B214**, 381 (1988);
   R. D. Carlitz, J. C. Collins, A. M. Mueller, Phys. Lett. **B214**, 229 (1988).
5. G. Altarelli, G. Parisi, Nucl. Phys. **B126**, 298 (1977).

6. P. Ratcliffe, Phys. Lett. **B192**, 180 (1987).
7. R. L. Jaffe, A. Manohar, Nucl. Phys. **B337**, 509 (1990).
8. G. Bodwin, J. Qui, Phys. Rev. **D41**, 2750 (1990).
9. S.D. Bass, N.N. Nikolaev, A.W. Thomas, Adelaide Univ. Preprint ADP-133-T8 (1990).
10. A.V. Manohar, Phys. Rev. Lett. **65**, 2511 (1990).
11. A. V. Efremov, J. Soffer, O. V. Teryaev, Nucl. Phys. **B346**, 97 (1990).
12. G. Altarelli, B. Lampe, Z. Phys. **C47**, 315 (1990);
    G. Altarelli, Preprint CERN TH-5675/90.
13. G.G. Ross and R.G. Roberts, Preprint RAL-90-062 august 1990.
14. S. Forte, Phys. Lett. **B224**, 189(1989); Nucl. Phys. **B311**, 1 (1990).
15. G. Veneziano, Mod. Phys. Lett. **A4**, 1605 (1989);
    G. M. Shore, G. Veneziano, Phys. Lett. **B244**, 75 (1990).
16. A. V. Efremov, J. Soffer, N. Törnqvist, Phys. Rev. Lett. **64**, 1495 (1990).
17. G. Veneziano, Nucl. Phys. **B159**, 213 (1979).
18. D. I. Diakonov, M. V. Eides, JETP **81**, 434 (1981).
19. N. Törnqvist, These proceedings.
20. O. Dumbrajs et al., Nucl. Phys. **B216**, 277 (1983).
21. B. Bagchi and A. Lahiri, Comment on ref.16 to be published in Phys. Rev. Lett..
22. N. Törnqvist and P. Zenczykowski, Phys. Rev. **D29**, 2139 (1984).
23. A.V. Efremov, J. Soffer and N. Törnqvist, Comment on ref.37 Preprint CPT-90/P.2451 submitted to Phys. Rev. Lett..
24. J.E. Mandula, Phys. Rev. Lett. **65**, 1403 (1990).
25. We are indebted to J. Mandula for some discussions on this question which led to the results given in eq.(11).
26. J. Ellis, M. Karliner and C.T. Sachrajda, Phys. Lett. **B231**, 497 (1989) and references therein.
27. G. Preparata, P.G. Ratcliffe and J. Soffer, Preprint Milano MITH 90/16 (submitted to Phys. Lett. B).

THE g_1 PROBLEM: MUCH ADO ABOUT NOTHING

Aneesh V. Manohar[*]

*Department of Physics, University of California, San Diego,
9500 Gilman Drive, La Jolla, CA 92093-0319.*

ABSTRACT

Implications of the EMC experiment and several other experiments for the strangeness content of the proton are presented. The gluon contribution to the proton spin is shown to be arbitrary due to factorisation ambiguities, so adding a gluon contribution cannot explain why Δs might be large. The polarized gluon distribution Δg is shown to be given by a gauge invariant but non-local operator. Other topics discussed include the gauge variation of the forward matrix element of K^μ, whether Δg can be measured by studying large transverse momentum jets, the angular momentum sum rule, and the implications of the renormalisation group invariance of $\Delta \Sigma + (\alpha_s/2\pi)\Delta g$ for quark models. Δs in the proton is analysed using the chiral quark model, and in the large N_c limit.

THE g_1 STRUCTURE FUNCTION AND A SUM RULE

The cross-section for deep inelastic lepton scattering from a spin-1/2 hadron target such as a nucleon can be written in terms of the hadronic tensor

$$W^{\mu\nu}(p,q,s) = \frac{1}{4\pi}\int d^4x\, e^{iq\cdot x}\, \langle p,s|\,[j^\mu(x),j^\nu(0)]\,|p,s\rangle\,, \tag{1}$$

where j^μ is the electromagnetic current, and s^μ is the spin four vector, $s\cdot p = 0$, normalised so that $s^2 = -M^2$. $W^{\mu\nu}$ can be decomposed into four linearly independent structure functions for a spin-1/2 target,

$$W_{\mu\nu} = -F_1 g_{\mu\nu} + \frac{F_2}{\nu}p_\mu p_\nu + \frac{ig_1}{\nu}\epsilon_{\mu\nu\lambda\sigma}q^\lambda s^\sigma + \frac{ig_2}{\nu^2}\epsilon_{\mu\nu\lambda\sigma}q^\lambda\left(p\cdot q\, s^\sigma - s\cdot q\, p^\sigma\right)\,, \tag{2}$$

where $\nu = p\cdot q$. With this normalisation, the structure functions $\{F_1, F_2, g_1, g_2\}$ scale (up to logarithms) in the deep inelastic limit. The European Muon Collaboration (EMC) measured the asymmetry

$$A_1 \equiv \frac{\mu^\uparrow p^\uparrow - \mu^\uparrow p^\downarrow}{\mu^\uparrow p^\uparrow + \mu^\uparrow p^\downarrow}\,,$$

in $\mu - p$ scattering, where the terms denote cross-sections and the arrows $\uparrow$ and $\downarrow$ denote polarisations along the beam direction. A_1 is related to g_1 by

$$g_1 \cong A_1 F_1/(1+R),$$

up to terms which are negligible in the deep inelastic limit. F_1 and $R = \sigma_L/\sigma_T$ are the conventional unpolarised target observables in deep inelastic muon scattering. The

---

[*] On leave from the Department of Physics, Massachusetts Institute of Technology, Cambridge, MA 02139.

EMC extrapolated their measured structure function from $x \approx 0.02$ down to $x = 0$, and published a result for the first moment of $g_1(x)$,[1]

$$\int_0^1 dx\, g_1^p(x) = 0.126 \pm 0.010 \pm 0.015 \ . \tag{3}$$

There is a sum rule for the first moment of $g_1(x)$ which can be derived using the operator product expansion and analyticity,[2]

$$2\int_0^1 dx\, g_1^p(x, Q^2) = \left(1 - \frac{\alpha_s(Q^2)}{\pi} + \mathcal{O}(\alpha_s^2)\right)\left\{\frac{1}{9}\left[4\Delta u(Q^2) + \Delta d(Q^2) + \Delta s(Q^2)\right]\right. \\ \left. - \frac{2\epsilon}{9}\left[\Delta u(Q^2) + \Delta d(Q^2) + \Delta s(Q^2)\right]\right\} + \mathcal{O}\left(\frac{\Lambda^2}{Q^2}\right), \tag{4}$$

where

$$2\Delta q(\mu^2)s_\alpha = \left.\langle p, s|\bar{q}\gamma_\alpha\gamma_5 q|p, s\rangle\right|_{\mu^2}, \tag{5}$$

and $|p, s\rangle$ is the proton state of momentum p and spin s. The label μ refers to the mass scale at which the axial vector current operator is renormalised. The power series in $\alpha_s(Q^2)$ in (4) represents a perturbative calculation of the coefficient function of the axial vector currents and the $\mathcal{O}(\Lambda^2/Q^2)$ terms are higher twist corrections. ϵ is the correction due to heavy quarks and radiative effects,[3] and for the EMC experiment, $\epsilon \approx 0.034$.

There is no gluon operator which contributes to the right hand side of the sum rule because there is no gauge invariant, twist two, dimension three, local gluon operator which can appear in the operator product expansion. Thus there is no term proportional to Δg on the right hand side of the sum rule. I will discuss the polarised gluon distribution and its possible contribution to the sum rule in greater detail later in this article.

The EMC result determines one linear combination of Δu, Δd, and Δs. Isospin invariance implies

$$(\Delta u - \Delta d)s_\mu = 2\langle p, s|A_\mu^3|p, s\rangle = g_A s_\mu, \tag{6}$$

where g_A is the axial vector coupling in neutron β-decay. $SU(3)$ symmetry is required to determine the other linearly independent combination,

$$(\Delta u + \Delta d - 2\Delta s)s_\mu = 2\sqrt{3}\langle p, s|A_\mu^8|p, s\rangle = (3F - D)s_\mu, \tag{7}$$

where F and D are the $SU(3)$ invariant amplitudes for the axial current in hyperon semileptonic decay. Fitting F and D to the hyperon semileptonic decays,[4] and using the errors determined from the fit gives

$$F = 0.47 \pm 0.04, \qquad D = 0.81 \pm 0.03. \tag{8}$$

These numbers differ from previous standard values because the experimental values for the hyperon lifetimes (especially the neutron lifetime) have changed significantly. Using (8) gives

$$\Delta u = 0.74 \pm 0.10,\ \Delta d = -0.54 \pm 0.10,\ \Delta s = -0.20 \pm 0.11, \tag{9}$$

(renormalised at $\mu^2 = Q^2_{\rm EMC}$), or equivalently

$$\Delta\Sigma\left(Q^2_{\rm EMC}\right) = 0.01 \pm 0.29, \qquad (10)$$

where $\Delta\Sigma = \Delta u + \Delta d + \Delta s$ is the singlet quark distribution. The analysis so far has assumed $SU(3)$ symmetry. There are large corrections to $SU(3)$ symmetry relations from the symmetry breaking quark mass matrix. To leading order, these corrections are of two types: a) chiral logarithmic corrections which are calculable using the lowest order $SU(3)$ symmetric lagrangian, and b) uncalculable corrections from higher dimension operators in chiral perturbation theory. The chiral logarithmic corrections to the axial vector currents in hyperon semileptonic decay and to the T^8 current have been computed.[5,6] The conclusion is that the $SU(3)$ breaking corrections are as large as the lowest order result, so that $SU(3)$ symmetry is badly broken. This has already been noted for the hyperon masses, and for hyperon nonleptonic decays.[5]

Much of the recent interest in the EMC result was due to the values for Δs and $\Delta\Sigma$ determined in (9) and (10). There is a large theoretical uncertainty in these values because of $SU(3)$ breaking effects. In addition, there may be additional uncertainty in the extrapolation of the measured value of $g_1(x)$ down to $x = 0$.[7] This makes the extraction of reliable values for Δq extremely questionable. Nevertheless, the questions originally raised by the EMC result about the role of strange quarks in the structure of the proton are still interesting, and should be investigated further.

IS THE PROTON STRANGE?

The strangeness content of the proton is not well understood at the present time. The EMC experiment has provided some indication that the matrix element $\langle p|\bar{s}\gamma_\mu\gamma_5 s|p\rangle$ in the proton may be non-zero. The same matrix element can also be determined from low-energy elastic neutrino-proton scattering. To discuss this and related experiments, it is convenient to introduce the proton form factors[3]

$$\langle p',s|\bar{s}\gamma^\mu s|p,s\rangle = F_1^{(s)}(\ell^2)\,\overline{\mathcal{U}}\gamma^\mu\mathcal{U} + iF_2^{(s)}(\ell^2)\,\frac{\overline{\mathcal{U}}\sigma^{\mu\nu}\ell_\nu\mathcal{U}}{2M_N}$$
$$\langle p',s|\bar{s}\gamma^\mu\gamma_5 s|p,s\rangle = G_1^{(s)}(\ell^2)\,\overline{\mathcal{U}}\gamma^\mu\gamma_5\mathcal{U} + G_3^{(s)}(\ell^2)\,\ell^\mu\overline{\mathcal{U}}\gamma_5\mathcal{U} \qquad (11)$$
$$\langle p',s|\tfrac{1}{2}(\bar{u}\gamma^\mu\gamma_5 u - \bar{d}\gamma^\mu\gamma_5 d)|p,s\rangle = G_1^{(3)}(\ell^2)\,\overline{\mathcal{U}}\gamma^\mu\gamma_5\mathcal{U} + G_3^{(3)}(\ell^2)\,\ell^\mu\overline{\mathcal{U}}\gamma_5\mathcal{U},$$

where $\ell = p' - p$, and $\mathcal{U}$ is a Dirac spinor. [Note that F and G defined in (11) are elastic form factors, and are not related to the deep inelastic scattering structure functions.] The axial vector coupling of the Z^0 boson is proportional to $\tfrac{1}{2}(\bar{u}\gamma^\mu\gamma_5 u - \bar{d}\gamma^\mu\gamma_5 d - \bar{s}\gamma^\mu\gamma_5 s)$, so that a measurement of the axial nucleon-Z^0 form factor determines $G_1^{(3)} - \tfrac{1}{2}G_1^{(s)}$. At zero momentum transfer, we know that

$$G_1^{(3)}(0) = \tfrac{1}{2}(\Delta u - \Delta d) = \tfrac{1}{2}g_A, \qquad (12)$$

which only requires using $SU(2)$ symmetry. Ahrens et al.[8] have measured the axial $Z^0 - N$ form factor at momentum transfers of order $\ell^2 = (0.4 - 1.05)$ GeV^2. Using their extrapolation to $\ell^2 = 0$, and including heavy quark and radiative corrections gives[3]

$$G_1^{(s)}(0) = \Delta s = -0.15 \pm 0.08. \qquad (13)$$

There is a large uncertainty in the extrapolation to zero momentum transfer, so the result (13) should not be taken too seriously. If the experiment can be repeated at lower ℓ^2, then this method for determining Δs will be more reliable than that using deep inelastic scattering, because we need only use $SU(2)$ symmetry rather than $SU(3)$ symmetry to extract Δs from the experimental measurement. In any case, the results of Ahrens et al. indicate that $G_1^{(s)}(\ell^2 \neq 0) \neq 0$.

The other strange elastic form factors are the vector form factors $F_1^{(s)}(\ell^2)$ and $F_2^{(s)}(\ell^2)$. The strangeness charge of the proton is zero, so $F_1^{(s)}(0) = 0$, but $F_1^{(s)}(\ell^2)$ need not vanish at non-zero ℓ^2. The strange vector form factors can be determined by doing parity violating electron scattering experiments at Bates and CEBAF.[9] The strange vector form factors can be extracted from the measurement using $SU(2)$ symmetry and the known electromagnetic form factors. The SAMPLE experiment at Bates should be able to determine $F_2^{(s)}$.[10] There is also a CEBAF proposal to measure $F_1^{(s)}$ at non-zero ℓ^2 by studying elastic $e-p$ scattering in the forward direction.[11] The theoretical predictions for $F_2^{(s)}(0)$ are[3] $F_2^{(s)}(0) = -0.12 \pm 0.2$ in the chiral quark model[12] and $F_2^{(s)} = -0.45$ in the Skyrme model. R.L. Jaffe has estimated the strange charge radius, $r_s^2 = 6dF_1^{(s)}/d\ell^2$, to be $r_s^2 = (0.11-0.22)$ fm^2 using vector meson dominance.[13]

The other evidence that there may be large strange quark matrix elements in the proton comes from the $\pi-N$ sigma term[14] which yields

$$\tfrac{1}{2}(m_u + m_d)\langle p|\,\overline{u}u + \overline{d}d\,|p\rangle \simeq 45-60 \text{ MeV}. \tag{14}$$

From this one can estimate the value of $\langle p|\,\overline{s}s\,|p\rangle$ by working to first order in $SU(3)$ symmetry breaking. The masses for the baryon octet B (a 3×3 traceless matrix) are given by

$$\mathcal{H} = m_0\,\text{Tr}\,\overline{B}B + d\,\text{Tr}\,\overline{B}\{M,B\} + f\,\text{Tr}\,\overline{B}[M,B] + s\,\text{Tr}\,M\,\,\text{Tr}\,\overline{B}B + \mathcal{O}(M^2), \tag{15}$$

where M is the quark mass matrix. From the measurement of the pion-nucleon sigma term (14) and the observed baryon masses, one can fit for d, f, s and m_0. Then

$$m_s \frac{\partial m_p}{\partial m_s} = m_s\langle p|\,\overline{s}s\,|p\rangle = m_s(d-f+s) = 334 \pm 132 \text{ MeV}, \tag{16}$$

where the quoted error reflects both experimental uncertainties and an estimate of the $\mathcal{O}(M^2)$ contributions. In other words, if the strange quark was massless, the proton would only weigh approximately 600 MeV. Some justification for the large value for $\langle p|\,\overline{s}s\,|p\rangle$ has been given in the context of both the Skyrme[15] and bag[16] models of the proton. There is some recent work[17] which suggests a possible breakdown of chiral perturbation theory, and therefore complicates the extraction of the sigma term. [Note that for a heavy quark such as the top quark, one expects $\langle p|\,m_t\overline{t}t\,|p\rangle = 2m_p/27 \approx 70$ MeV, using the scale anomaly.[18] This is because $\langle p|\,\overline{t}t\,|t\rangle$ falls like $1/m_t$, so that $\langle p|\,m_t\overline{t}t\,|p\rangle$ approaches a finite value for large m_t, provided we keep the bare QCD coupling fixed. The naive argument that the proton wavefunction should not contain any t quarks, because it can lower its energy by removing them is not correct. There is a direct contribution to the proton energy from $\overline{t}t$ pairs which increases the energy, and an indirect contribution from the t quark loop which lowers Λ_{QCD} and reduces the energy. The minimum energy occurs for a non-zero value of $\langle p|\,m_t\overline{t}t\,|p\rangle$. Thus a "naive" expectation for $\langle p|\,m_s\overline{s}s\,|p\rangle$ should be 70 MeV rather than zero.]

Finally, there is evidence from charm production in neutrino scattering[19] that the spin averaged parton distribution $s(x) + \bar{s}(x)$ in the nucleon is small. In particular, the momentum fraction carried by strange quarks is less than 0.026, *i.e.*

$$\langle p| \bar{s}\gamma^{(\mu} D^{\nu)} s |p\rangle = (0.026 \pm 0.006)(2p^\mu p^\nu).$$

Thus one does not expect there to be a large strange quark contribution (polarised or unpolarised) at large values of x. However, the momentum fraction carried by strange quarks is equal to

$$\int_0^1 dx\, x(s(x) + \bar{s}(x))$$

so there is no constraint on

$$\Delta s = \int_0^1 (s_\uparrow(x) - s_\downarrow(x) + \bar{s}_\uparrow(x) - \bar{s}_\downarrow(x)),$$

since Δs can get significant contributions at small values of x.

The strange quark content of the proton is not yet well determined. There are some experiments that suggest that the strangeness is small; others suggest that it is large. It is important to keep in mind that the experiments measure different s quark operators, and that there is no contradiction between the various results. One expects that the strangeness content of the proton should be small if the s quark is treated as a heavy quark. On the other hand, for a light s quark, one expects there to be some s quarks in the proton. Unfortunately, in the real world, the s quark mass has an intermediate value, and cannot be treated in either limit. This makes calculations more difficult, but also more interesting.

THE PARTON MODEL AND FACTORISATION

There has been some controversy about the role of gluons in the sum rule for the g_1 structure function. In the operator product expansion analysis, there is no gluon operator that can contribute to the first moment of g_1, as mentioned earlier. To better appreciate the role of gluons, it is useful to rederive the sum rule directly using factorisation in QCD, without resorting to the operator product expansion. This is the method of Efremov and Teryaev (ET),[20] Altarelli and Ross (AR),[21] and Carlitz, Collins and Mueller (CCM).[22]

The g_1 structure function for γp scattering can be written in factorised form as the convolution of a hard cross-section for γ-parton scattering with parton distribution functions:

$$g_1^{\gamma p} = \hat{g}_1^{\gamma q} \otimes \Delta q(x) + \hat{g}_1^{\gamma g} \otimes \Delta g(x) \qquad (17)$$

where $\Delta q(x)$ is the polarised quark distribution $q_\uparrow(x) - q_\downarrow(x) + \bar{q}_\uparrow(x) - \bar{q}_\downarrow(x)$, and $\Delta g(x)$ is the polarised gluon distribution $g_\uparrow(x) - g_\downarrow(x)$ in the proton, and a $\hat{}$ denotes a hard cross-section. [For simplicity, the formulæ will be written for a single quark flavour of unit charge. The generalisation to the case of three flavours with fractional charges is trivial.] The hard cross-sections are computable order by order in QCD perturbation theory, and depend on the particular scattering process considered. The parton distribution functions cannot be computed using perturbative QCD, because they are sensitive to infrared effects. They are, however, universal, so the same distribution function can be used in different processes. For example, one can use the distribution functions measured in Drell-Yan to compute the deep inelastic scattering cross-section. There are ambiguities in the factorisation procedure, because one can redefine the hard cross-sections and the parton distributions such that the physical cross-sections are kept fixed. Thus one needs a well-defined prescription to decide which quantities are assigned to the parton densities, and which are assigned to the hard cross-section. A particularly simple method for making this separation was developed for unpolarised cross-sections by Collins and Soper.[23] It is straightforward to generalise their formalism to the polarised case. One defines polarised quark and gluon distributions for a target A in terms of non-local light-cone correlation functions,[24]

$$\Delta q/A(x) = \frac{1}{4\pi} \int d\xi^- e^{-ix\xi^- P^+} \langle A, P| \bar{\psi}_a \left(0, \xi^-, 0_\perp\right) \gamma^+ \gamma_5 W^a{}_b \psi^b \left(0, 0, 0_\perp\right) \\ + \bar{\psi}_a \left(0, 0, 0_\perp\right) \gamma^+ \gamma_5 W^{\dagger a}{}_b \psi^b \left(0, \xi^-, 0_\perp\right) |A, P\rangle, \quad (18)$$

$$\Delta g/A(x) = \frac{i}{4\pi x P^+} \int d\xi^- e^{-ix\xi^- P^+} \langle A, P| G_a^{+\alpha} \left(0, \xi^-, 0_\perp\right) W^a{}_b \tilde{G}_\alpha^{+b} \left(0, 0, 0_\perp\right) \\ - G_a^{+\alpha} \left(0, 0, 0_\perp\right) W^{\dagger a}{}_b \tilde{G}_\alpha^{+b} \left(0, \xi^-, 0_\perp\right) |A, P\rangle. \quad (19)$$

where W is the Wilson line operator for the fundamental representation in (18), and for the adjoint representation in (19).

$$W^a{}_b = P \exp \left\{ ig \int_0^{\xi^-} dy^- A^+(0, y^-, 0_\perp) \right\}^a{}_b. \quad (20)$$

These non-local correlation functions need to be renormalised, and we will choose dimensional regularisation and $\overline{MS}$ subtraction.

To compute the hard cross-section, one applies factorisation to γ-parton scattering, instead of to γ-hadron scattering. As an example, the photon-gluon scattering cross-section can be written as

$$g_1^{\gamma g} = \hat{g}_1^{\gamma q} \otimes \Delta q/g(x) + \hat{g}_1^{\gamma g} \otimes \Delta g/g(x) \quad (21)$$

where $\Delta q/g(x)$ and $\Delta g/g(x)$ are the parton densities in a gluon target. $g_1^{\gamma q}$ can be computed in perturbative QCD. The parton densities in a gluon target can also be computed in perturbative QCD by using (18) and (19) with the target A taken to be a gluon. This enables us to determine the hard cross-section using (21). Both these calculations will in general, be infrared sensitive. However, all the infrared dependence cancels when they are substituted into (21) to determine the hard cross-section, so that the hard-cross section can be reliably computed using QCD perturbation theory. To compute the gluon contribution to $g_1^{\gamma p}$ at lowest non-trivial order, one needs to compute $\hat{g}_1^{\gamma g}$ to order α_s using

$$\hat{g}_1^{\gamma g} = g_1^{\gamma g} - \Delta q/g(x) + \mathcal{O}\left(\frac{\alpha_s}{2\pi}\right)^2, \quad (22)$$

since $\Delta q/g$ is $\mathcal{O}(\alpha_s)$, and at lowest order $\Delta g/g(x) = \delta(x-1)$, $\hat{g}_1^{\gamma-q} = \frac{1}{2}\delta(x-1)$. [The gluon target has been chosen to have helicity $+1$.] The gluon-photon scattering graph gives[22]

$$g_1^{\gamma g}(x) = \frac{\alpha_s}{4\pi} \left\{ [2x-1] \log \frac{Q^2(1-x)}{m^2 x - p^2 x^2(1-x)} + 3 - 4x + \frac{p^2 x(1-x)}{m^2 - p^2 x(1-x)} \right\}. \quad (23)$$

The graph is infrared divergent, and has been regulated by giving the quark a mass m, and the gluon a momentum p. The polarised quark density in a helicity +1 gluon is given by[25]

$$\Delta q/g(x) = \frac{\alpha_s}{2\pi}\left\{(2x-1)\log\frac{\mu^2}{m^2-p^2x(1-x)} + \frac{p^2x(1-x)}{m^2-p^2x(1-x)}\right\}, \quad (24)$$

where the infrared divergences have again been regulated by giving the quark a mass m, and the gluon a momentum p. The ultraviolet divergence has been regulated by dimensional regularisation and $\overline{MS}$ subtraction. The hard photon-gluon cross-section is obtained from (22),

$$\hat{g}_1^{\gamma g}(x) = \frac{\alpha_s}{4\pi}\left\{(2x-1)\log\frac{Q^2(1-x)}{\mu^2 x} + 3 - 4x\right\}. \quad (25)$$

Note that the infrared divergences have cancelled, as they should. Taking moments of (17) undoes the convolution, so that

$$M_n(g_1^{\gamma p}) = M_n(\hat{g}_1^{\gamma q})M_n(\Delta q(x)) + M_n(\hat{g}_1^{\gamma g})M_n(\Delta g(x)) \quad (26)$$

The gluon contribution to the first moment of g_1 is given by $M_1(\hat{g}_1^{\gamma g})$. The first moment of (25) is zero, so there is no hard gluon contribution to the first moment of g_1. The same result has been obtained by Bodwin and Qiu.[26] Bodwin and Qiu have also shown that any factorisation scheme that respects gauge invariance, Lorentz invariance, and certain analyticity properties gives zero for the first moment of $\hat{g}_1^{\gamma g}$.

K^μ AND Δg

The parton model distribution functions defined in the previous section can be compared with the those obtained using the operator product expansion. The odd moments of the quark and gluon distributions eq.(18) and (19), can be computed explicitly,

$$M_n(\Delta q) = \frac{1}{2}\left(\frac{\sqrt{2}}{M}\right)^n \langle A|\overline{\psi}\left(i\partial^+\right)^{n-1}\gamma^+\gamma_5\psi|A\rangle, \quad n\geq 1, \; n \text{ odd},$$

$$M_n(\Delta g) = \frac{i}{2}\left(\frac{\sqrt{2}}{M}\right)^n \langle A|G^{+\alpha}\left(i\partial^+\right)^{n-2}\tilde{G}_\alpha^+|A\rangle, \quad n\geq 3, \; n \text{ odd}.$$

These moments agree with those obtained by defining quark and gluon distributions in terms of twist two operators in the operator product expansion. The operator product expansion only involves local operators, so one can determine M_3, M_5, etc. of the gluon distribution in terms of matrix elements of local operators. There is no gluon operator corresponding to the first moment of the polarised gluon distribution. This does not mean that the first moment vanishes; instead it can be determined in terms of the higher moments,

$$\begin{aligned}
M_1(\Delta g) &= \int_0^1 \Delta g(x)dx = \int_0^1 \frac{x^2\Delta g(x)dx}{1-(1-x^2)} \\
&= \int_0^1 dx\, x^2\Delta g(x)[1+(1-x^2)+(1-x^2)^2+\ldots] \\
&= (M_3) + (M_3 - M_5) + (M_3 - 2M_5 + M_7) + \ldots \\
&= \lim_{R\to\infty}\sum_{r=0}^{R}\sum_{k=0}^{r}(-1)^k \frac{r!}{k!(r-k)!}M_{3+2k}.
\end{aligned} \quad (27)$$

The first moment is given by (27) if the limit exits.[27] Equivalently, using (18), and taking the first moment implies that

$$\Delta g = M_1\left(\Delta g(x)\right) = \int_0^1 dx\ \Delta g(x)$$
$$= \frac{1}{4\sqrt{2}M}\int_{-\infty}^{\infty} d\xi^-\ \epsilon(\xi^-) \times \langle A|\, G^{+\alpha}\left(\xi^-\right)\tilde{G}_\alpha^+(0) - G^{+\alpha}(0)\tilde{G}_\alpha^+\left(\xi^-\right) |A\rangle, \qquad (28)$$

where $\epsilon(z) = 1$ if $z > 1$, and -1 if $z < 1$. Thus the first moment of the polarised gluon distribution is given by the matrix element of a gauge invariant but non-local gluon operator. It has been suggested that Δg is given by the proton matrix element of K^μ,

$$K^\mu = \epsilon^{\mu\nu\alpha\beta}\ \text{Tr}\ A_\nu \left(F_{\alpha\beta} - \tfrac{2}{3}A_\alpha A_\beta\right),$$

and that this matrix element is gauge invariant because the gauge variation of K^μ is a total derivative, and vanishes at zero momentum transfer.[21] The gauge variation of forward matrix elements of K^μ can be studied in two dimensional massless quantum electrodynamics (the Schwinger model). This theory has anomalies, winding number in the gauge sector, confinement, and no $U(1)$ Goldstone boson, and thus in many respects resembles QCD. It has the advantage over QCD in that it is exactly soluble. Using the exact solution, one can easily show that forward matrix elements of K^μ explicitly depend on the choice of gauge.[28]

REDEFINITIONS AND Q^2 EVOLUTION

The sum rule for the g_1 structure function has no gluon contribution if one defines the parton densities using light-cone correlation functions, or using the operator product expansion. As mentioned before, there are factorisation ambiguities, so one can redefine the parton densities and the hard cross-sections such that the physical cross-section is kept invariant. For example, one can make the redefinitions

$$\Delta q_\lambda = \Delta q + \lambda \frac{\alpha_s}{2\pi}\Delta g, \quad \hat{g}_{1\lambda}^{\gamma g}(x) = \hat{g}_1^{\gamma g}(x) - \frac{\alpha_s}{4\pi}\lambda\delta(x), \qquad (29)$$

for arbitrary λ. The redefinition advocated by ET, AR, and CCM is $\lambda = 1$. One must keep in mind that if one makes the redefinition (29) for parton densities, then a similar redefinition must be made in other processes. For example, the axial Z^0-nucleon coupling at zero momentum is now proportional to

$$\Delta u_\lambda - \Delta d_\lambda - \Delta s_\lambda + \lambda \frac{\alpha_s}{2\pi}\Delta g \qquad (30)$$

so that the Z^0 now couples directly to polarised gluons at zero momentum transfer.[29] Clearly, making a redefinition does not affect any experimental predictions. The simplest choice is to use $\lambda = 0$, so that the Z does not couple directly to gluons.

The choice $\lambda = 1$ has been advocated instead, since distributions $\Delta q_{\lambda=1}$ are renormalisation group invariant, and therefore $\Delta q_{\lambda=1}$ should match directly to the quark distributions in a quark model description of the proton. To see why this argument is incorrect, consider the renormalisation group equations for the polarised quark and gluon distributions[3]

$$\mu\frac{d}{d\mu}\begin{pmatrix}\Delta\Sigma \\ -N_f\frac{\alpha_s}{2\pi}\Delta g\end{pmatrix} = \begin{pmatrix}\gamma_A & 0 \\ \gamma_A & 0\end{pmatrix}\begin{pmatrix}\Delta\Sigma \\ -N_f\frac{\alpha_s}{2\pi}\Delta g\end{pmatrix} \qquad (31)$$

where $\Delta\Sigma$ is the singlet quark distribution, and γ_A is the anomalous dimension of the axial current[30,31]

$$\gamma_A = N_f \left(\frac{\alpha_s}{\pi}\right)^2,$$

and N_f is the number of quark flavours. The non-singlet distributions have vanishing anomalous dimension, because the divergence of the non-singlet axial current is proportional to a dimension three operator, so that the symmetry is softly broken. Thus

$$\mu \frac{d}{d\mu}\left(\Delta q + \frac{\alpha_s}{2\pi}\Delta g\right) = \mu \frac{d}{d\mu}\Delta q_{\lambda=1} = 0. \tag{32}$$

The original derivation of (31) assumed that the gluon distribution Δg was given by the local operator K^μ. We have seen that Δg is actually given by the matrix element of a non-local operator (28). The non-local operator differs from K^μ by a surface term, which does not affect the renormalisation group equations, so that (31) is still correct.

To understand why (32) does not allow one to match $\Delta q_{\lambda=1}$ directly onto a quark model, consider the renormalisation of the vector and axial non-singlet currents,

$$J_V^{a\nu} = \bar{q}\, T^a \gamma^\nu q, \qquad J_A^{a\nu} = \bar{q}\, T^a \gamma^\nu \gamma_5 q,$$

which satisfy the renormalisation group equations

$$\mu \frac{d}{d\mu} J_{V,A}^{a\nu} = 0.$$

The baryon matrix elements of the vector and axial currents at zero momentum transfer are given by

$$\begin{aligned}\langle B| J_V^{a\nu} |B\rangle &= \operatorname{Tr} \bar{B}\gamma^\nu [T^a, B], \\ \langle B| J_A^{a\nu} |B\rangle &= F\operatorname{Tr} \bar{B}\gamma^\nu\gamma_5 [T^a, B] + D\operatorname{Tr} \bar{B}\gamma^\nu\gamma_5 \{T^a, B\}.\end{aligned} \tag{33}$$

The zero momentum vector current is an $SU(3)$ generator, which acts on baryons by commutation since the baryons transform as the adjoint representation of $SU(3)$, i.e. $F = 1$, $D = 0$. This implies that the strange charge of the proton must vanish, $F_1^{(s)}(0) = 0$. The axial current, on the other hand, has $F \neq 1$, and $D \neq 0$. Like the vector current, the axial current is also not renormalised, but the axial current can have non-trivial values for F and D because the chiral $SU(3)$ symmetry is spontaneously broken, and the baryons do not tranform as a representation of $SU(3)_L \otimes SU(3)_R$. Thus one cannot use the non-renormalisation of the axial current to conclude that $F = 1$, $D = 0$. Similarly $\Delta s_{\lambda=1}$ having no anomalous dimension does not imply that $\Delta s_{\lambda=1} = 0$ in the proton, because the strange axial charge is spontaneously broken. Another example is the operator $m_s \bar{s}s$, which is also renormalisation group invariant,

$$\mu \frac{d}{d\mu} m_s \bar{s}s = 0. \tag{34}$$

However, we have already seen that $\langle p| m_s \bar{s}s |p\rangle \neq 0$.

Δg AND LARGE TRANSVERSE MOMENTUM JETS

CCM[22] have suggested $\Delta q_{\lambda=1}$ is the cross-section corresponding to one-jet events, and that $(\alpha_s/2\pi)\Delta g$ is the contribution from two-jet events in deep inelastic lepton-proton scattering. Thus one might be able to measure the two contributions separately by identifying jets in the final state. They propose to distinguish jets using a cutoff on transverse momentum. There are possible problems with gauge invariance in using a transverse momentum cutoff, as have been noted by Bodwin and Qiu.[26] Instead of using transverse momentum as a cutoff, let us use cuts that are relevant for an experimental separation of the final state into one-jet and two-jet events. The cuts I will use are that each jet has some minimum energy E_{min}, that the opening angle between jets is greater than a minimum angle θ_{min}, and that Q^2 is greater than Q^2_{min} to avoid contamination by higher twist effects. In this case, one can show that the one and two jet events cannot be separated in a fixed target experiment, because the two jets are collinear. In a colliding beam experiment, it is possible to make a separation into two-jet and three-jet events (there is now an additional target jet), but that the the three jet event rate is not given by $(\alpha_s/2\pi)\Delta g$, but by $\int d\xi \Gamma_\xi \Delta g(\xi)$, where Γ_ξ is a function which depends on the precise experimental cuts.[32] Thus the polarised gluon distribution can, in principle, be measured by studying three-jet events. However $\Delta q_{\lambda=1}$ is not relevant for this measurement.

THE ANGULAR MOMENTUM SUM RULE

Translational invariance implies that the the stress tensor $T_{\mu\nu}$ is conserved, $\partial_\mu T^{\mu\nu} = 0$. $T_{\mu\nu}$ can be chosen to be symmetric and gauge invariant, for example by writing the Lagrangian in a background metric, and varying with respect to $g^{\mu\nu}$. The current associated with Lorentz transformations is a rank-3 tensor constructed entirely from $T_{\mu\nu}$,

$$M^{\mu\nu\lambda} \equiv x^\nu T^{\mu\lambda} - x^\lambda T^{\mu\nu}. \tag{35}$$

$M^{\mu\nu\lambda}$ is conserved, $\partial_\mu M^{\mu\nu\lambda} = 0$, because $T^{\mu\nu}$ is symmetric and conserved. $M^{\mu\nu\lambda}$ is gauge invariant, and has no totally antisymmetric part,

$$\epsilon_{\alpha\mu\nu\lambda} M^{\mu\nu\lambda} = 0$$

or equivalently

$$M^{\mu\nu\lambda} + M^{\lambda\mu\nu} + M^{\nu\lambda\mu} = 0.$$

The nucleon matrix element of $M^{\mu\nu\lambda}$ is[4]

$$\mathcal{M}^{\mu\nu\lambda}(p,0,s) = \left[2p^\mu \left(ip^\lambda \partial^\nu - ip^\nu \partial^\lambda \right) \right.$$
$$\left. + \frac{1}{2} \left(2p^\mu \epsilon^{\lambda\nu\beta\sigma} - p^\nu \epsilon^{\mu\lambda\beta\sigma} - p^\lambda \epsilon^{\nu\mu\beta\sigma} \right) p_\beta s_\sigma \right] (2\pi)^4 \delta^4(0). \tag{36}$$

The nucleon has spin-1/2 because the coefficient in front of the second term is 1/2, which is the angular momentum sum rule. A more useful form of the sum rule is obtained if we take the nucleon to be in an eigenstate of helicity $+1/2$ and choose its momentum to define the $\hat{e}_3$-axis, then $p_0^\mu = (p^0, 0, 0, p^3)$ and $s_0^\mu = (p^3, 0, 0, p^0)$. Substituting in (36) we obtain the sum rule,

$$\mathcal{M}^{012}(p_0, 0, s_0) = \frac{1}{2}(2\pi)^4 \delta^4(0) 2E. \tag{37}$$

The angular momentum tensor for QCD is

$$M^{\mu\nu\lambda} = \frac{i}{4}\bar{\psi}x^\nu\left(\gamma^\mu \overleftrightarrow{D}^\lambda + \gamma^\lambda \overleftrightarrow{D}^\mu\right)\psi + \frac{i}{4}\bar{\psi}x^\nu\left(\gamma^\mu \overleftrightarrow{\partial}^\nu + \gamma^\nu \overleftrightarrow{\partial}^\mu\right)\psi \qquad (38)$$
$$+ 2\,\text{Tr}\left(x^\nu G^{\mu\alpha}G^\lambda_\alpha - x^\lambda G^{\mu\alpha}G^\nu_\alpha - \tfrac{1}{4}G^{\alpha\beta}G_{\alpha\beta}\left(x^\nu g^{\mu\lambda} - x^\lambda g^{\mu\nu}\right)\right).$$

Clearly, there is no obvious breakup of the tensor into pieces that one can identify as being due to the quark spin, gluon spin, quark orbital angular momentum, and gluon orbital angular momentum. One can redefine the angular momentum tensor by adding a superpotential,

$$M'^{\mu\nu\lambda} = \mathcal{M}^{\mu\nu\lambda} + \partial_\alpha N^{[\mu\alpha][\nu\lambda]} \qquad (39)$$

where $N^{[\mu\alpha][\nu\lambda]}$ is an arbitrary tensor antisymmetric in $[\mu\alpha]$ and $[\nu\lambda]$. The superpotential does not affect the conserved charges such as the total angular momentum. By choosing a suitable superpotential, one obtains[4]

$$M'^{\mu\nu\lambda}_{\text{QCD}} = \frac{i}{2}\bar{\psi}\gamma^\mu\left(x^\lambda\partial^\nu - x^\nu\partial^\lambda\right)\psi + \frac{1}{2}\epsilon^{\mu\nu\lambda\sigma}\bar{\psi}\gamma_\sigma\gamma_5\psi$$
$$- 2\,\text{Tr}\left\{G^{\mu\alpha}\left(x^\nu\partial^\lambda - x^\lambda\partial^\nu\right)A_\alpha\right\} + 2\,\text{Tr}\left\{G^{\mu\lambda}A^\nu + G^{\nu\mu}A^\lambda\right\} \qquad (40)$$
$$- \frac{1}{2}\,\text{Tr}\,G^2\left(x^\nu g^{\mu\lambda} - x^\lambda g^{\mu\nu}\right).$$

The terms in (40) correspond successively to the orbital and spin angular momentum of the quarks and gluons. Note that with the exception of $\epsilon^{\mu\nu\lambda\sigma}\bar{\psi}\gamma_\sigma\gamma_5\psi$, the terms in (40) are not separately gauge invariant, and mix under renormalisation. It is worth noting that the term corresponding to the gluon spin is not the same as either K^μ or the non-local gluon operator (28). One can apply the angular momentum sum rule (37) to the proton using (40). The quark spin term gives a contribution $\Delta\Sigma(2\pi)^4\delta^4(0)2E$. The other terms are not simply related to quantities measurable in scattering experiments, so the angular momentum sum rule is not very useful for analysing the EMC results. There is no direct connection between the EMC results and the spin of the proton, because the EMC measurement determines the axial charge of the proton, not its angular momentum.

MODELS[33]

Physicists were disturbed by the notion of a large strange quark component to the proton, because it conflicts with the simple quark model picture of the proton as a bound state of u and d quarks. However, the quark model picture also conflicts with the measurements of the spin-independent structure functions F_1 and F_2. The simple quark model picture suggests that the u and d quark distributions are of the form $\delta(x - \tfrac{1}{3})$, smeared out by a suitable quark model wave function, and that $u(x) = 2d(x)$, both of which do not agree with the data. The key to resolving these contradictions is to remember that the constituent quarks of the quark model are not the same as the current quarks of QCD. I will use the uppercase letters U, D, S to denote the constituent quarks, and lowercase letters u, d, s to denote the current quarks, which are the fields which appear in the QCD lagrangian. The proton is made of two U quarks, and one D quark. One can write a parton distribution for the constituent quarks, $q_Q(x)$, which is the probability of finding a current quark q with momentum fraction x in a constituent quark Q. The proton parton distributions $u(x)$ and $d(x)$ are given by (neglecting smearing effects due to the constituent quark wavefunction)

$$u(x) = 2u_U(x) + u_D(x), \qquad d(x) = 2d_U(x) + d_D(x). \tag{41}$$

Isospin invariance implies $u_U(x) = d_D(x)$, and $u_D(x) = d_U(x)$. Thus we can write

$$d(x) = u_U(x) + 2u_D(x). \tag{42}$$

$u(x)$ is no longer equal to $2d(x)$, because $u_D(x) \neq 0$, i.e. because the constituent D quark contains u quarks, and $\{u(x), d(x)\}$ are no longer smeared out δ-functions.

Instead of working with the parton distribution functions for the strange quarks, I will concentrate only their first moment, i.e. on the matrix elements of the axial current. It is now easy to see how

$$\langle p|\,\bar{s}\gamma^\mu\gamma_5 s\,|p\rangle \neq 0.$$

The flavour octet axial currents in QCD can be expressed in terms of the constituent quarks as

$$\bar{q}\gamma^\mu\gamma_5 T^a q = g_A^{(8)}\,\overline{Q}\gamma^\mu\gamma_5 T^a Q + \ldots, \tag{43}$$

where ... represents higher dimension operators involving the Q fields, and $g_A^{(8)}$ is a non-perturbative renormalisation constant between current and constituent quarks. Similarly, for the singlet current

$$\bar{q}\gamma^\mu\gamma_5 q = g_A^{(0)}\,\overline{Q}\gamma^\mu\gamma_5 Q + \ldots. \tag{44}$$

By combining these two equations, it is easy to see that

$$\bar{s}\gamma^\mu\gamma_5 s = \tfrac{1}{3}\left(g_A^{(0)} - g_A^{(8)}\right)\left[\overline{U}\gamma^\mu\gamma_5 U + \overline{D}\gamma^\mu\gamma_5 D\right] + \tfrac{1}{3}\left(g_A^{(0)} + 2g_A^{(8)}\right)\overline{S}\gamma^\mu\gamma_5 S. \tag{45}$$

Using the non-relativistic wave function for the proton:

$$\langle p,s|\,\overline{U}\gamma^\mu\gamma_5 U\,|p,s\rangle = \tfrac{4}{3}s^\mu, \qquad \langle p,s|\,\overline{D}\gamma^\mu\gamma_5 D\,|p,s\rangle = -\tfrac{1}{3}s^\mu, \tag{46}$$

so that

$$\langle p,s|\,\bar{s}\gamma^\mu\gamma_5 s\,|p,s\rangle = \tfrac{1}{3}\left(g_A^{(0)} - g_A^{(8)}\right)s^\mu. \tag{47}$$

The singlet and octet currents are renormalised differently in QCD, because the singlet current can mix with gluons. Thus if $g_A^{(0)} \neq g_A^{(8)}$, one can have many strange (s) quarks in the proton, with the proton being made of only up (U) and down (D) quarks. A large value for Δs then implies that $g_A^{(0)}$ is very different from $g_A^{(8)}$, i.e. quark annihilation effects are important in the axial vector (1^{-+}) channel in QCD. We already know that these effects are important in the pseudoscalar (0^{-+}) channel, because of the large $\eta' - \eta$ mass difference.

The large annihilation effects seem to directly contradict Zweig's rule,[34] which says that quark annihilation effects are small. Zweig's rule is based on the spectrum and decays of the baryons and mesons, which are made of constituent quarks. Thus Zweigs's rule really says that annihilation effects are small for the constituent (U, D, S) quarks. This arises naturally in a chiral quark model for constituent quarks, where the constituent quarks are weakly interacting.[12] The distinction between current quarks and constituent quarks is also important to keep in mind when proposing further tests of the idea that there are large strange quark matrix elements in the proton. The proton contains many $s\bar{s}$ pairs, but very few $S\overline{S}$ pairs. Thus ϕ production from protons will not be large because a ϕ is a $S\overline{S}$ state, not a $s\bar{s}$ state.

LARGE N_c

There has been some discussion in the literature about $\Delta\Sigma$ and Δs in the limit of a large number of colours. It can be shown that in the $N_c \to \infty$ limit,

$$\Delta u - \Delta d \sim \mathcal{O}(N_c), \quad \Delta u + \Delta d \sim \mathcal{O}(1), \quad \Delta\Sigma \sim \mathcal{O}(1), \quad \Delta s \sim \mathcal{O}(1/N_c). \qquad (48)$$

The suppression of $\Delta\Sigma$ relative to $\Delta u - \Delta d$ is a trivial effect. It arises because the N_c quarks in the baryon have their spins combined to form a state with spin-1/2. For $\Delta\Sigma$, the N_c terms from the N_c quarks cancel to leave an effect of order one. For $\Delta u - \Delta d$, the minus signs from Δd cancel the minus signs for spin down d quarks, so that the N_c quarks can produce an effect of order N_c. The result (48) is different from the Skyrme model calculation of Δs and $\Delta\Sigma$[35] because in the Skyrme model, one does not take the limit $N_c \to \infty$. The Clebsch-Gordan coefficients used in calculating Skyrmion expectation values are for the **8** of flavour $SU(3)$, rather than for the flavour representation of a baryon made up of N_c quarks. The true $N_c \to \infty$ limit in the Skyrme model gives (48). The interesting results from EMC were that $\Delta\Sigma$ was small, and that Δs was large. Large N_c cannot explain this pattern.

There is also no relation between the $\eta' - N$ coupling and $\Delta\Sigma$. The analog of the Goldberger-Treiman relation for the η' is not valid, because the η' is not a Goldstone boson, *i.e.* there are large variations in hadronic form factors on the scale of the η' mass.

CONCLUSIONS

The EMC experiment has revived interest in the flavour structure of hadrons. There is as yet no reliable evidence for non-zero $\langle p|\bar{s}\gamma_\mu\gamma_5 s|p\rangle$ because of theoretical uncertainties in F and D, and in the extrapolations used by EMC, and by Ahrens *et al.*. Understanding the strangeness of the proton is theoretically interesting, because there is no good reason why all s-quark operators in the proton must vanish, since the s quark is not particularly heavy compared to the strong interaction mass scale. There have been preliminary lattice investigations of both Δg[36] and $\langle p|\bar{s}\gamma_\mu\gamma_5 s|p\rangle$,[37] and more accurate results on both quantities will be obtained soon. There will also be new experimental results in the near future that should provide additional information about s-quarks in the proton. Finally, the suggestion that there may be a large gluon contribution to the proton spin does nothing to help understand the EMC results or the non-zero value of $\langle p|\bar{s}\gamma_\mu\gamma_5 s|p\rangle$.

ACKNOWLEDGEMENTS

I would like to thank R.L. Jaffe for suggesting the title. The work described here was done in collaboration with R.L. Jaffe, E. Jenkins, and D.B. Kaplan. I would also like to thank R. Carlitz, J. Collins, S. Ellis, H. Georgi, J. Kuti, and J. Mandula for helpful discussions. This research was supported in part by a grant from the Alfred P. Sloan Foundation, by a National Science Foundation Presidential Young Investigator award #PHY-8958081, and by the Department of Energy under grant #DE-FG03-90ER40546.

REFERENCES

1. J. Ashman, *et al.*, Phys. Lett. **206B**, 364 (1988);
   J. Ashman, *et al.*, Nucl. Phys. **B328**, 1 (1989).

2. J. Kuti and V.F. Weisskopf, Phys. Rev. **D4**, 3418 (1971);
   J. Ellis and R.L. Jaffe, Phys. Rev. **D9**, 1444 (1974). .
3. D.B. Kaplan and A.V. Manohar, Nucl. Phys. **B310**, 527 (1988).
4. R.L. Jaffe and A.V. Manohar, Nucl. Phys. **B337**, 509 (1990).
5. J. Bijnens, H. Sonoda, and M.B. Wise, Nucl. Phys. **B261**, 261 (1985).
6. E. Jenkins and A.V. Manohar, UCSD/PTH 90-23.
7. F.E. Close and R.G. Roberts, Phys. Rev. Lett. **60**, 1471 (1988).
8. L.A. Ahrens, *et al.*, Phys. Rev. **D35**, 785 (1987).
9. R.D. McKeown, Phys. Lett. **219B**, 140 (1989);
   D.H. Beck, Phys. Rev. **D39**, 3248 (1989);
   See E.J. Beise and R.D. McKeown, Caltech preprint OAP-707, October 1990, for a review of some of the possible experiments.
10. Bates proposal #89-06, R.D. McKeown and D.H. Beck, contact people.
11. CEBAF proposal #89-23, R. Carlini, contact person.
12. A.V. Manohar and H. Georgi, Nucl. Phys. **B234**, 189 (1984).
13. R.L. Jaffe, Phys. Lett. **229B**, 275 (1989).
14. T. P. Cheng and R. Dashen, Phys. Rev. Lett. **26**, 594 (1971), Phys. Rev. **D13**, 216 (1976);
    C. A. Dominguez and P. Langacker, Phys. Rev. **D24**, 1905 (1981).
15. J. F. Donoghue and C. R. Nappi, Phys. Lett. **168B**, 105 (1986).
16. G. E. Brown, K. Kubodera and M. Rho, Phys. Lett. **192B**, 273 (1987);
    see also R. L. Jaffe, Phys. Rev. **D21**, 3215 (1980), C. L. Korpa and R. L. Jaffe, Comm. Nucl. Part. Phys. **17**, 163 (1987).
17. J. Gasser, H. Leutwyler, and M.E. Sainio, BUTP-90/31.
18. M.A. Shifman, A.I. Vainshtein, and V.I. Zakharov, Phys. Lett. **78B**, 443 (1978).
19. H. Abramowicz, *et al.*, Z. Phys. **C15**, 19 (1982);
    K. Lang, *et al.*, in 1985 Intl. Symp. on Lepton and Photon Interactions at High Energies, ed. M. Konuma and K. Kakahashi, (Kyoto, 1986); For a review, see F. Sciulli, *ibid.*
20. A.V. Efremov and O.V. Teryaev, Dubna preprint, JINR EZ-88-297 (1988).
21. G. Altarelli and G.G. Ross, Phys. Lett. **212B**, 39 (1988).
22. R.D. Carlitz, J.C. Collins, and A.H. Mueller, Phys. Lett. **214B**, 229 (1988).
23. J.C. Collins and D.E. Soper, Nucl. Phys. **B194**, 445 (1982).
24. A.V. Manohar, Phys. Rev. Lett. **65**, 2511 (1990).
25. A.V. Manohar, UCSD/PTH 90-20.
26. G. Bodwin and J. Qiu, Phys. Rev. **D41**, 2755 (1990).
27. This is a simple generalisation of a result of F. Hausdorff, Math. Z. **9**, 74 (1921) for the classical moment problem.
28. A.V. Manohar, UCSD/PTH 90-31.
29. I would like to thank P.A. Souder for this remark.

30. S. Adler, in Lectures on Elementary Particle Physics and Quantum Field Theory, ed. S. Deser, M. Grisaru, and H. Pendleton (MIT Press, Cambridge, 1970).
31. J. Kodaira, Nucl. Phys. **B165**, 129 (1979).
32. A.V. Manohar, UCSD/PTH 90-24.
33. This section is taken from A.V. Manohar, in Parity Violation in Electron Scattering, ed. E.J. Beise and R.D. McKeown, (World Scientific, Singapore 1990), p. 20; See [3] and [4] for a more detailed discussion.
34. G. Zweig, CERN preprints 401, 402, (1964), unpublished.
35. S. Brodsky, J. Ellis, and M. Karliner, Phys. Lett. **206B**, 309 (1988).
36. J. Mandula, Phys. Rev. Lett. **65**, 1403 (1990).
37. A.D. Patel, unpublished.

What Have We Learned From Global QCD Analysis of Unpolarized Parton Distributions?

Wu-Ki Tung

Illinois Institute of Technology[1], Chicago Illinois 60616
and
Fermi National Accelerator Laboratory, P.O. Box 500, Batavia, Illinois 60510

Abstract

This review of unpolarized parton distributions in the perturbative QCD framework examines contemporary issues on next-to-leading order studies, assesses the current status of global analysis of these distributions, and discusses the wide scope of physical processes relevant for improving the second-generation analyses and for resolving the remaining open questions. Subtlies and pitfalls are delineated – with a view of extracting pertinent lessions for the emerging study of polarized parton distributions.

1 Introduction

The parton model has its foundation in perturbative Quantum Chromodynamics (QCD). It furnishes a comprehensive framework for describing general high energy processes in current and planned accelerators and colliders. The basic formula – the Factorization Theorem – relates a *measurable* physical cross-section to a sum over relevant *fundamental partonic* hard cross-sections (which can be calculated perturbatively) convoluted with corresponding *parton distributions* (which can, in principle, be extracted from a set of standard experiments at moderately high energies). With proper attention to their definition and a consistent convention, these distribution functions are *universal* – i.e., they are independent of the physical process to which they are applied.

In leading-order (LO), the QCD parton framework reproduces original parton model results[1], with scale-dependent parton distributions. Since the early 70's, this simple model has enjoyed spectacular successes in unifying the phenomenology of all sorts of high energy processes to about the 10–15% level within currently available x range, modulo some overall "K-factors" for processes such as lepton-pair production. Furthermore, it has been used as an indispensable tool to make projections for future physics at much higher energies and small x.[2] In these applications, the well-known leading order parton parametrizations[2,3] played an essential role.

In recent years the use of QCD parton model has developed to a stage at which much improved knowledge of parton distribution functions are clearly required along three distinct fronts:

(i) In physical processes which provide precise tests of the Standard Model, the precision of experimental measurements has improved dramatically on the one

---
[1]Permanent Address

hand, and the relevant hard cross-sections have been calculated to the next-to-leading order (NLO) and beyond on the other. In order to make real progress, it is crucial to know the parton distributions to a comparable degree of accuracy. This, in turn, prescribes the use of NLO evolution of the distributions and requires a much more detailed comparison with experimental data in the extraction of these distributions, especially for the sea-quarks which are not well determined because of their relative small size. Relevant processes in this category are: electron, muon & neutrino deep inelastic scattering (DIS); lepton-pair production (LPP or DY), and (W, Z, γ) production. It has become increasingly clear that *the lack of detailed knowledge of the parton distributions often constitutes the largest source of uncertainty in precise tests of the Standard Model* in these processes (e.g. the determination of the Weinberg angle in DIS).

(ii) In the study of jet physics and associated production of (W, Z, γ) with jets, which are important on their own right as well as sources of significant background for "new physics", reliable knowledge of the gluon distribution is crucial to the predictive power of the QCD parton model. But the gluon distribution is, so far, not very well determined.

(iii) Physical processes at future colliders (SSC, LHC, ...) involve partons at very small x, well beyond the currently measurable range (around $0.03 < x < 0.75$). In order to make quantitative predictions, it is important: (a) to gain more theoretical insight on the small-x behavior of parton distributions and on the interface of small-x physics to "soft" (e.g. Reggeon) physics; and (b) to determine phenomenologically the range of possible small-x extrapolation of parton distributions consistent with currently available data.

This review cannot cover the full landscape of unpolarized parton distributions which now touches practically all areas of high energy physics. In the next section, I shall outline the NLO QCD parton formalism and use the examples of the gluon and the sea-quarks to illustrate non-trivial features which must be taken into account in modern quantitative applications. Although these examples are fairly simple conceptually, they have largely been overlooked by practitioners of QCD phenomenology, resulting in frequent misunderstanding and sometimes misuse of the formalism. In Sect. 3, I shall survey the existing parton distribution parametrizations and summarize the relevant issues confronting "second generation" global analyses of parton distributions. And in Sect. 4, I shall comment on issues pertaining to the proper use of these newer distributions. For specifics on recent developments, the reader can consult the Proceedings of two recent workshops devoted to this field, [4] and [5].[2]

2 QCD Parton Model in Next-to-Leading Order

For the sake of simplicity, I shall use a generic lepton-hadron scattering process as an explicit example. All the issues I discuss also apply to hadron-hadron scattering. The generic process is of the form: $\ell + H \longrightarrow C + X$ where C either represents an identified final-state particle with specific attributes (such as

---

[2]The present survey is adapted and updated from the review given by this author in [4].

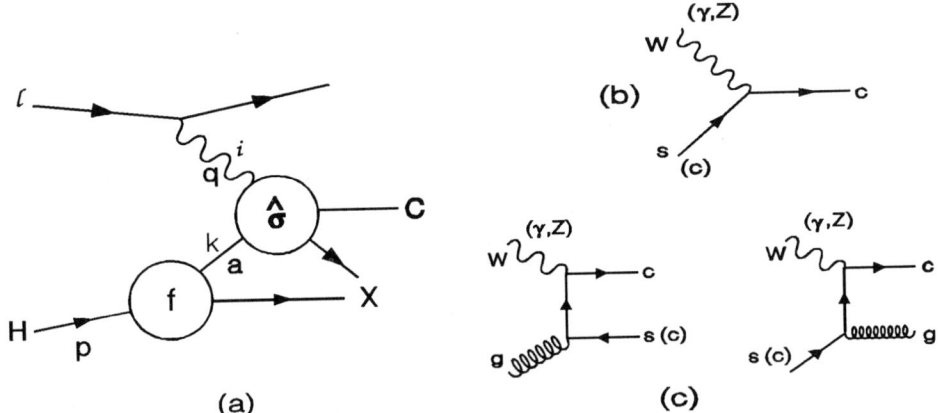

Figure 1: (a) The QCD parton picture and the Factorization Theorem; (b) Example of LO hard scattering; and (c) The corresponding NLO terms.

heavy mass or large transverse momentum) or is null in the case of total inclusive scattering. The "master equation" of the QCD Parton Model is the *factorization formula*[6] which reads:

$$\sigma^i_{H \to C}(q,p) = \sum_a f^a_H(\xi,\mu) \otimes \hat{\sigma}^i_{a \to C}(q,k,\mu) \qquad (1)$$

where, as illustrated in Fig. 1a, H is the target hadron label; a is the parton label; i is the electroweak vector boson helicity label; (q,p,k) are the momenta of the vector boson, the hadron, and the parton respectively; μ is a renormalization scale; and ξ is the fractional momentum carried by the parton with respect to the hadron. The symbol $\otimes$ denotes a convolution (over the variable ξ) of the parton distribution function f^a_H and the hard vector-boson-parton cross-section $\hat{\sigma}^i_{a \to C}$.

The hard cross-section $\hat{\sigma}^i_{a \to C}$ can be calculated in perturbative QCD. For our example, the first two terms in the series arise from Feynman diagrams of the kind shown in Fig. 1b,c, and take the form:

$$\hat{\sigma}^i(\xi, Q/\mu, \alpha_s(\mu)) = \hat{\sigma}^i_0 \, \delta(1-\xi) + \alpha_s(\mu) \, \hat{\sigma}^i_1(\xi, Q/\mu) + O(\alpha_s^2) \qquad (2)$$

where we have suppressed the initial parton label a. The LO $\hat{\sigma}^i_0$ (*cf.* Fig. 1b) is a simple constant proportional to the square of the electro-weak coupling of the parton. To calculate the NLO hard cross-section $\hat{\sigma}^i_1$ (*cf.* Fig. 1c) one encounters divergences (arising from integration over final state momenta) which must be subtracted in order to yield meaningful answers.

The subtraction, in effect, removes that part of the NLO contribution pertaining to almost on-the-mass-shell and collinear parton lines which is already included in the LO term by virtue of the use of QCD-evolved parton distributions. Since the subtraction, hence the hard cross-section $\hat{\sigma}$, is <u>renormalization scheme</u> and <u>renormalization scale</u> *dependent* while the physical cross-section on the left-hand side of Eq. (1) must be *independent* of these theoretical artifices, *the parton distribution functions f_H^a must be scheme-dependent objects* to match the definition of $\hat{\sigma}$. We shall demonstrate that *the scheme-dependence of the gluon and sea-quark distributions can be very substantial* – contrary to conventional expectations. This will lead to important phenomenological consequences.

2.1 Scheme-Dependence of Gluon and Sea-quark Distributions

We illustrate the scheme dependence of parton distributions by comparing two commonly used schemes, bearing in mind that the choice of scheme is in principle unlimited.

The $\overline{MS}$ scheme is defined by an "universal" subtraction prescription to facilitate perturbative calculations *independent of any physical process*. It is used by most theorists in the calculation of hard matrix elements. The $\overline{MS}$ parton distributions are guaranteed by their definition to satisfy the momentum sum rule. In this scheme, the NLO formula for the F_2 structure function of virtual γ deep inelastic scattering reads:

$$F_2^\gamma(x, Q) = f_{\overline{MS}}^q \otimes \left[C_{2,q}^{(0)} + \alpha_s\, C_{2,q}^{(1)\overline{MS}} \right] + f_{\overline{MS}}^G \otimes \alpha_s\, C_{2,G}^{(1)\overline{MS}} + O(\alpha_s^2) \quad (3)$$

where $C^{(i)}, i = 0, 1$ are the standard hard matrix elements in LO and NLO, often called the Wilson coefficients.[7] On the right-hand side of this equation there is an implicit sum over the quark flavor index.

The "DIS" scheme[8], on the other hand, was defined *specifically* to make the relation between the parton distributions and the deep inelastic scattering structure function F_2^γ simple. This is achieved by absorbing all the NLO terms on the right-hand side of Eq. (3) into the *definition* of f_{DIS}^q:

$$F_2^\gamma(x, Q) \equiv f_{DIS}^q \otimes C_{2,q}^{(0)} + O(\alpha_s^2) \quad (4)$$

Comparing the two equations, we find the difference between the quark distributions in the DIS scheme and the $\overline{MS}$ scheme to be:

$$f_{DIS}^q(x, Q) - f_{\overline{MS}}^q(x, Q) = \alpha_s(f^q \otimes C_{2,q}^{(1)\overline{MS}} + f^G \otimes C_{2,G}^{(1)\overline{MS}}) + O(\alpha_s^2) \quad (5)$$

No explicit label is given to the parton distributions on the right-hand side since these terms are of one order higher in α_s, thus either scheme should do. Eq. (4) does not say anything about the gluon distribution in the DIS scheme. However, if the momentum sum rule is to be preserved, the second moment of the gluon distribution in this scheme will be indirectly determined by those of the quarks. To complete the definition, it is conventional to generalize the condition on the second moment to all moments of the gluon distribution.[9][14] We then obtain:

$$f_{\overline{MS}}^G(x, Q) - f_{DIS}^G(x, Q) = \alpha_s(f^{q_s} \otimes C_{2,q}^{(1)\overline{MS}} + f^G \otimes C_{2,G}^{(1)\overline{MS}}) + O(\alpha_s^2) \quad (6)$$

where q_s denotes the singlet quark distribution.

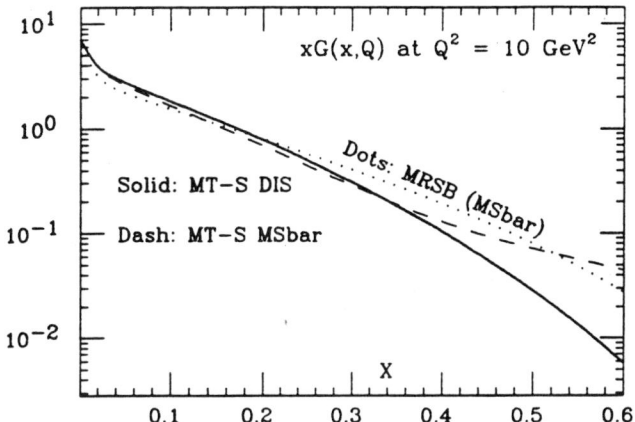

Figure 2: Comparison of the same MT gluon distribution in two different schemes, as well as with a different distribution (MRSB).

These equations allow us to convert parton distributions from one scheme to the other. As shown in Eq. (5) and Eq. (6), the difference of the same distribution in the two schemes is *nominally* of order α_s (rhs) compared to the individual distributions (lhs). Thus, little attention has been given to the scheme dependence of the distributions in practical applications – indeed it is not uncommon to discuss the gluon and quark distributions *as if* they are *unique* and *physical*. Let us see how far this customary practice can lead us astray.

The Shape of the Gluon Distribution: We first consider the question of "hard" vs. "soft" gluons which is, so far, an unsettled issue. We would like to show that this question itself is very scheme-dependent, hence has no content unless a consistent scheme is specified. Assuming that f^G is *soft* in one of the schemes (say f^G_{DIS}) – for example, it might behave like $(1-x)^\eta$ with $\eta \geq 6$,[10] then it approaches zero very fast as $x \longrightarrow 1$. However, since the first term on the right-hand side involves the *hard* valence quark distribution, it will be fairly "hard" – say, behaving like $(1-x)^\eta$ with $\eta \sim 4$.[15] As a consequence, the same distribution in the other scheme ($f^G_{\overline{MS}}$ in this example) will necessarily be hard! Thus *the "hardness" or "softness" of the gluon distribution is a very scheme-dependent concept*, which does not necessarily have an independent meaning. To see this point in concrete terms, we show in Fig. 2 a comparison of the gluon distribution at $Q^2 = 10 GeV^2$ from one of the MT fits[9] expressed in the DIS scheme (solid line), from the same fit in the $\overline{MS}$ scheme (dashed), and from the MRSB set (dotted) which is in the $\overline{MS}$ scheme. The enormous difference between the first two curves at large x – one can clearly be characterized as "soft", the other "hard" – arises entirely from the transformation from one scheme to the other; they actually represent the same gluon distribution! This plot underlines the necessity to specifying the defining scheme of the gluon distribution in any meaningful discussion about "soft" or "hard" gluons.

The conversion of the gluon distribution from one scheme to another necessarily turns a "soft" gluon distribution into a harder distribution, since the redefinition involves re-interpreting the gluons radiating off the (hard) valence quarks. In this connection, we note that: it is natural to expect a soft gluon distribution in the DIS scheme, as the contribution from the gluon to F_2 is absorbed into that from the quarks by fiat; whereas in the $\overline{MS}$ scheme, the gluons radiating off the valence are indeed counted as gluon partons. We add that the numerical difference between the distributions in the two renormalization schemes diminishes with increasing Q, becoming insignificant beyond $Q^2 = 100 GeV^2$ or so.

The Size of the Sea Quark Distributions: A similar situation exists for the sea-quark distributions *over the entire range of x*. For this case we refer to Eq. (5) and take f^q to be f^{sea}. It is well-known that, for small and moderate values of Q, the gluon distribution is much larger numerically than the sea-quark distributions. For instance, in terms of the fractional momentum carried by the partons, the ratio of these distributions is around 0.50 : 0.03 – a factor of greater than 10. Hence, the second term on the right-hand side of the equation can easily be of the same order of magnitude as the individual terms on the left-hand side even if it is formally of order α_s. In other words, *the size of the sea-quark distributions can depend substantially on the scheme in which they are defined*; and *it is not very meaningful to talk about a* LO *sea-quark distribution* since its definition is always coupled to the much bigger gluon distribution.[11]

Comments on the DIS-scheme: The widely used DIS-scheme was defined to render simple the formula for F_2, *and only F_2*. Even in the same scheme, however, the other deep inelastic scattering structure functions – F_1, F_3 or $F_{\text{left}}, F_{\text{right}}$ – *do* contain non-trivial NLO contributions from both quarks and gluons. Also, the simplification only applies to the *total inclusive* structure function F_2. In practice, one is often interested in semi-inclusive processes such as the production of jets at a given transverse momentum or the production of heavy flavors (charm, bottom, etc.). In fact, the scattering of the (electroweak) vector boson with gluons in the hadron (a NLO hard process) is *mainly responsible* for producing final state jets at non-vanishing transverse momentum and final state heavy quark flavors (by pair-production), especially at energies not too far above threshold. Thus, although it is theoretically allowed to absorb the entire NLO contribution to F_2 into the definition of the DIS scheme quark distributions (which implicitly assumes collinear, on-the-mass-shell partons), this convenience for the total inclusive process comes clearly at the expense of *over-subtraction* (since the NLO diagrams contain non-collinear, off-the-mass-shell quark configurations as well). Hence, in applications to semi-inclusive processes, the use of DIS scheme distributions requires some care, and may lead to counter-intuitive results. The use of this definition is partially responsible for the "large" NLO corrections found for certain processes, such as vector boson production.

2.2 Non-trivial Order of Magnitude Estimates in QCD

Traditionally, in applying the perturbative QCD formalism to physical processes, the various terms which contributes to the right-hand side of Eq. (1) are classified as LO, NLO, etc., according to the perturbation expansion of the hard cross-section $\hat{\sigma}^i_{a \to C}$ only. However, we have just pointed out circumstances under which terms of different orders will mix because the distribution functions $f_H^a(\xi,\mu)$ on the right-hand side can differ significantly in order of magnitude. In particular, we already mentioned the large gluon to sea-quark ratio, compensating an explicit power of α_s.[3]

Thus, for the purpose of order of magnitude estimates, the parton distributions can be classified into two distinct classes: (i) those of order 1 – the gluon and the "valence" quarks (u, d) (to be denoted by f^{large} below); and (ii) those effectively of order α_s – the active sea quarks (to be called f^{small}). In this subsection, we describe phenomenological consequences of this observation in physically interesting processes. Taking into account the order of magnitudes of *both* the parton distributions *and* the hard cross-section, the perturbative QCD formula Eq. (1) can be reorganized, schematically, as follows,

$$\sigma_{phys} = f^{\text{large}} \otimes \hat{\sigma}_{LO}^l + \left[f^{\text{small}} \otimes \hat{\sigma}_{LO}^s + f^{\text{large}} \otimes \hat{\sigma}_{NLO}^l \right]$$
$$+ \text{numerically smaller terms} \qquad (7)$$

where the labels (l, s) on $\hat{\sigma}$ are flavor indices matching those on the parton distributions. This organization gives a more realistic ordering of the terms than the conventional one at non-asymptotic energies: the two terms inside the brackets are numerically comparable.

This point becomes *crucial* when the leading term on the right-hand side is absent or suppressed (because either $\hat{\sigma}_{LO}^l$ vanishes or contains a suppression factor due to the electro-weak coupling). Then the traditional LO analysis which only keeps the $f^{\text{small}} \otimes \hat{\sigma}_{LO}^s$ term becomes totally inadequate, since the "NLO term" in the same square bracket is of the same numerical order. A case in point is charm production in neutrino deep inelastic scattering: Here the d-quark contribution ($f^{\text{large}}\hat{\sigma}_{LO}^l$) is suppressed by a small weak coupling. Thus, all existing theoretical and experimental analyses of this process have focused on the "LO" scattering of the weak W-boson on the strange quark in the hadron target ($f^{\text{small}}\hat{\sigma}_{LO}^s$, cf. Fig. 1b). The above discussion clearly suggests that the "NLO" contribution from the scattering of W on the gluon ($f^{\text{large}}\hat{\sigma}_{NLO}^l$, Fig. 1c) can be just as important. The results of a recent calculation confirm this;[11] a representative figure is shown in Fig. 3. The gluon contribution is seen to be around 50% of the "LO" term, hence must be included in a proper QCD formulation of the problem.

Another approach sometimes found in the literature concerning this class of problems, especially involving heavy flavor production in neutral-current processes, is to invoke *only* the gluon contribution (the so-called "gluon-fusion" mechanism)[12][4] – just the opposite of the quarks-only approach. Although this may make sense in some restricted kinematic region, notably just above the threshold of producing the heavy flavor pair, the LO quark scattering diagram must also play an important role, and eventually become the dominant mechanism with

---

[3]This is easily understood: Since the sea-quarks qualitatively arise from the splitting of the gluon; they contain an *implicit* power of α_s with respect to the gluon, at least for moderate values of Q.

[4]Recent calculations of heavy quark production at HERA have all been based on this mechanism.

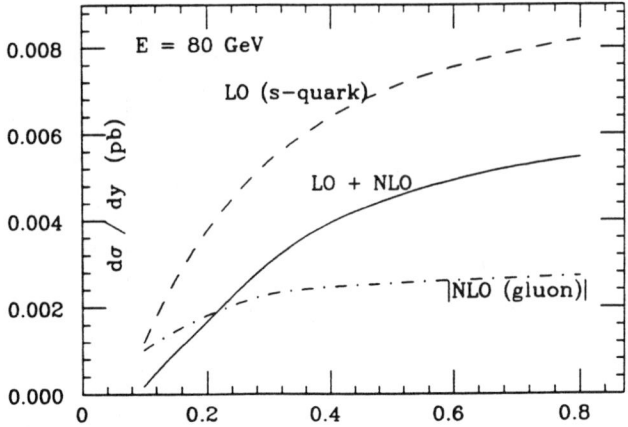

Figure 3: Charm production in deep inelastic changed-current scattering – "LO" cross-section compared to the "NLO" result including the gluon contribution.

increasing energy. The above discussion should make it clear that a consistent treatment has to include both mechanisms if it is to be quantitatively reliable over the entire energy range.

We summarize the key points underlying the topics discussed in the last two subsections. Within the QCD parton model, the contributions from the sea quarks and from gluons are always inextricably intertwined. In spite of the conventional designations of LO and NLO respectively, they can be numerically comparable. The precise division between the two mechanisms is tied intimately to the choice of renormalization scheme used during the calculation. Although it is theoretically possible to minimize the contribution from one or the other mechanism to *one given quantity* (e.g. F_2) by a specific choice of scheme, both terms must be included in the analysis of all other physical quantities. In order to achieve consistent results, the choice of scheme must be specified explicitly in all these applications.

3 Overview of Global Analyses of Parton Distributions

The global analysis of parton distributions involves the quantitative comparison of experimental data from a wide range of physical processes with the QCD parton model master equation, Eq. (1), for the purpose of extracting a set of universal parton distribution functions. These can then be used in other applications: to make "predictions" as well as to provide stringent tests of the self-consistency of the perturbative QCD framework itself or of the Standard Model in general. Since any compelling indications of inconsistency of the SM are signs of "new physics", and since even direct search for new physics must rely heavily on understanding of the background from conventional physics, the systematic analysis of parton distributions is intimately tied to all these ventures. Thus, *contemporary global analyses of parton distributions must incorporate all relevant modern high statistics experimental results and apply the* NLO QCD *formalism in a consistent manner* as described in Sec. 2. In the following, we: (i) briefly review and assess the existing parton distribution parametrizations; (ii) highlight the relevant phenomenological issues for modern quantitative global analysis; and (iii) summarize current efforts on global analyses and list the open questions.

3.1 Review of Parton Parametrizations

The list of widely used parton distributions is a long one. Prominent among these are the pioneering works of Feynman-Field and Buras-Gaemers; followed by the widely used distributions of Gluck-Hoffmann-Reya, Duke-Owens, and Eichten-Hinchliffe-Lane-Quigg.[2,3] These are all based on leading order QCD-evolved distributions extracted by comparison with data existing up to about 1983. In recent years, many second generation high statistics experiments have become available and more refined global analyses have been carried out in NLO by Martin-Roberts-Stirling[13], Diemoz-Ferroni-Longo-Martinelli[14], Aurenche-Baier-Fontannaz-Owens-Werlen[15], and Morfin-Tung[9], reflecting the needs of the current time. Table I lists most of the currently used parton distributions and the experimental data on which the analyses were based.

	D-O[3]	EHLQ[2]	(H)MRS[13]	DFLM[14]	M-T[9]
ν-DIS	CDHS	CDHS	CDHSW,(CCFRR)	CHARM	CDHSW,(CCFR)
μ-DIS	EMC	—	EMC,BCDMS	—	EMC,BCDMS
D-Y	E288,ISR	—	(E288),(E605)	—	E288,E605
Dir-γ	—	—	WA70	—	—

Table I: Parton Distribution sets and data used.
References to the experiments are: BCDMS,[16] CDHS,[17] CDHSW,[18] CCFR,[19] CCFRR,[20] E288,[21] E605,[22] EMC,[23] WA70.[24] Parentheses around entries indicate that the corresponding data were only used partially.

The experimental developments which have the most significant impact on recent analyses as compared to the previous LO ones are: (i) results of the high statistics CDHS[17] neutrino experiment, on which most earlier analyses were heavily dependent, has since been considerably revised and supplanted by the new CDHSW[18] results; (ii) the very accurate new data on muon scattering from the BCDMS[16] collaboration has become available. The disagreement between the BCDMS and the earlier EMC[23] results has been a source of much uncertainty and discussion for the past two years. However, recent comprehensive studies, including the introduction of the new analysis of the SLAC-MIT experiments[25] as an independent check on the normalization as well as a recent reanalysis of the EMC[26] data, has gone a long way to resolve the discrepancy.[27]

These recent developments cause the predictions of the earlier parton distributions on deep inelastic scattering –the main source of information on these distributions – to disagree with the best current data up to 15-20%. Differences of this size correspond to many standard deviations in these high statistics experiments. There is no room here to make a detailed comparison. (See [28][29][30].) To illustrate the general state of affairs, we show one typical plot on the comparison of current data with the results calculated from representative parton distribution

114 Unpolarized Parton Distributions

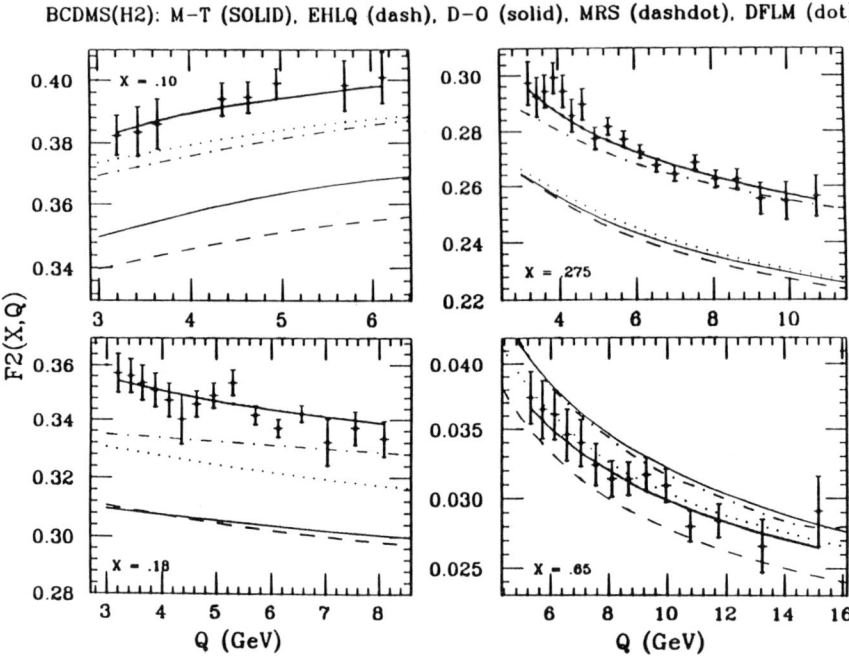

Figure 4: Comparison of a subset of BCDMS F_2 data[16] on hydrogen with calculated results from representative parton distribution sets.

sets in Fig. 4. This figure clearly demonstrates the glaring discrepancy between recent data and the popular classical parton distributions. It underlines the necessity for using up-to-date parton distributions in quantitative QCD analyses where accuracy is intended.

3.2 Relevant Issues for Quantitative Global Analysis

We shall now briefly summarize the important experimental and theoretical considerations which affect a serious quantitative global QCD analysis of parton distributions. A systematic discussion can be found in the report of the Structure Function and Parton Distribution group in the 1988 Snowmass Proceedings.[28]

On the experimental side, in additional to the choice of physical processes and experiments to fit, all the following factors can affect the consistency and the correctness of the results: (i) the selection of data within a given experiment, according to kinematic cuts in Q^2, W or other variables. (How do the results depend on the values of these cuts?) (ii) Are systematic errors included in the fitting procedure? This is, in fact, a critical issue for the second generation analysis since: (a) errors on current high statistics experimental data are dominated by systematic errors; and (b) when data from several experiments are used in a chi-square or likelihood analysis, the fitting procedure is simply meaningless without including the systematic errors. However, most existing global analysis *do not* include systematic errors.[5] (iii) Do the different experimental data sets apply the same "corrections" (e.g. "slow-rescaling", "isoscalar", etc.) to their data analyses; and, if not, how should one handle the differences?[28] Unfortunately, serious differences in applying "corrections" do occur in published experimental analyses of related measurements, and these differences are too often overlooked in phenomenological global analysis. (iv) Finally, distinct physical quantities measured in the same experiments can have correlated errors. A proper fit to the data must take into account the correlations. No global analysis to date has attempted to incorporate this in a systematic way for all data sets.

On the theoretical side, the parton distributions and QCD parameters one obtains from global fits can depend (in addition to the choice of LO or NLO formalism) on: (i) the functional form used for the initial distributions, especially if it happens to be too restrictive; (ii) the number of parameters which are allowed to vary when the fit to data is made. Unfortunately, there is no proven way to assure a correct choice on either of these considerations. A reasonable choice when fitting a given set of data may not remain so when additional experiments and/or physical processes are incorporated. Finally, results on these global analyses in the very small x region – a region of much interest in applications to "predict" the high energy behavior of standard and new physics – are heavily dependent on whether the small-x *extrapolation* is fixed by an assumed functional form or is characterized by parameters to be fit to existing data.

These issues should be of concern to users of parton distributions in quantitative analyses, as the reliability of the distributions can have an impact on the basic physics one is trying to extract from the measured quantities.

3.3 Current Status of Global Analyses and Open Issues

There are two active programs on global analysis of parton distributions based on current data.[13,9] They have recently been summarized.[31][32] They both fit the data listed in Table 1, although they do use different procedures, analysis criteria, and parametrizations. Each group found a range of possible behavior for most distributions, especially for the gluon and the sea quarks. This makes direct comparison of the distributions difficult. (Interested readers should consult the detailed accounts in the original papers.) It also underlines the fact that, in spite of past efforts and the current detailed work, there remain many *open questions* which will require a combined effort of refined and expanded experiments, further theoretical clarifications, and continued phenomenological analysis to be resolved. The list of open questions and their potential resolution includes:

⋄ *The gluon distribution*: What is the proper or optimal definition? (Cf. Sec. 2.1 and [33]) How can it be determined in an unambiguous way? Current data on DIS do not determine the gluon distribution well because of the lack of high-Q information on the longitudinal structure function. The expected measurement

---

[5]This is ionic as most experiments devote more effort on understanding the systematic errors than on anything else.

of this key function at HERA will be most useful. Lepton-pair production data imposes better constraints (through combined effects of sea-quarks and gluons)[9]. The "dedicated study" of the gluon distribution based on direct-photon production shows a great deal of promise,[15][34] but is also subject to a number of experimental limitations (limited kinematic range, large errors, isolation criteria, etc.) and theoretical uncertainties (choice of scales, photon fragmentation functions, bremsstrahlung contributions, etc.). Some of these complications have been clarified recently,[35] and new data from fixed-target and collider experiments are expected in the near future.[36] Are there other comparable or better ways to determine the gluon distribution?

⋄ *The sea-quark distributions*: In view of the discussion of significant scheme-dependence of these distributions given in Sec. 2.1 and Sec. 2.2, is there a natural choice of scheme that can be generally agreed upon? What is the dependence of $f_{sea}^{q^i}$ on the flavor label i? Is it SU(3)-symmetric, SU(2)-symmetric, or totally non-symmetric? Current data on charm production (dimuon final state) in DIS have been interpreted to indicate a non-SU(3)-symmetric sea.[37] However, the discussion of Sec. 2.2 suggests that the interpretation, based on a LO picture, needs to be reassessed. There is also recent suggestion that the $\bar{u}$ and $\bar{d}$ distributions may also be different.[38] Is this really the case? All these questions can only be settled by more detailed experiments and more comprehensive theoretical analysis. The next-to-leading order QCD tools are now available, they need to be systematically applied to yield consistent results.[11]

⋄ *Small-x behavior of the parton distributions*: In the familiar perturbative QCD formalism, the Q-dependence of parton distributions $f^i(x,Q)$ is governed by the evolution equation with calculable kernels; however, the x-dependence beyond currently measurable range is unknown except for some qualitative guidelines based on Regge-type of arguments at some unspecified scale. Much attention has been given to developing theoretical tools to extend our understanding of the parton model into the small-x region.[39] There are several distinct aspects to the "small-x problem": (i) the resummation of large $(\alpha_s \log(1/x))^n$ terms for fixed Q; (ii) the region of large $\log(1/x)$ and large $\log Q$; and (iii) the region of saturation of parton densities and the breakdown of the parton picture as we know it. Promising recent progress have been made and they are reviewed in two recent workshops.[4,5]

From the phenomenological point of view, one can ask what reasonable constraints on the small-x extrapolation of the parton distributions can be obtained by global fits to current data. Shown in Fig. 5a are plots of the gluon distribution and the predicted structure function F_2 extending to very small x from two sets of NLO parton distributions which both fit current data well.[9] The difference between the extrapolated results is seen to be quite large. This difference can be resolved by direct measurement of F_2 (and hopefully $G(x,Q)$) at HERA as well as by some suitable hadron-hadron collision process such as lepton-pair production (DY) at the colliders. Fig. 5b illustrates this point by showing the anticipated rapidity distribution of the lepton-pair at the Tevatron for $Q = 20 GeV$ using several choices of parton distributions, including the two mentioned above. We see that both the shape and the normalization of the y-dependence are quite sensitive to the uncertain small-x behavior of the parton distributions. Similar observation has been made recently by the MRS group concerning W-production at SSC.[40]

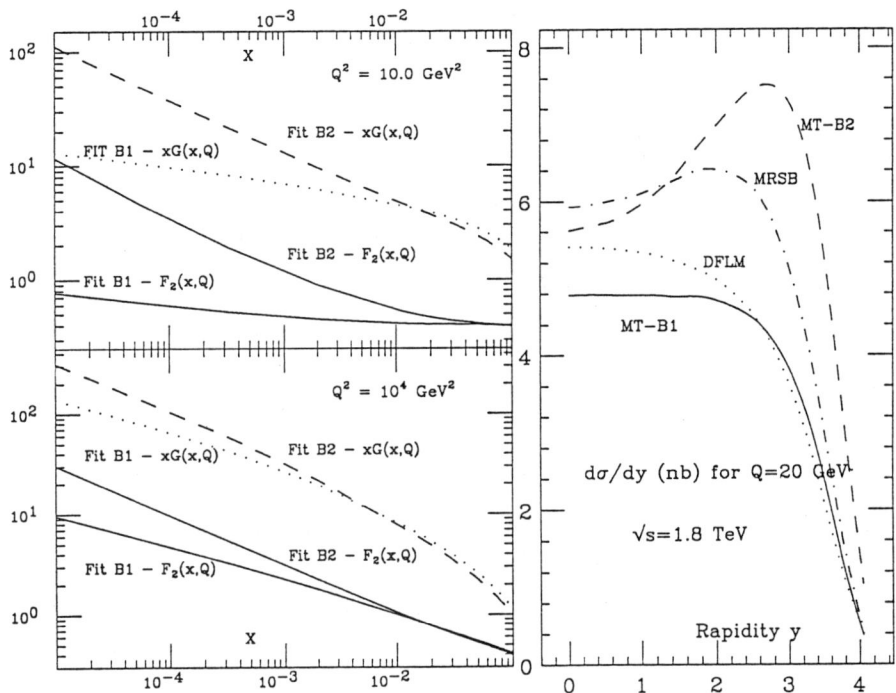

Figure 5: (a) Small-x extrapolation of $G(x,Q)$ and $F_2(x,Q)$ based on two global fits to current data.[9]; (b) Rapidity distribution of Drell-Yan pairs predicted by some representative parton distributions.[9]

In order to address the open questions described above, we need to go beyond the traditional reliance on DIS and lepton-pair production processes. While important progress will continue to be made in these areas, especially with the exciting expansion of experimental range offered by HERA, the coming of age of fixed-target direct photon experiments and quantitative measurements of an ever wider range of processes at the hadron colliders have opened up many more possibilities of determining the remaining uncertainties of parton distributions and testing the consistency of QCD. Foremost among these are W-, Z-production cross-sections (including rapidity and transverse momentum distributions), heavy flavor production, jet production (with or without associated vector bosons), as well as DY and direct-photon processes.

4 Some Remarks on the Use of Parton Distributions

As the parton model advances from the original "naive" genre to a comprehensive "QCD-based" framework, many subtle and unfamiliar features arise as we have described in the previous sections. This section focuses on issues pertaining to the proper use of the parton distributions in light of these advances.

⋄ *Leading Order Parton Distributions:* The LO QCD parton model is still a very powerful and simple framework which describes a wide range of physical processes to within 10–20% accuracy. In these applications, it is sufficient to use LO parton distributions. In fact, it is preferable to do so, than to use the NLO distributions as the latter are extracted by comparing complete NLO formulas with experiment. (A built-in error is always incurred by the mixed use of a LO hard cross-section with NLO parton distributions.) It is important, however, that the LO distributions used do accurately reflect existing data where applicable. As discussed in Sec. 3.1, the most often used first generation parton distributions, unfortunately, strongly disagree with the high precision DIS and DY data due to significant recent experimental advances. A LO parton distribution set which fits recent data has been included in one of the new analysis.[9]

⋄ *Next-to-leading Distributions and Scheme Dependence:* For applications which require more accuracy, the NLO formula is needed. In applying the NLO QCD formalism, it is important to make sure that the hard cross-section formula and the parton distributions are both *defined in the same renormalization scheme* (cf. Sec. 2.1). The often found practice of applying all available distributions, irrespective to the order and to the scheme, to the same hard cross–section formulas in "comparison of experiments with QCD" in the literature is clearly ill-founded.

Although the outcome of a given NLO global analysis of parton distributions can, in principle, be presented in any scheme, most published distribution sets are given only in the scheme used in the original analysis. It is therefore, incumbent on the user to make the necessary adjustment in order to bring about consistency. Among the current NLO parton distributions, the MRS[13] ones are in the $\overline{MS}$ scheme, the DFLM[14] ones are in the DIS scheme, whereas the MT[9] distributions are given in *both* schemes.

⋄ *Scale Dependence:* All QCD "predictions" are subjected to an uncertainty associated with the choice of renormalization and factorization scales. This uncertainty diminishes at very high energies or, in principle, when more and more higher order terms are included. At current energies and to NLO only, this scale-dependence can be substantial for certain processes. There is, so far, no clear consensus among theorists on whether there is some sensible method for making an intelligent (if not "correct") choice of scale for a given situation. Because of the practical importance of this issue, there will be continued discussion and debate about this topic.

⋄ *"Theoretical Uncertainty" due to Parton Distributions:* A common *myth* in contemporary phenomenology is to apply a variety of parton distributions to a given physical process and then cite the range of results obtained as the "theoretical uncertainty" associated with parton distributions. There is no justification for such a practice, since: (i) the LO and NLO distributions are not designed to be used the same way; and (ii) some of these distributions are already known to disagree strongly with current data, as mentioned above. In fact, as shown in Fig. 4, in many cases the deviations from the correct results are all to the *same* direction for most of these distributions (rather than "bracketing" the right answer).

Since existing data do not completely determine all the parton distributions, it is, of course, important to assess the range of uncertainties due to current lack of knowledge, both in performing precision tests and in making predictions for the future. One can do this properly by obtaining a range of parton distributions in the *same* global analysis with allowance made for the uncertain features remaining. This approach has been used in characterizing the small-x extrapolation of parton distributions in recent studies.[13,9] It can also be applied to other features such as the flavor dependence of the sea quarks.[9]

◊ *Comparison of Parton Distribution Sets:* The various current parton distribution sets differ considerably in: (i) the choice of input experimental data (cf. Table I); (ii) the treatment of experimental errors and corrections (cf. Sec. 3.2); and (iii) the selection of functional forms for the input distributions and the number of free parameters. (cf. Sec. 3.2) Many implicit assumptions ride on the choice made in (iii) which users do not see explicitly. For precision applications, these factors must be examined carefully before definite conclusions are drawn. Critical comparison of different sets of distributions have been discussed at various workshops but turns out to be quite difficult to carry out due to factors listed above.

5 Lessons?

Perturbative QCD stands as one of the main pillars of contemporary high energy physics. It plays a central role in pushing quantitative tests of the Standard Model to ever increasing accuracy (hopefully, even to its eventual limit); in studying the signals and backgrounds for new physics; and in providing hints for attempts to formulate non-perturbative solutions to strong interaction physics. Most applications of QCD are based on the Factorization Theorems which have a simple parton model interpretation and rely on the use of the universal parton distributions to relate measurable quantities to the fundamental processes of the underlying theory. This framework applies to the unpolarized world as well as the polarized world.

We have emphasized the importance of *precise definitions* for parton distribution functions in quantitative QCD analysis. Interestingly, this point is perhaps better understood in the world of polarized partons than in the more "mature" unpolarized one, thanks to the much publicized controversies surrounding the possible interpretation of the EMC spin measurement.[41] The matter of precise definition of the polarized quark and gluon distributions will be discussed extensively in this Workshop.

This survey also underscored the unfortunate consequences of the lack of full attention to the subtleties and pitfalls of NLO QCD in the literature on unpolarized QCD physics. These pitfalls could be avoided by good communication between the theorists, phenomenologists and experimentalists. However, there seems to have developed a *polarization* in the *world of unpolarized parton physics* due to less-than-effective communication between the three groups. The legacy of sloppy use of the parton model, by now deeply ingrained in some sectors of this community, confuses the issue in many contemporary discussions and impedes real progress. Let us hope the emerging *physics community of polarized partons* shall remain *unpolarized* so that the underlying physics can be revealed most expeditiously in what would be an extremely challenging and arduous venture under the best circumstances.

Acknowledgement: I would like to thank my collaborators M. Aivazis, J. Morfin, F. Olness and numerous other colleagues, in particular John Collins, for many useful discussions.

References

[1] R.P. Feynman, *Photon-Hadron Interactions*, Benjamin, (1972).

[2] E. Eichten *et al*, *Rev. Mod. Phys.*, **56**, 579 (1984) and Erratum **58** 1065 (1986).

[3] M. Glueck *et al*, *Z. Phys.* **C13**, 119 (1982); D. Duke and J. Owens, *Phys. Rev.* **D30**, 49 (1984).

[4] *Proceedings of Workshop on Hadron Structure Functions and Parton Distributions*, Fermilab, April 1990, Ed. D. Geesaman *et al*, World Scientific Pub. (1990).

[5] *Proceedings of Workshop on Parton Distribution Functions at Small-x*, DESY, May 1990, Ed. J. Bartel, North Holland Pub. (to be published)

[6] For a recent review and original references, see J. Collins, D. Soper and G. Sterman in *Perturbative Quantum Chromodynamics*, A. Mueller, Ed., World Scientific Pub., 1989.

[7] Cf. E.G. Floratos *et al*, *Nucl. Phys.* **B192**, 417 (1981) and references cited therein.

[8] G. Altarelli, R.K.Ellis, & G.Martinelli, *Nucl. Phys.* **B143**, 521 (1978); and *ibid.* **B157**, 461 (1979).

[9] J.G. Morfin and W.K. Tung, *Z. Phys.* C. (to be published), Fermilab and IIT preprint.

[10] See, *e.g.* Eichten *et al*, Ref.2; Diemoz *et al*, Ref.14.

[11] M.G. Aivazis, F. Olness & Wu-Ki Tung, *Phys. Rev. Lett.* **65**, 2339 (1990); and to be published.

[12] J. Leveille, T. Weiler, *Nucl. Phys.* **B147**, 147 (1979); T. Weiler, *Phys. Rev. Lett.* **44**, 304 (1980); M. Arneodo, *et al, Z. Phys.* **C35**, 1 (1987).

[13] A.D. Martin, R.G. Roberts & W.J. Stirling, *Phys. Rev.* **D37**, 1161 (1988), *Mod. Phys. Lett.* **A4**, 1135 (1989); P.N. Harriman, A.D. Martin, R.G. Roberts & W.J. Stirling, *Phys. Rev.* **D42**, 798 (1990).

[14] M. Diemoz *et al, Z. Phys.* **C39**, 21 (1988).

[15] P. Aurenche *et al, Phys. Rev.* **D39**, 3275 (1989).

[16] BCDMS Collaboration (A.C. Benvenuti,*et al.*), *Phys. Lett.* **B223**, 485 (1989); *Phys.Lett.*, **B237**, 592, 1990; *Phys.Lett.*,**B237**, 599, 1990.

[17] H. Abramowicz *et al, Z. Phys.* **C17**, 283 (1984); *Z. Phys.* **C25**, 29 (1984); *Z. Phys.* **C35**, 443 (1984).

[18] J.P.Berge *et al*, Preprint CERN-EP/89-103 (1989).

[19] Private communications from Sanjib Mishra.

[20] D.B. MacFarlane *et al, Z. Phys.* **C26**, 1 (1984).

[21] A.S.Ito *et al, Phys. Rev.* **D23**, 604 (1981).

[22] C.N. Brown *et al, Phys. Rev. Lett.* **63**, 2637 (1989).

[23] J.J. Aubert *et al, Nucl. Phys.* **B293**, 740 (1987).

[24] M. Bonesini *et al, Z. Phys.* **C38**, 371 (1988).

[25] A. Bodek, in Ref. 4.

[26] K. Bazizi, S.J. Wimpenny, and T. Sloan, contribution to *Proceedings of the 25th International Conference on High Energy Physics*, Singapore, 1990 (to be published).

[27] A. Milsztajn, in Ref. 4; and A. Milsztajn, A. Staude, K. Teichert, M. Virchaus and R. Voss, CERN-footnotesize PPE/90-135.

[28] Wu-Ki Tung *et al*, in *Proceeding of the 1988 Summer Study on High Energy Physics in the 1990's*, S. Jensen, Ed., World Scientific, (1990).

[29] K. Charchula *et al*, DESY report DESY 90-019 (1990).

[30] T. Sjostrand, summary talk at the LHC Workshop, Aachen, 1990.

[31] See A. Martin and R. Roberts, Ref. 4.

[32] J. Morfin, Ref. 4.

[33] K. Ellis, Ref. 4.

[34] J. Owens, in Ref. 4.

[35] E. Berger and J.W. Qiu *Phys. Lett.* **248B**, 471 (1990); J. Owens and J.W. Qiu, in Proceedings of Snowmass 90 (to be published).

[36] See reviews by C. Bromberg, R.M. Harris, L. Camilleri, and G.F. Egan in Ref. 4.

[37] CDHS: H. Abramowicz, *et al*, *Phys. Rev. Lett.* **57**, 298 (1986); *Z. Phys.*, **C28**, 51 (1985); CHARM: J.V. Allaby, *et al*, *Z. Phys.* **C36**, 611 (1987); CCFR: K. Lang *et al*, *Z. Phys.* **C33**, 483 (1987); S.R. Mishra *et al*, in *Proceedings of 14th Rencontres de Moriond*, Mar. 1889.

[38] Recent discussions on this subject are motivated by results reported by the footnotesize NMC experiment at the PANIC XII (MIT, 1990) and XXVth International HEP Conferences (Singapore, 1990), See, for instance: G. Preparata *et al*, Marseille Preprint CPT-90/PE.2417; S. Kumano, Indiana Preprint IU/NTC 90-14.

[39] For a comprehensive review of pioneering works in this field, see: L. Gribov, E. Levin, and M. Ryskin, *Phys. Rep.* **100**, 1 (1983). See also recent reviews by E. Levin in Ref.4 and Ref.5.

[40] J. Stirling, Ref. 5.

[41] For reviews, see: K. Rith in Ref. 4; T.P. Cheng, Proceedings of DPF90 meeting, Rice University, 1990; A. Mueller, report to the PANIC XII conference, MIT, 1990 (to be published).

POLARIZATION PROSPECTS FOR KEK

Kenichi Imai
Department of Physics, Kyoto University, Kyoto 606 Japan

ABSTRACT

A polarized collider has been proposed as an polarization option of the "PS Collider" which is a future project of the Proton Synchrotron at KEK. A brief description on the current plan about the collider is given and the possible physics with this machine is discussed.

PS COLLIDER

At KEK, we have two accelerator complexes, the TRISTAN and the 12GeV Proton Synchrotron (PS). The TRISTAN is e^+e^- collider at 30 x 30 GeV. Recently the longitudinal polarization of about 40% has been observed unexpectedly. At the PS, extensive efforts have been made to accelerate a polarized proton beam. So far the polarized proton of about 10^9/spill can be accelerated up to 4 GeV with the polarization of about 40%.[1] Since we have now both polarized proton and electron (positron), we can think of various options as a future polarization prospect such as a polarized ep collider. But here we discuss a polarized pp collider as our future plan since the polarized pp collider seems to be a good project on the bases of both physics and feasibility.

The "PS Collider" was proposed as a heavy ion collider which provides the collision of gold beams at the energy of 5 - 7 GeV/nucleon each in order to achieve the highest baryon density for the study of quark-gluon plasma. The design study of the PS-collider was made recently and the details are given in the design report.[2] This PS-collider can be used for the collision of polarized protons or deuterons as a "polarized collider" with some modifications to the original design. The PS-collider is a rather modest plan of the extension of the Proton Synchrotron. The two rings for the collision will be located in the existing experimental hall. as shown in Fig.1. The polarized protons are extracted from the synchrotron at low energy (1.8 GeV) to avoid the depolarization and injected into the collision rings where the protons are accelerated up to 19 GeV at maximum. The dipole magnets are superconducting magnets of which maximum field is about 6.5T. There are two collision points in the ring.

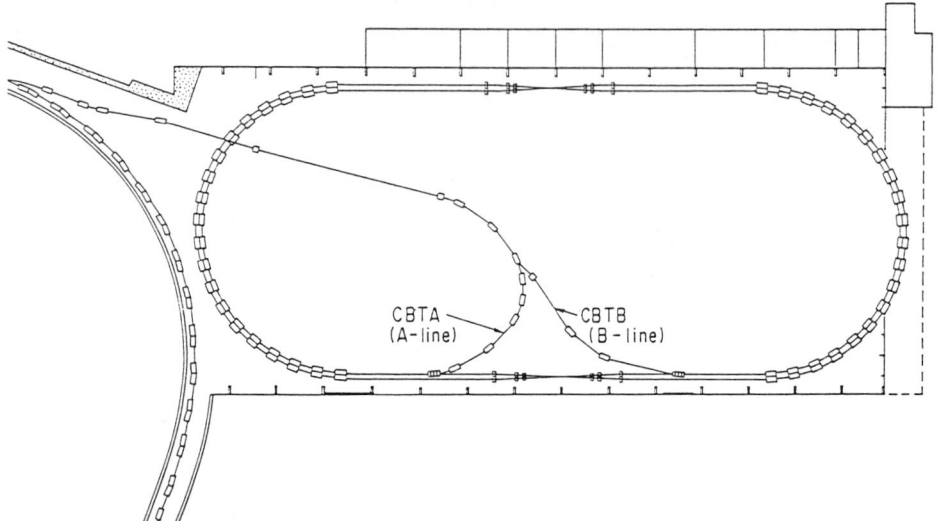

Fig.1 The PS collider. The polarized beam is extracted from the existing Proton Synchrotron and injected into the collider rings.

In order to keep the polarization and to get the longitudinal polarization at the collision point, the Siberian Snake method will be applied. Several schemes of the Snakes have been considered. One of the scheme is shown in Fig.2 schematically. In this scheme, two snake magnets which rotate the spin by 90° each and one snake which rotates the spin by 180° are employed to obtain the longitudinal polarization at one crossing point and the vertical polarization at the other straight section. Although the longitudinal spin asymmetry is important at high energy, the vertical polarization is necessary not only for physics experiments but to monitor the beam polarization. The design study for the scheme to handle the spin is still in progress. The expected luminosity is 10^{27}-10^{30} cm^{-2} sec^{-1}. The beam polarization of about 60% is anticipated by the various improvements of the ion source and the booster synchrotron. The energy is variable up to 19 x 19 GeV.

With this machine we can study the polarization phenomena at higher energy region than the present FNAL polarized proton beam of which energy corresponds to 10 x10 GeV in the case of the collider. The advantage of the collider for the polarization physics is not only that we can get higher energy but also we can use pure polarized proton beam and target which is very important for the measurement of inclusive reactions,

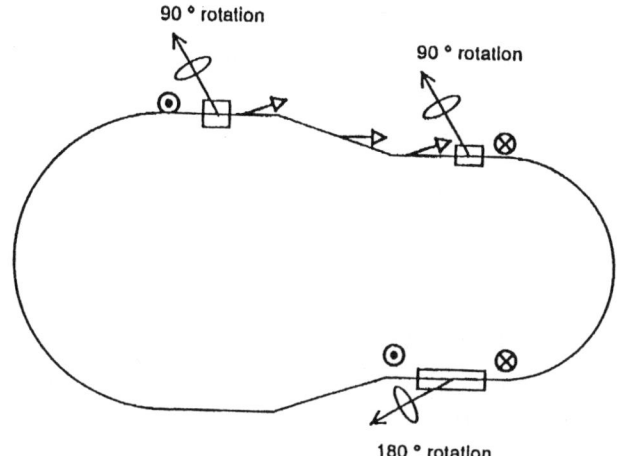

Fig.2 A scheme of the Snakes to provide longitudinal polarization. ⌀ indicates the axis of ths spin precession in the snake.

because the effective nucleon polarization of the widely used polarized target such as butanole is very low and nuclear A-dependence of the various reactions are not well understood which is important to deduce the correct values for the nucleon-nucleon scattering. A typical collider detector has 4π coverage of the acceptance. Another advantage is to get better statistics and cover the full range of x_F. Moreover if we employ a general purpose detector, we can measure various reactions simultaneously.

PHYSICS OF POLARIZED COLLIDER

There must be many interesting physics which can be studied with this polarized collider, and they are discussed in this workshop. Here we want to focus on the particular topics, namely the "spin crisis",as an example of physics which can be done with this machine. It is widely accepted that the polarized gluon inside a proton may solve the spin crisis[3] and polarized proton-proton scattering is important in order to measure the gluon polarization just like polarized lepton-nucleon scattering for the measurement of the quark polarization. To do this, several processes have been suggested such as single and double jet production, direct γ, Drell-Yan, heavy quark, heavy quarkonium production etc. Among them two reactions are considered to be useful in this energy region to measure the gluon polarization, namely longitudinal spin parallel-antiparallel asymmetry,

A_{LL}, of the direct γ production and the charmonium production.

For the proton-proton scattering, the dominant elementary process of the direct g production is expected to be a gluon Compton process, $q + G \to q + \gamma$. If this assumption is correct, the A_{LL} for the proton-proton system can be roughly described as a product of the gluon polarization, $\Delta G(x)/G(x)$, quark polarization, $\Delta q(x)$, and the asymmetry of the elementary process, $a_{LL}(q+G \to q+\gamma)$. Therefore $\Delta G(x)/G(x)$ can be obtained from the measurement of the A_{LL}, since the a_{LL} can be calculated from the perturbative QCD and $\Delta q(x)$ has been determined by deep-inelastic polarized lepton scatterings. The cross section of the direct γ production in the proton-proton scattering was measured at ISR which is more or less similar machine as the one discussed here[4]. Thus the detection of the direct γ is feasible and actually it is easier than fixed target experiments. Because the laboratory energy of π^0 is lower for the collider and hence the separation of the two γ is larger, it is easier to distinguish the direct γ and two.γ from π^0 which is a main background at high P_T region. According to a simple estimate, we expect to determine the A_{LL} at the accuracy of 6% at $P_T = 5$ GeV/c by assuming one month of running at the luminosity of 10^{30} cm^{-2} sec^{-1}.

Another process to measure the gluon polarization is the charmonium production, because the dominant process to produce the heavy quarkonium in the proton-proton collision is expected to be gluon-fusion process. Among various charmonium states, J/φ, χ_{c1} and χ_{c2} are the only states which can be practically detected. The χ_{c1} and χ_{c2} can be identified through their radiative decay to the J/φ, where the branching ratio is 27.3% and 13.5% respectively, by measuring a lepton pair and γ in the final state. Because of the small branching ratio of the radiative decay (0.66%), χ_{c0} is very difficult to detect. If the elementary process is the gluon-gluon fusion, the χ_{c1} can not be produced and the asymmetry for the charmonium production can be simply written as,

$$A_{LL}(x_F) = a_{LL} \Delta G(x_1)/G(x_1) \Delta G(x_2)/G(x_2)$$

,where the a_{LL} is the asymmetry of the elementary process and $\Delta G(x_1)/G(x_1)$ is the gluon polarization (spin structure function of gluon) of 1st proton. The a_{LL} for the χ_{c0} and χ_{c2} are expected to be +1 and -1, respectively. Therefore we can determine $\Delta G(x)/G(x)$ by the measurement of the A_{LL} for the production of the χ_{c2} under this assumption. This assumption can be checked by the production ratio of χ_{c1} and χ_{c2}. The dominance of the χ_{c2} production in the proton-proton scattering was claimed although the statistics was rather poor[5]. By measuring the angular distribution of the decay, the alignment of χ_{c2} and J/φ can be also obtained. The assumption can be further checked by the measurement of γ directions which determines the a_{LL} experimentally[6]. Thus the advantage of the χ_{ci} measurement is that the assumption to determine the gluon spin can be confirmed simultaneously.

For the J/φ production, intermediate states are important and so the A_{LL} can not be so simple as the one for the χ_{c2}. But the model calculations have been made for the A_{LL} which is not small in this energy region[7] and also we can expect much better accuracy for the measurement. The charmonium production , as a whole, is important process for the determination of the gluon polarization.

Experimentally, J/φ and $\chi_{c1,2}$ were already detected at ISR although χ_{c1} and χ_{c2} could not be distinguished because of the energy resolution of the calorimeter used before. We need a better calorimeter to distinguish and it is possible if ,for example, pure CsI is employed. In order to obtain the x-dependence of ΔG(x)/G(x), X_F and s dependence of the A_{LL} must be measured and it is quite possible for the polarized collider. According to a simple estimate, we expect to detect 50k χ_{c2} events/month in total. The detector could be a typical collider detector which can detect electrons, gamma, hadrons and muons. In order to do the physics mentioned here, we will need a finely segmented electro-magnetic calorimeter of very good energy resolution.

In summary, we have proposed a polarized collider as a future extension of the Proton Synchrotron at KEK. It will provide a unique opportunity for high energy spin physics and be useful to solve the "spin crisis". The energy domain of this machine is complementary to the RHIC and SSC.

I would like to acknowledge Drs H.Sato, Y.Mori, T.Toyama, H.En'yo and S.Yamashita for valuable discussions about the polarized collider at KEK.

REFERENCES
1. H.Sato et al., Nucl. Instr. & Methods A272, 617 (1989).
2. J.Chiba et al., KEK Report 90-13 (1990).
3. G.Altarelli and W.J.Stirling, Particle World 1, 40 (1989).

4. M.Diakonou et al., Phys. Lett. 87B 292 (1979) and T.Akesson et al., Phys. Lett. 158B, 282 (1985).
5. D.A.Bauer et al., Phys. Rev. Lett. 54, 753 (1985).
6. J.L.Crtes and B.Pire, Phys. Rev. D38, 3586 (1988).
7. A.P.Contogouris, S.Papadopoulos and B.Kamal, Phys. Lett. 246, 523 (1990).

RECENT RESULTS AND FUTURE PROSPECTS FOR THE
POLARIZED BEAM PROGRAM AT FERMILAB
(For the Fermilab E-704 Collaboration[*])

A. Yokosawa
High Energy Physics Division, Argonne National Laboratory
Argonne, Illinois 60439 USA

ABSTRACT

We summarize activities concerning the Fermilab polarized beams. They include a brief description of the polarized-beam facility, measurements of beam polarization by polarimeters (Fermilab E-581), asymmetry measurements in the π^0 production at high $p_\perp$ and in the Λ (Σ^0), $\pi^\pm$, π^0 production at large x_F, and $\Delta\sigma_L$(pp, $\bar{p}$p) measurements (Fermilab E-704). In future we plan to investigate the proton-spin crisis by determining the gluon spin distribution in inclusive production of direct gamma, $\chi 2$, and J/ψ.

INTRODUCTION

About ten years ago, we wrote several physics proposals in order to justify the construction of a polarized-beam line at Fermilab. They are as follows:

P-581 Construction of a Polarized Proton Beam Facility in the Meson Area
P-674 Asymmetries in Inclusive Pion and Kaon Production at Large-x with a Polarized Beam
P-675 Asymmetry Measurements for Dimuon Production in the J/ψ Mass Region
P-676 An Experiment to Measure $\Delta\sigma^{Tot}$ in p-p and $\bar{p}$-p Scattering Between 100 and 500 GeV/c
P-677 A Study of the Spin Dependence in the Inclusive Production of Lambda Particles with the Polarized Beam at Fermilab
P-678 Proposal to Study the Spin Dependence in Inclusive π^0 and Direct Gamma Production at High $p_\perp$ with the Polarized Proton Beam Facility at Fermilab
P-682 Study of the p_T Dependence of $\pi^\pm$ Inclusive Production with a Polarized Proton Beam and Target
P-688 Nuclear-Size Dependence of Single-Spin Asymmetries in High-p_T Hadron Production
P-689 Measurement of the Asymmetry in Calorimeter Triggered High-$p_\perp$ Events Using a Polarized Proton Beam and Target

Among these an integration program was suggested as an initial experiment by the Fermilab management. Results were E-581, measurements of the beam polarization and E-704 integrating P-674, 676, 677, and 678.

The Fermilab polarized-beam facility at 200 GeV/c became operational in 1987. At that time, we mainly concentrated on the measurements of beam polarization by polarimeters and on the studies of beam polarization monitors. Recently we carried out the polarized-beam program designated as E-704 by an international collaboration.

We summarize these activities including the discussions of some preliminary data on $p_\perp$ and x_F dependence.

POLARIZED BEAM FACILITY

The Fermilab polarized-beam facility[1] was operated during the past TeV-II (fixed target) period which ended in February 1988.

An extracted beam from the Tevatron is delivered through the MP primary-beam line to the Meson Detector Building where a 0.73-interaction-length Be target is utilized to produce Λ and $\bar{\Lambda}$ at $\theta_{c.m.} \approx 0°$. Protons and antiprotons from the Λ and $\bar{\Lambda}$ decays respectively are brought to a final target position in the MP hall through the MP secondary beam (200 GeV/c) line.

Polarized protons from the virtual sources as shown in Fig. 1 are focussed in the tagging section, where both the momentum and polarization are selected.[1]

The typical beam flux ($\Delta p/p = \pm 5\%$) for 3×10^{12} incident protons per 20-sec spill at 200 GeV/c were: (P_{av} is average polarization)

	Tagged Beam $P_{av} = 45\%$	Total Protons (antiprotons)	Background π's
Protons	1.0×10^7	2.0×10^7	2.0×10^6
Antiprotons	5.0×10^5	1.0×10^6	5.0×10^6

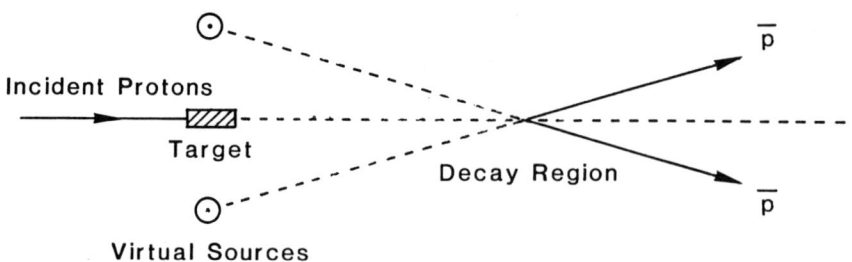

Figure 1 Virtual sources (top view).

PRIMAKOFF-EFFECT MEASUREMENT

The asymmetry of the nuclear coherent Coulomb π^0 production process ("Primakoff process"), was measured[2] for the first time with the use of the polarized-proton beam. The apparatus consisted of a lead-glass calorimeter for π^0 detection and a magnetic spectrometer for the scattered protons. A large asymmetry in the region of $|t'| < 0.001$ $(GeV/c)^2$ and $1.36 < M(\pi^0 p) < 1.52$ GeV/c^2 was observed for the reaction $p + Pb \to p + \pi^0 + Pb$, where the Coulomb process is predominant. The expected null asymmetry was observed in the larger $|t'|$ region where the diffractive-dissociation process is predominant.

The observed ϕ-angle dependence of the coherent π^0 production process may be expressed as $1 + (f\, T(\theta)\, P_B) \cos \phi$, where $T(\theta)$ is the analyzing power (target azimuthal asymmetry) for photoproduction of π^0 from a polarized proton target at c.m. polar angle θ, ϕ is the azimuthal angle, P_B is the transverse polarization, and the parameter f is a dilution factor caused by the diffractive dissociation. The raw asymmetry at ϕ is obtained as

$$A(\phi) = \left(N^\uparrow(\phi) - N^\downarrow(\phi)\right) / \left(N^\uparrow(\phi) + N^\downarrow(\phi)\right) = f\, T(\theta)\, P_B \cos \phi = \varepsilon \cos \phi,$$

where $N^\uparrow(\phi)$ and $N^\downarrow(\phi)$ are the number of events at ϕ for the up and down spin direction of the incident proton, respectively.

The measured asymmetry for the Coulomb process is consistent with the analyzing power (about -70%) of the π^0 production process deduced from existing low-energy $\gamma + p \to \pi^0 + p$ data. The results demonstrate that the Primakoff process is useful for the measurement of proton and antiproton polarization at high energy.

COULOMB-NUCLEAR INTERFERENCE MEASUREMENTS

The analyzing power, A_N, of proton-proton, proton-hydrocarbon, and antiproton-hydrocarbon scattering in the Coulomb-nuclear region was measured with use of the polarized-proton and polarized-antiproton beams.[3] For the elastic scattering at small $|t|$, a set of scintillation counters was utilized to detect the recoil proton which stops within a very short range in the scintillator. The results at $|t| \sim 0.003$ $(GeV/c)^2$ show the value $A_N = (2.4 \pm 0.9)\%$ with the polarized-proton beam, and $A_N = (-4.6 \pm 1.9)\%$ with the polarized-antiproton beam both on a hydrocarbon target, and also $A_N = (4.5 \pm 2.8)\%$ of proton-proton scattering. These results are consistent with predictions[4-6] based on Coulomb-nuclear interference. Recently A_N measurements were repeated with much higher statistics than those mentioned above.

SINGLE-SPIN ASYMMETRY IN $p^\uparrow p \to \pi^\circ X$ AND $\bar{p}^\uparrow p \to \pi^\circ X$ AT HIGH $p_\perp$

This experiment is recently completed at incident proton momentum of 200 GeV/c. Preliminary data of the $p^\uparrow p$ reaction show that the asymmetry values (A_N) at $x_F \approx 0$ are approximately zero (or small negative) up to $p_\perp$ = 3.5 GeV/c and then begin to rise to ~ + 40% in the region of $p_\perp$ = 4 to 5 GeV/c as shown in Fig. 2. At lower energies as seen in the BNL[7] ($p^\uparrow p \to \pi^+ X$), CERN[8] ($pp^\uparrow \to \pi^\circ X$),

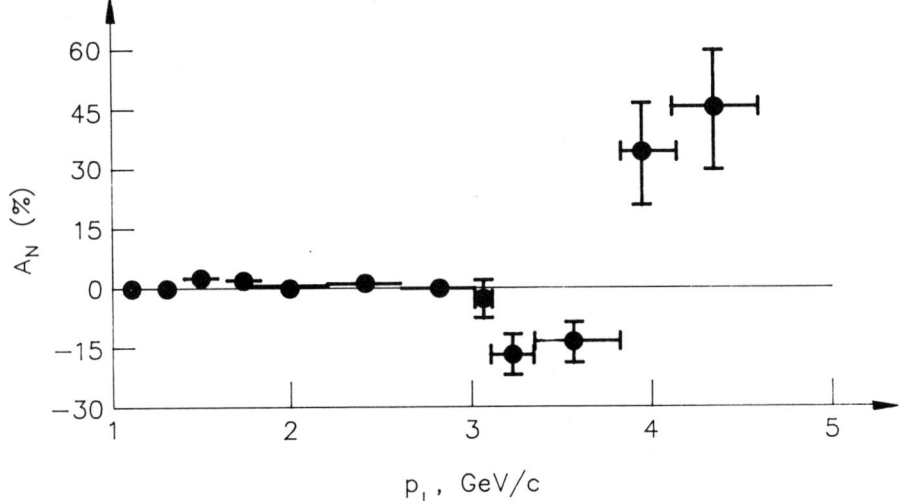

Figure 2 $p_\perp$ dependence of A_N at $x_F \approx 0$ (preliminary data).

Serpukhov[9] ($\pi^- p^\uparrow \to \pi^\circ X$) data, this rapid rise from zero to large positive values,* was also observed as shown in Fig. 3 although none of the data exceeded $p_\perp$ = 3 GeV/c. A new finding is that all the A_N data of π° or π^+ production at $x_F \approx 0$ show the large positive asymmetries begin at $x_\perp$ = 0.4 in the region $\sqrt{s}$ = 5 to 20 GeV as shown in Fig. 4.

---

* Data taken with polarized targets (Refs. 8 and 9) are normalized to the measurements with polarized beams, that is, the sign of A_N ($hp^\uparrow \to hX$, where h represents hadron) is reversed. Note that A_N ($p^\uparrow h \to hX$) = $-A_N$ ($hp^\uparrow \to hX$) at $x_F \approx 0$.

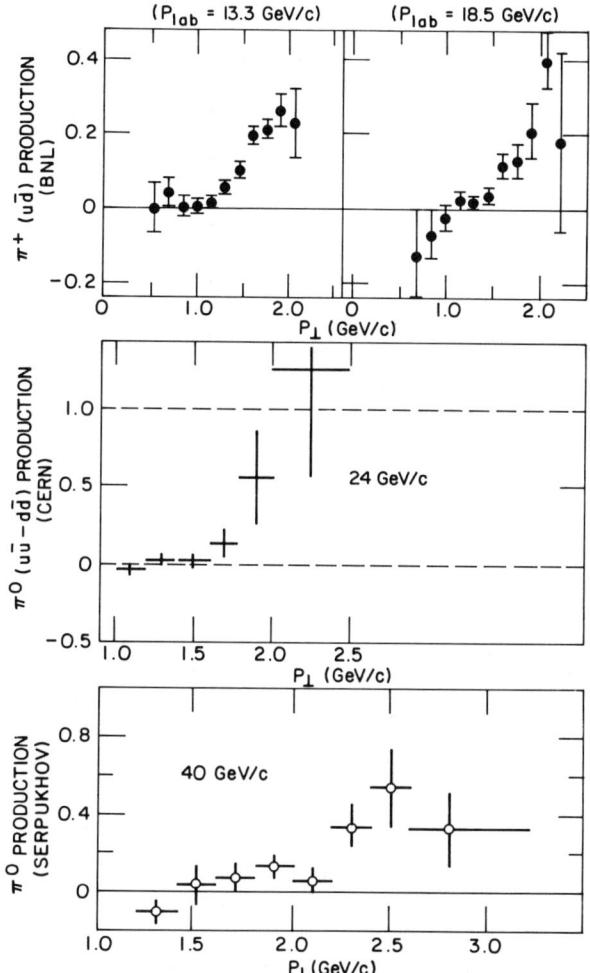

Figure 3 Asymmetry $A_N(\%)$ vs. $p_\perp$ at p_{lab} = 1.33 to 40 GeV/c

This is strong indication that we are indeed observing asymmetries caused by hard scattering. We note that the common crossing point $x_\perp$ = 0.4 was pointed out in the Ref. 9 in the region $\sqrt{s}$ = 5 to 8.5 GeV/c. Theoretically single-spin asymmetries are discussed within the context of the QCD hard-scattering model.[10] By knowing the quark content of π^+ = $u\bar{d}$ and π^0 = $(u\bar{u} - d\bar{d})/\sqrt{2}$, polarized u quark in the polarized proton beam is considered to be the career of the spin information.

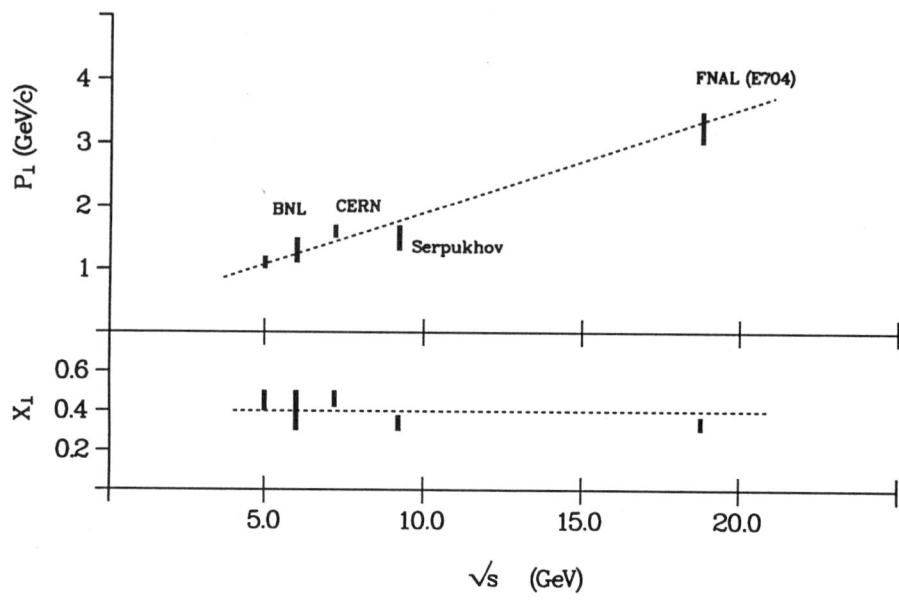

Figure 4 Onset of structure in A_N vs. $\sqrt{s}$.

Single-spin asymmetry in $\bar{p}^\uparrow p \to \pi^0 X$ shows a similar $p_\perp$ dependence as the $p^\uparrow p$ case. However, data are limited only up to $p_\perp$ = 3.5 GeV/c.

x_F DEPENDENCE OF SINGLE-SPIN ASYMMETRY IN $p^\uparrow p \to \pi^0 X$ AND $\bar{p}^\uparrow p \to \pi^0 X$

Measurements on the x_F dependence at 200 GeV/c covering $p_\perp$ up to 2 GeV/c were recently completed. Asymmetry values in the $p^\uparrow p$ reaction are consistent with zero up to x_F = 0.3 to 0.4, and then linearly increase to + 20% near x_F = 1.0 as shown in Fig. 5. The data suggest an influence of polarized u quarks at large x_F. Also they are consistent with earlier data[11] taken at $\langle x_F \rangle$ = 0.52. This is the first x_F dependence data ever obtained in the π production. It is interesting to notice that our data resembled x_F dependence for Λ polarization in $pp \to \Lambda^\uparrow X$ where polarized s quarks were considered[12] to be responsible for high polarization.

Single-spin asymmetry in $\bar{p}^\uparrow p \to \pi^0 X$ shows a similar x_F dependence as $p^\uparrow p$ case.

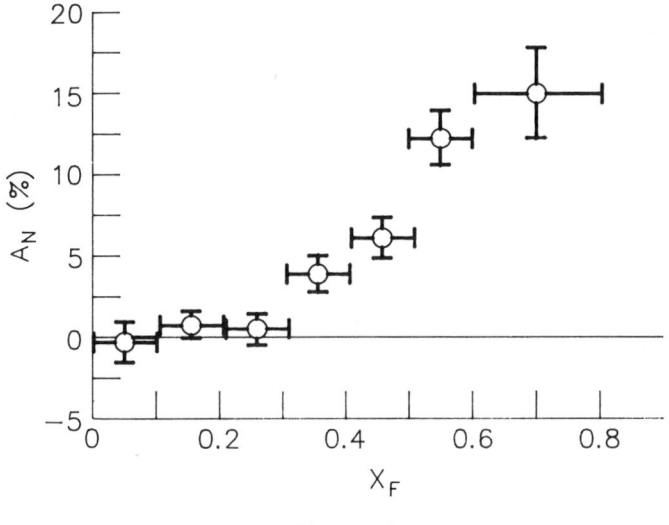

Figure 5

x_F dependence of A_N at $p_\perp$ = 0.5 to 2.0 GeV/c (preliminary data).

A SHORT SUMMARY ON $x_\perp$ AND x_F DEPENDENCE

There are two distinct common phenomena observed in the $x_\perp$ and x_F dependence of the one-spin asymmetries. The asymmetry values are zero or small for both $x_\perp$ and x_F < 0.3 to 0.4. Then there is a rise from zero to large positive values for $x_\perp$ and x_F > 0.3 to 0.4. It will be interesting to find out if these two phenomena are related. One strong hint is that polarized u quarks seem responsible to the rise in A_N at high $x_\perp$ and high x_F. It is interesting to investigate if these phenomena may be related to the origin of proton spin. High-$p_\perp$ and high -x_F scattering phenomena were interpreted as an indication for the existence of rotating color charges in polarized protons.[13]

$\Delta\sigma_L$ (pp and $\bar{p}$p), AND DOUBLE-SPIN ASYMMETRY IN $p^\uparrow p^\uparrow \to \pi^\circ X$ AND $\bar{p}^\uparrow p^\uparrow \to \pi^\circ X$

Difference in total cross sections for pure spin states, $\Delta\sigma_L$ (pp and $\bar{p}$p), was simultaneously measured at 200 GeV/c with the π° production up to $p_\perp$ = 3 GeV/c. Data are currently being analyzed.

$\pi^\pm$ AND HYPERON PRODUCTION ON HYDROGEN TARGET WITH POLARIZED BEAM

Measurements of $p^\uparrow p \to (\pi^\pm, \Lambda^\circ, \Sigma^\circ) + X$ at 200 GeV/c were completed and data are currently being analyzed.

FUTURE EXPERIMENTS

We have proposed the following simultaneous measurements using a polarized beam and a polarized target.

i) <u>Spin dependence in direct-gamma production at high $p_\perp$.</u>
To understand the basic question of the origin of proton spin, we may be able to determine the gluon spin distribution in the proton by measuring the spin correlation parameter A_{LL} in the direct-γ production at high $p_\perp$, with longitudinally polarized protons on longitudinally polarized target nucleons.

$$A_{LL} = (1/P_B P_T) \frac{N(\overrightarrow{\rightarrow}) - N(\overrightarrow{\leftarrow})}{N(\overrightarrow{\rightarrow}) + N(\overrightarrow{\leftarrow})} ,$$

where P_B is beam polarization, P_T is target polarization, and arrows indicate the spin direction in the laboratory system.
The QCD Compton effect, "gluon + quark → gamma + quark", is expected to be the dominant mechanism for direct-γ production at large $p_\perp$. The parameter A_{LL} is approximately proportional to the gluon polarization.[14,15]

Our plan is to carry out the A_{LL} measurements up to $p_\perp$ = 5 GeV/c with reasonable statistical accuracy. Our main detectors consist of lead-glass counters and proportional wire chambers.

ii) <u>Spin dependence of χ2 (3555) production.</u>
We plan to measure the spin dependence of χ2 (3555) production, which will be <u>simultaneously</u> carried out with the above mentioned direct-gamma measurements. The double-spin asymmetry, A_{LL}, in $p^\uparrow p^\uparrow \to$ χ2 (3555) → J/ψ + γ is also expected to provide a means to study the spin dependent gluon structure function. The 15% decay branching ratio of χ2 (3555) to J/ψ + γ allows us to analyze the helicity of the charmonium state.

There is general agreement[16] theoretically that the χ2 (3555) state is mainly produced by gluon-gluon fusion as shown below and there are promising experimental results[17] suggesting that simple gluon fusion is sufficient to account for the χ2 (3555) production in proton interactions at 200 and 250 GeV/c. The measured two-spin asymmetries (as defined below) give information on the initial gluon polarization, which can be used to reconstruct the gluon spin distribution in the polarized proton.[18] By considering the fusion process, if the initial helicity state is (+-), that is $(\overrightarrow{\leftarrow})$, the J_z = 2 and this state produces χ2.

To be exact, the observable A_{LL} is related to the distribution function of a polarized gluon in a polarized proton expressed as $G_+(x)$ and $G_-(x)$ with same- and opposite-sign helicities respectively.[18]

$$A_{LL}(x_F) = (1-R)/(1+R)\left[\frac{\Delta G}{G}(x_1) \cdot \frac{\Delta G}{G}(x_2)\right],$$

where x_1, x_2, and x_F are the longitudinal-momentum fraction of gluons, R is the ratio of matrix elements f_+ (f_-) which are the squared matrix elements for the production of $\chi 2$ out of two gluons with same- (opposite-) sign helicities, and $\Delta G/G(x) \equiv (G_+(x) - G_-(x))/(G_+(x) + G_-(x))$.

Theoretical predictions were recently made on $A_{LL}(x_F)$ in both the J/ψ and χ production based on a perturbative QCD approach.[19,20]

REFERENCES

§   Work supported in part by the U.S. Department of Energy, Division of High Energy Physics, Contract W-31-109-ENG-38.

*   The E-581/704 Collaboration: Argonne (US), Fermilab (US), Univ. of Iowa (US), Kyoto-Kyoto Sangyo-Kyoto Education-Hiroshima (Japan), LAPP (France), Los Alamos (US), Northwestern Univ. (US), Osaka-Okayama (Japan), Rice Univ. (US), CEN Saclay (France), IHEP Serpukhov (USSR), INFN Trieste- Messina-Udine (Italy), Univ. Occup. Envior. Health (Japan).

1) D. Grosnick et al., Nucl. Instrum. and Meth. A290, 269 (1990).
2) D. C. Carey et al., Phys. Rev. Lett. 64, 357 (1990).
3) N. Akchurin et al., Phys. Lett. B229, 299 (1989).
4) B. Z. Kopeliovich and L. I. Lapidus, Yad. Fig. 19, 218 (1974); (Sov. J. Nucl. Phys. 19, 114 (1974)).
5) N. H. Buttimore, E. Gotsman, Am. E. Leader, Phys. Rev. D18, 694 (1978).
6) N. H. Buttimore, Proc. 6th Int. Symp. on High Energy Spin Phys. Marseille, J. Soffer, ed., Journal de Physique, 46, C2 (1985) p. 643.
7) S. Sanoff et al., Phys. Rev. Lett. 64, 995 (1990).
8) J. Antille et al., Phys. Lett. B94, 523 (1980).
9) V. D. Apokin et al., Phys. Lett. B243, 461 (1990).
10) D. Sivers, Phys. Rev. D41, 83 (1990); Phys. Rev., to appear.
11) B. E. Bonner et al., Phys. Rev. Lett. 61, 1918 (1988).
12) T. deGrand and H. I. Miettinen, Phys. Rev. D24, 2419 (1981); Phys. Rev. D31 (E), 661 (1985).
13) Liang Zuo-tang and Meng Ta-chung, Phys. Rev. 42, 2380 (1990); Meng Ta-Chung, Proc. 25th Rencontre de Moriond, Mar. '90, Les Arc, France.
14) M. B. Einhorn and J. Soffer, Nucl. Phys. 274, 714 (1986); N. S. Craigie, K. Hidaka, M. Jacob, and F. M. Renard, Phys. Reports 99, 143 (1983).

15) E. L. Berger and J. W. Qiu, Phys. Rev. D40, 788 (1988); D40 (RC) 3128, (1989).
16) S. D. Ellis et al., Phys. Rev. Lett. 36, 1263 (1976); B. L. Ioffe, Phys. Rev. Lett. 39, 1589 (1977); C. E. Carlson et al., Phys. Rev. D18, 760 (1978); H. Fritzsch, Phys. Lett. 67B, 217 (1977); M. Gluck et al., Phys. Rev. D17, 2324 (1978); L. M. Jones et al., Phys. Rev. D17, 1782 (1978); V. Berger et al., Z. Phys. C6, 169 (1980); J. H. Kuhn, Phys. Lett. 89B, 385 (1980); Y. Afek et al., Phys. Rev. D22, 86 (1980).
17) D. A. Bauer et al., Phys. Rev. Lett., 54, 753 (1985); Y. Lemoigne et al., Phys. Lett. 113B, 509 (1982).
18) J. L. Cortes and B. Pire, Phys. Rev. 38 (RC), 3586 (1988).
19) A. P. Contogouris, S. Papadopoulos, and B. Kamal, Phys. Lett. B246, 523 (1990).
20) M. A. Doncheski and R. Robinett, Phys. Lett. B248, 188 (1990).

AN OVERVIEW OF THE RHIC PROJECT

T. LUDLAM
Brookhaven National Laboratory, Upton, New York 11973[†]

INTRODUCTION: THE RHIC LAYOUT

The Relativistic Heavy Ion Collider (RHIC) project at Brookhaven has now gotten the go–ahead for construction funding by the U.S. Congress, and will soon become the flagship facility for quark matter research with high energy heavy ion collisions. This field of research, virtually non–existent as recently as ten years ago, has spawned a search for new discoveries in high–temperature, high–density nuclear matter produced in collisions of heavy nuclei at the highest attainable energies.[1] Brookhaven and CERN have been in the forefront of this research, utilizing existing proton accelerators to produce high energy ion beams for fixed target experiments in which the c.m. collision energies are many GeV per nucleon.[2] The RHIC project, which is scheduled to become operational in 1997, will provide a dedicated facility for these experiments, with an energy increase of more than an order of magnitude.[3] With its two rings of superconducting magnets, RHIC will be able to collide proton beams as well as the higher mass nuclear species. Of special interest for this meeting is the potential for high luminosity proton-proton collisions with polarized beams. The quantitative details of this potential are spelled out by S.–Y. Lee in another contribution to these proceedings.[4]

The complex of accelerators which will make up the RHIC facility is shown in Fig. 1. Ion beams originating in the Tandem Van de Graaff accelerator are transported through a long transfer line and injected into the Alternating Gradient Synchrotron (AGS). For the fixed–target experimental program, these fully–stripped ion beams are accelerated in the AGS to an energy of $29\left(\frac{Z}{A}\right)$ GeV/amu, where $\frac{Z}{A}$ is the charge–to–mass ratio. At present the heaviest ions that can be accelerated in this way are silicon nuclei (A = 28), reaching an energy of 14.5 GeV/amu.

The AGS Booster Synchrotron[5] is nearing completion as a construction project, and will be commissioned in 1991. The advent of the Booster will allow fully-stripped ions up to gold (A = 197) to be accelerated with high intensity. (Heavier elements, up to Uranium, will be accessible with some degradation in performance.) The Tandem–Booster–AGS complex then becomes the injector for RHIC. The collider will consist of two rings of superconducting magnets for accelerating, storing and colliding beams of heavy ions at energies up to 100 GeV/amu. Much of the "footprint" seen for RHIC rings in Fig. 1 already exists — the tunnel and four of the six collision halls were built for the partially completed CBA project early in the 1980's.

[†]Work performed under Contract No. DE–AC02–76CH00016 with the United States Department of Energy.

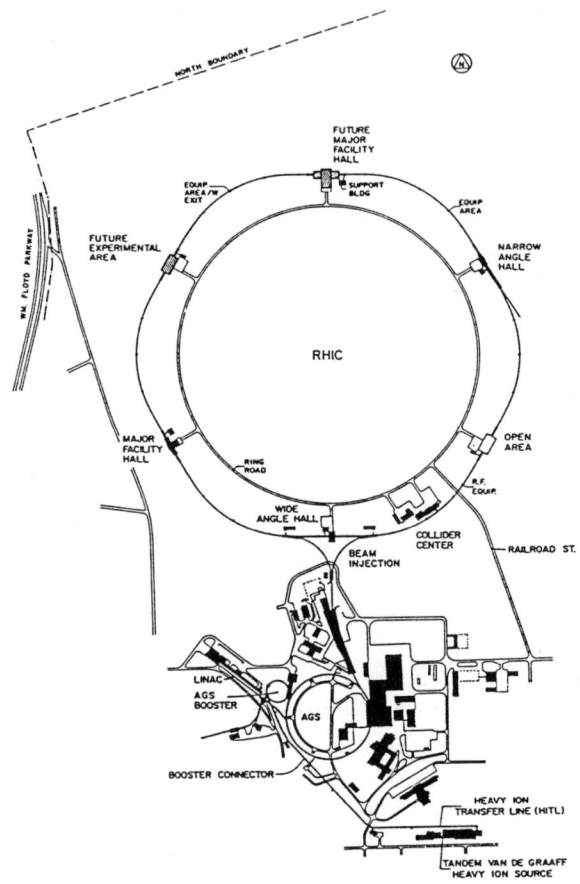

Fig. 1 Site map of present and future accelerators at Brookhaven. The Tandem Van de Graaff and the AGS with its linac injector are existing machines. The Booster Synchrotron for pre- -acceleration before injection into the AGS is nearing completion. The RHIC colliding beams accelerator to the north of the AGS complex is beginning construction in 1991.

THE RHIC DESIGN

The basic parameters of the RHIC facility are illustrated in Fig. 2. The design calls for a top beam energy of 100 GeV/nucleon for ions of mass A = 200, and the acceleration of ion masses spanning the full periodic table. The complete accelerator complex, consisting of Tandem, Booster, AGS and RHIC will provide c.m. collision energies for gold beams ranging from AGS energy up to 100 + 100 GeV/amu. This energy range is covered with no inaccessible gaps, and adequate beam intensities throughout. The energy range is covered in three segments: As shown in Fig. 2, the range between fixed target AGS experiments and high energy collider operation can be spanned by using one of the RHIC beams striking a fixed target. For this operation an internal gas jet target would be used.

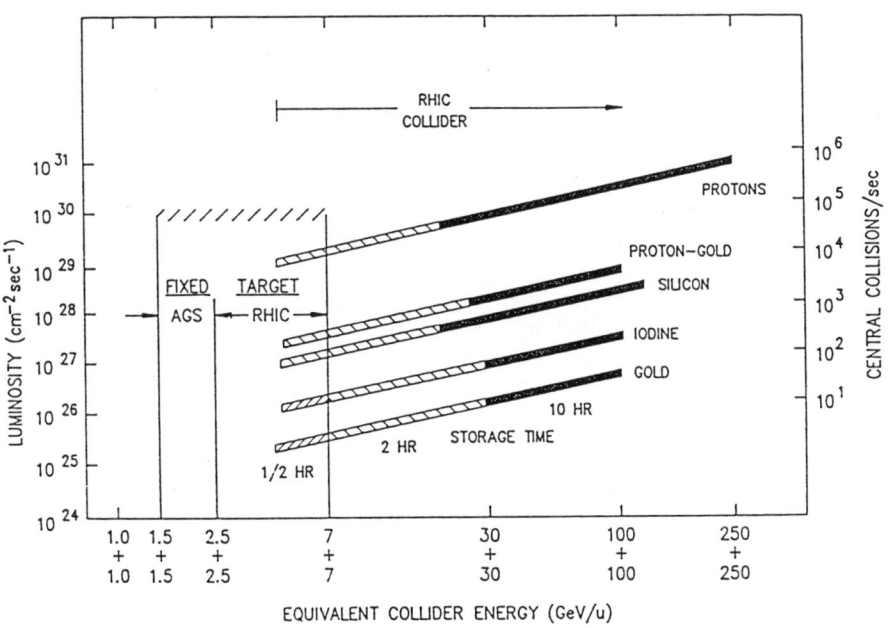

Fig. 2 The design luminosity, for various ion masses, as a function of collision energy over the full range accessible with AGS and RHIC. On the left-hand scale, central collisions correspond to impact parameter less than 1 fermi.

The layout of the RHIC collider is shown schematically in Fig. 3. The circumference of the collider is 3833 meters. It consists of two accelerator rings with six crossing regions (insertions) where the counter-rotating beams are brought into collisions and experiments carried out. Particle bunches accelerated in the AGS to its top energy (28 GeV for protons; 11 GeV/amu for gold) are transferred to the collider, with single bunches of ions injected 57 times into each ring in boxcar fashion. Filling time per ring will be about one minute. For gold, as an example, there will be $\sim 1.1 \times 10^9$ ions/bunch, or 6×10^{10} ions in 57 bunches in each ring. For the lightest ions, hydrogen and deuterium, approximately 10^{11} ions/bunch can be stored in the machine. Acceleration will take approximately 60 seconds. Bending and focussing of the ion beams is achieved with superconducting magnets. Given that the machine will be built in the existing CBA tunnel, a cost optimization is achieved by filling the circumference with relatively low field magnets. The maximum energy of 100 GeV/amu for gold ions (250 GeV for protons) is reached with a magnetic field of 3.5 Tesla. Maximum operational flexibility is obtained with the magnets of each ring in separate vacuum vessels, with the beams in the arcs separated by 90 cm.

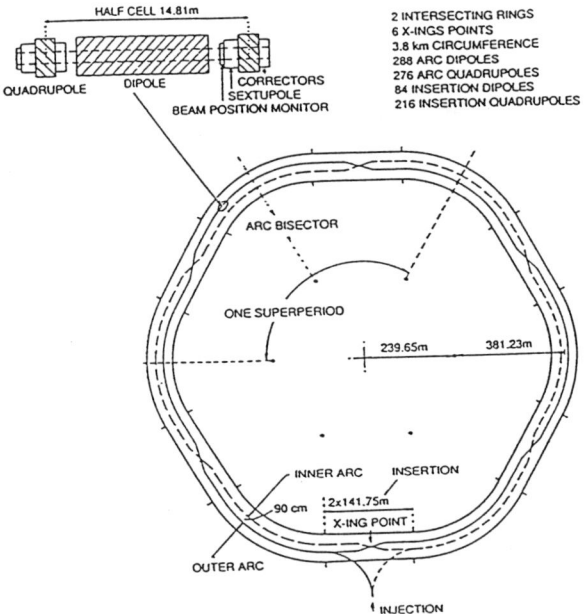

Fig. 3 Layout of the storage rings for the RHIC collider.

The six beam crossing regions, four of which will be available for first-round experiments, are designed to accommodate a range of detector configurations to fulfill the needs of experiments. The free space available for experimental equipment in each crossing region is 9 meters on either side of the intersecting point. For head-on collisions with gold ion beams at top energy, a luminosity of $2 \times 10^{26} \text{cm}^{-2} \text{sec}^{-1}$ over a 10 hour beam lifetime is expected. The r.m.s. length of luminous interaction region where the beams cross is 22.5 cm. For proton–proton collisions the design luminosity is $1 \times 10^{31} \text{cm}^{-2} \text{sec}^{-1}$ Collisions of unequal species, e.g., protons in one beam and gold in the other, will be possible as well. The design of the machine is such that the luminosity values shown in Fig. 2 can, in principle, be increased by roughly a factor of ten soon after the machine becomes operational. If the experimental program warrants it, this would be an early upgrade of the machine performance.

THE PRESENT STATUS

The RHIC project, under the auspices of the U.S. Department of Energy, will begin in 1991 and will cost a total of 397 million dollars for construction. The scheduled completion date is April, 1997. Included in the total cost of the project are funds amounting to approximately 97 million dollars which will be used for the construction of first-round detectors. These figures include the cost of inflation.

As noted above, a large fraction of the RHIC facility already exists. For the injector complex, the Tandem Van de Graaff, AGS, and heavy ion transfer line are already operational; the Booster Synchrotron is under construction. Most of the conventional construction for the collider is complete, including the ring tunnel, main service building and experimental halls for four of the six intersection regions. In addition, the liquid helium refrigerator, capable of cooling all of the superconducting magnets in the collider has been completed (as part of the CBA project) and successfully tested. The refrigerator has a capacity of 25 kilowatts at a temperature of 4.3K. The estimated heat load for RHIC is $\sim$ 10 kilowatts at 4.6K.

One of the most important elements of the RHIC proposal is the design of the superconducting magnets for the accelerator rings. These magnets are the largest component of the cost of the machine, and their fabrication and installation is the major determinant of the construction schedule. The design of these magnets is based on the cosine theta coil structure, which has been adopted for the Tevatron, HERA and SSC magnets as well. The RHIC dipole magnets are designed to operate at a relatively low field (3.5T), and thus the coil can be wound in a single layer of superconductor. Figure 5 shows a magnet assembly, consisting of a dipole, quadrupole and corrector coils, mounted in a cryostat. These magnets are fully designed, and have been the major component of an intensive RHIC R&D program which began in 1987. Full-size, "machine quality" prototypes of dipoles, quadrupoles, sextupoles and the correction coil package have been built and successfully tested. In the case of the dipoles, the largest component of the magnet system, three of the prototypes built thus far have been assembled by an industrial manufacturer. All of the prototype dipole magnets (8 have been tested so far) have reached field strengths of approximately 4.6 T, or 35% higher than the operating field of RHIC, with virtually no training. The RHIC magnets have been designed to be amendable to fabrication in quantity in industrial facilities, and it is planned that a significant fraction of the magnets for RHIC will be built by commercial industry.

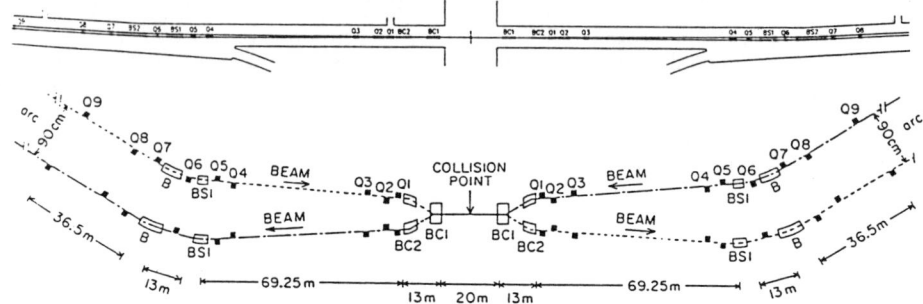

Fig. 4 The magnet lattice in the intersection regions of RHIC. The upper drawing is to scale, and shows the 6 o'clock region where the injection lines from the AGS enter the RHIC tunnel. The lower drawing shows an expanded view of the magnet layout.

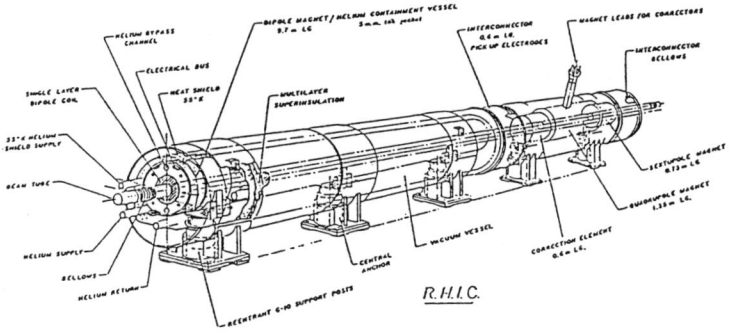

Fig. 5 RHIC magnet assembly: The drawing shows a half-cell of the arc magnet lattice, including a dipole, corrector package, quadrupole and sextupole magnets enclosed in their cryostat.

The magnet R&D program is nearing completion, with tests now in progress of a full cell of the machine magnet lattice (see Fig. 3), which consists of a string of eight magnet elements, with their cryogenic and electrical connections, operating as a system as they will in the final machine.

EXPERIMENTS AND DETECTORS

Of the six crossing regions built into the RHIC rings, those at the 2, 4, 6 and 8 o'clock positions (see Fig. 1) have completed experimental halls, including support buildings and (except in the 4 o'clock "open area") crane coverage. The RHIC plan calls for mounting experiments initially in these four areas, leaving the remaining two unfinished until some later time. If a fifth experimental area were to be made available for the first–round research program – say, for a dedicated polarized proton experiment – a decision would have to be made by the summer of 1992. From the point of view of the machine operation, the heavy ion program is expected to run about 40 weeks per year. This would leave approximately two months for dedicated polarized proton running, if desired.

Detector designs for the recording of events at RHIC, with sensitivity corresponding to the expected signals for new phenomena in nuclear matter under extreme thermodynamic conditions, will require the extension of known techniques for particle detection beyond the present ranges of application in elementary particle and nuclear physics. With beam energies up to 100 + 100 GeV/amu and ion masses up to ≥ 200, the total energy in each collision can reach up to 40 TeV in the center–of–mass: a range far beyond that of any present accelerator or any existing detector system. Unlike the proposed SSC and LHC colliders, which will accelerate elementary particles to such an energy and produce hundreds of very high energy particles in the final state, the most interesting events at the RHIC collider are expected to produce tens of thousands of final–state particles in each collision, with proportionately less energy carried away by each particle. Thus, while the basic detector technology will have much in common with the detector systems developed for colliding beams of elementary particles, the design of detector systems for the heavy ion collider must address a different set of problems.

This past April Brookhaven issued a call for Letters of Intent for RHIC experiments. These Letters, due on September 28, 1990, were meant to serve as early indicators of community interest in specific areas of physics research at RHIC and the manner in which groups in the U.S. and abroad are coming together to form collaborations. The Letters were also to help the Laboratory assess the need for experiment–specific R&D funds and to make decisions regarding the allocation of these funds. Nine Letters of Intent were received. One of these proposes a measurement of the total and elastic proton-proton cross sections at RHIC. All of the others focus on the physics of ultra-relativistic nucleus- -nucleus collisions. Represented on these Letters were some 50 institutions, worldwide, and about 300 scientists.

A wide variety of detector systems is described in these Letters. Some are designed to look rather generally at the details of particle production in the violent and complex interactions expected from RHIC collisions. Others focus in a more specialized way on the predicted signals of new phenomena, in particular the formation of deconfined hadronic matter in the environment of extremely high temperature and energy density which these collisions are expected to produce. Many of these ideas aired at a workshop at Brookhaven in July, prior to the submission of the Letters of Intent.[6]

With several well-specified concepts now in hand for large detectors at RHIC, attention in the user community has turned to the need for R&D on detector techniques. Past workshops have pointed out a number of specific areas where detector technology must be advanced beyond the present state of the art in order to realize the needed capability for experiments at RHIC, and the recently received letters of intent provide a clear picture of the needs and priorities for detector development. Beginning in 1989, a substantial portion of the RHIC R&D funding was devoted to detector development, and this will continue as the Project moves to the construction phase, and an international community of users begins the work of preparing proposals that will lead to detailed designs for experiments.

REFERENCES

1. For a recent overview of the field see, *Proc. Seventh Int. Conf. in Ultra-Relativistic Nucleus–Nucleus Collisions*, G. Baym et al., eds., Nucl. Phys. A498, (1989).
2. A current summary of results is given in, *Proc. Workshop on Heavy Ion Physics at the AGS*, O. Hansen, ed., Brookhaven National Laboratory Report BNL–44911 (1990).
3. *Conceptual Design of the Relativistic Heavy Ion Collider RHIC*, Brookhaven National Laboratory Report BNL–52195 (May 1989).
4. S.-Y. Lee, *Prospects for Polarization at RHIC and SSC*, these proceedings.
5. W.T. Weng, *Progress and Status of the AGS Booster Project*, Proc. 14th Int. Conf. on High Energy Particle Accelerators, Particle Accelerators 27, 259 (1990).
6. *Proc. Fourth Workshop on Experiments and Detectors for a Relativistic Heavy Ion Collider*, M. Fatyga and B. Moskowitz, eds., Brookhaven National Laboratory Report BNL–52262 (1990).

REMARKS ON A POLARIZED RHIC*

G. Bunce
Brookhaven National Laboratory, Upton, New York 11973

ABSTRACT

A polarized proton program at RHIC, with the same luminosity as for unpolarized protons or L = 2 x 10^{32}cm^{-2}sec^{-1}, is a very attractive possibility. Solutions for technical difficulties seem to be mostly in hand. As a benchmark, the error on a 2-spin asymmetry measurement for jets at $p_\perp$ = 50 GeV/c, for one $\Delta p_\perp$ = 1 GeV/c bin, Δy = 1, $\Delta \phi$ = 2π is ΔA_{LL} = .005 for one month of running. This level of sensitivity offers a remarkable discovery potential.

In these remarks I will discuss the steps necessary to arrive at polarized protons in RHIC, and develop benchmarks for sensitivity for measuring spin asymmetries for direct photons, jets, W$^\pm$, and Drell-Yan production. Closing remarks include possibilities for detectors, a discussion of the focus of the spin program, how a RHIC spin program fits in with other plans, and suggestions on our next step.

Steps to achieve polarized protons at RHIC.

It is very difficult to accelerate polarized protons: the proton anomalous magnetic moment is large and its spin interacts with the focusing fields and with imperfections in the guide field of accelerators. At certain energies spin resonances occur which would depolarize the beam if rather elaborate precautions are not taken. At the ZGS (weak focusing) and then AGS (strong focusing), these problems have been overcome, with polarized protons accelerated to about 25 GeV/c. To arrive at high luminosity polarized protons in RHIC, many additional innovations are needed. Fortunately solutions are known for each of the technical problems anticipated for RHIC, and a key element, the Siberian Snake, has been thoroughly tested in a beautiful series of experiments at Indiana University. Alan Krisch described this work in his talk. Solutions for several other important issues were described by S.Y. Lee. I would like to remind you of the problems and solutions that are proposed.

The present AGS polarized proton source is more than adequate for obtaining high luminosity in RHIC. The normal AGS polarized intensity now is 10^{10} protons per pulse. The booster/accumulator, now nearly complete, will allow both the accumulation of many

---
* Work performed under the auspices of the U.S. Dept. of Energy.

pulses of polarized protons from the Linac, and to compress the stored protons into a single AGS bucket. The expected intensity gain from accumulation is a factor x20; the gain from filling each AGS/RHIC bucket in this way is x12. The required number of protons per RHIC bucket needed to achieve a 2 x 10^{32}cm^{-2}sec^{-1} luminosity is only 10^{11} bucket, well below the anticipated polarized proton intensity using the present AGS source. Therefore, the polarized proton luminosity should be the same as for unpolarized protons. This is a crucial point.

At present the spin resonances, which would depolarize the beam, are overcome in the AGS by correcting the fields at individual imperfection resonances (about 30 of these are encountered over the acceleration cycle) and by quickly shifting the betatron tune for intrinsic resonances (4 of these). The setup for running is time-consuming and painful. A proposed new device, called a partial snake (invented about 3 years ago by T. Roser of Michigan), will eliminate all the imperfection resonances, resulting in a stable and straightforward polarized proton injector for RHIC. This device has been tested at Indiana--see A. Krisch's talk at this meeting. L. Ratner of BNL has proposed a design for the AGS snake, based on a snake study group which met at BNL last spring.

The next difficulty is to transfer the polarized protons into RHIC. The transfer line changes level and bends in the horizontal plane 90°. S.Y. Lee and E. Courant studied this and found a neat solution: at one momentum, 24.8 GeV/c, the line is transparent to the proton spin. This momentum is below an important resonance in the AGS and is also above the RHIC transition energy. For details of this and the snake requirements for RHIC, see the talk by S.Y. Lee.

In RHIC two full Siberian snakes are needed for each ring. The snakes are a series of magnets which precess the spin through 180° and leave the beam in its proper orbit. They have the effect of cancelling all spin resonances for each 2 orbits. The snakes can precess the spin through a plane including the longitudinal direction, or in a plane transverse to the beam. By having one of each type of snake the spin is stable (and vertical if injected vertically) throughout RHIC. Therefore transverse spin experiments (a major program as advocated by Sivers, Artru and Ralston) could be done at any interaction region. S.Y. Lee has developed a way to split each snake on either side of an interaction region (see his talk). In this way the snake may be used as a spin rotator to give longitudinal polarization at the interaction region. This will work for one experiment, but others desiring longitudinal spin would need to install spin rotators in each beam, the equivalent of 2 additional snakes per experiment.

When one performs spin experiments, it is always attractive to reverse the spin as frequently as possible. In RHIC the situation is ideal: the spin for each bunch can be set independently, so that alternating bunches can have opposite spin. An empty bucket can be left to code the position of the alternating bunches. Thus, the spin can be reversed each bunch crossing, which offers exquisite control of systematic effects.

Finally, it is important to have high energy polarimeters. Since the RHIC energy is the same as the present Fermilab polarized beam, versions of the polarimeters described by D. Underwood and A. Yokosawa should work well.

We anticipate a luminosity of $L = 2 \times 10^{32} cm^{-2} sec^{-1}$ for polarized protons in RHIC with $P = .7$, and a maximum $\sqrt{s} = 500$ GeV. This assumes 57 bunches per ring, which can be increased by a factor 2 in an upgrade; a factor of 4 is also available by using 2×10^{11} protons per bunch although there are machine implications and the number of interactions per crossing would then be 8. L.G. Ratner has estimated hardware costs: about \$200K for the partial snake for the AGS, and \$5M for the 4 RHIC snakes and polarimeters. Other RHIC costs, including the development of an intersection region and low-β quads, need to be evaluated.

There is also interest in using polarized deuterons in RHIC. Deuterons have a much smaller anomalous magnetic moment, $G_D = -.143$ versus $G_p = 1.793$. Snakes which are built to precess proton spins through 180° will not work for deuterons. However, the spin resonances for deuterons are infrequent. S.Y. Lee and L. Ratner are planning to study the acceleration of polarized deuterons.

A final comment on the machine: since RHIC uses bunched beam, the two beams must have the same velocity to stay in phase at crossings. Therefore, it will not be possible to use asymmetric energies for the 2 beams.

Benchmarks for sensitivity.

A group met over the Friday lunch break and agreed on several parameters to use to estimate rates and sensitivity. We use $L = 2 \times 10^{32} cm^{-2} sec^{-1}$ and 10^6 seconds/month; $P_1 = P_2 = .7$; $\Delta A_{LL}/NN = \frac{1}{P_1 P_2} \frac{1}{\sqrt{N_{events}}}$. For $\Delta A_{LL}/NN = .05$, 1600 events are required, implying a sensitivity to a cross section of $\Delta\sigma = 10^{-35} cm^2$. We then developed the sensitivity for direct γ, jets, W production, and Drell-Yan pairs.

For direct γ we used a cross section from a calculation by Soffer et al.,

For $\sqrt{s} = 300$, $P_\perp = 20$, $\left.\dfrac{d\sigma}{dP_\perp dy}\right|_{y=0} = 2 \times 10^{-35}$

Therefore for $\Delta P_\perp = 1$, $\Delta y = 1$, $\Delta\phi = 2\pi$ we get at $P_\perp = 20$, $\Delta A_{LL} = .03/\text{GeV}$ for one month of running for direct γ.

For jets we have used the UA1/2 cross sections.

For $\sqrt{s} = 540$, $P_\perp = 50$, $\left.\dfrac{d\sigma}{dP_\perp dy}\right|_{y=0} = 10^{-33}$

Therefore for $\Delta P_\perp = 1$, $\Delta y = 1$, $\Delta\phi = 2\pi$, we have at $P_\perp = 50$, $\Delta A_{LL} = .005/\text{GeV}$ for one month of running for jets.

For W production we use

$$\sigma_T = 10^{-33} \text{ for } \sqrt{s} = 500,$$

or 2×10^5 Ws for one month of running in a 4π detector. For a 10% branching ratio,

for W, B.R. = .1, $\Delta A_L = .01$ for a one month run.

For example, a decay mode with a 10% branching fraction can be searched for using parity violations at a level of .01 in one month.

For Drell-Yan pairs with $M_{l^+l^-} \geq 10$ GeV, $\Delta\sigma_T = 0.35\ \mu b$, giving $\Delta A_{LL} = 0.008$ in one month for a 4π detector.

It should be emphasized that these rates do not include detection efficiency, but, in the case of direct γs and jets the rates are calculated on a per bin basis with $\Delta p_\perp = 1$.

Additional Remarks

Detectors. There are several proposed RHIC detectors which can be considered for spin experiments (see T. Ludlam's talk) also. The difficulty here is that RHIC detectors will need to handle low luminosity and very high multiplicity whereas a spin detector would be faced with the opposite, so this possibility must be studied. It has also been suggested that on the RHIC time scale (1997), several existing collider detectors may be available. Again, pp luminosity and e^+e^- luminosity are much lower than the RHIC luminosity.

Focus. In forming a RHIC proposal we will need to emphasize certain "expected" physics studies. A clear example from this meeting is the measurement of the gluon asymmetry through direct photon production. I would just like to remark here that the RHIC sensitivity for many processes, for example

$$\Delta A_{LL} \text{ (jets, } P_\perp = 50) = .005/\text{GeV}$$

$$\Delta A_L \text{ (W} \to \text{X, B.R.} = .1) = .01$$

$$\Delta A_{LL} (\gamma, P_\perp = 20) = .03/\text{GeV},$$

for a one month run offer a remarkable window for discovery. In fact, in spin physics surprises are the norm: it has been quite rare for results from spin experiments to fit within the standard model of the day. The EMC result is only one recent example.

The next step. Here I would argue that RHIC is ideal as the next step in spin experiments. Technically, it introduces the use of both partial and full snakes (the SSC requires 26 snakes per ring and snakes in the booster rings as well). From a physics perspective, the lower energy offers parton processes at relatively high quark momentum function x, where the SLAC data show the quarks to be clearly polarized.

Our next step. Technically, the partial snake in the AGS should be supported and go forward. We need clear physics arguments--I expect that these will evolve from this workshop and the RHIC sensitivity. We need to form a RHIC spin collaboration and develop a proposal.

Workshop Talks

Jets in polarized pp collisions

J. Ph. Guillet

CERN, CH-1211 Geneva 23, Switzerland

Abstract

We give some results on jet production for future polarized hadronic collider.

1 Introduction

Since the jets give the largest number of large P_t events for the new colliders and since they are sensitive to the parton dynamic (QCD) (gluon distribution, trilinear gluon vertices,...) their study is worth while being undertaken. But, on the other side, the jet cross-section is plagued by a big systematical error ($\sim 50\%$) which may be improved in the future (?). Note that, in this respect, an asymetry is a good observable for it is a ratio of two cross-sections; so we can expect that the systematical error is reduced.

First let set our notations. For a reaction with two polarized beams : $p(\lambda_1) p(\lambda_2) \rightarrow$ one or two jet $+X$ where λ is the longitudinal polarization of the proton ($\lambda = +1$ or -1), we define two types of "polarized" cross-section:

- the parity non-violating combination:

$$\Delta\sigma = \sigma_{++} - \sigma_{+-} - \sigma_{-+} + \sigma_{--},$$

- the parity violating combination:

$$\Delta_p\sigma = \sigma_{++} + \sigma_{+-} - \sigma_{-+} - \sigma_{--}.$$

The unpolarized cross-section is given by:

$$\sigma = \sigma_{++} + \sigma_{+-} + \sigma_{-+} + \sigma_{--}.$$

The plan of the paper is as follows. First, a brief recall of the jets in the unpolarized case is made. Secondly, we discuss seveal ways to measure the spin dependent gluon distribution with jets. And, thirdly, some results on mass-spectroscopy with two jets are given.

2 Results in the unpolarized case

The situation is more comfortable for theorists because the complete next-to-leading order has been computed (order α_s^3) [1] [2] [3]. Basically, the cross-section can be written as:

$$E\frac{d\sigma}{d^3\vec{P}} = \sum_{i,j} \int dx_a dx_b F_i(x_a, M^2) F_j(x_b, M^2) E\frac{d\hat{\sigma}_{ij}}{d^3\vec{P}}. \quad (1)$$

with

$$-O(\alpha_s^2): E\frac{d\hat{\sigma}_{ij}}{d^3\vec{P}} = f_{ij}(x_a, x_b, \alpha_s(\mu))$$
with $x_a x_b S + x_a T + x_b U = 0$

$$-O(\alpha_s^3): E\frac{d\hat{\sigma}_{ij}}{d^3\vec{P}} = g_{ij}(x_a, x_b, \alpha_s(\mu); \mu, M, R)$$
with $x_a x_b S + x_a T + x_b U \neq 0$

The improvements are the following.

- A reduction of the scale dependence (for instance the cross-section changes by a factor of 2 (resp. 20%) at order α_s^2 (resp. α_s^3) when the scales vary between $P_t/4$ and $2P_t$ for $P_t = 100$ Gev, $\sqrt{S} = 1.8$ Tev)

- At order α_s^3, there is a dependence on the jet definition (jet cone size) which is clearly more satisfactory.

Unfortunately, in the polarized case, $O(\alpha_s^3)$ has not been computed yet. So we will work with the leading order ($O(\alpha_s^2)$).

3 Polarized gluon distribution ΔG

Since the EMC experiment has extracted their data on the polarized proton structure function $g_1^p(x, Q^2)$, a large number of theoretical works have been done in order to match theory and experiment, leading to very different polarized gluon distribution. For example:

- using the Skyrm model [4], one can have that $\int_0^1 dx \Delta G(x, Q_0^2) \simeq 0$ where Q_0 is of order of 1 Gev;

- using another model related to the anomaly of the axial-vector current [5], this leads to $\int_0^1 dx \Delta G(x, Q_0^2) \simeq 4 - 5$.

Since our goal is not to discuss the theory of the polarized EMC effect, we let the reader to see the contributions on this subject in this conference.

In order to fix our ideas and to test the sensitivity of the different observables against ΔG, we describe the polarized distributions we will use after. Two sets of polarized distributions has been chosen:

- a standard set [7] with a $\int \Delta G \simeq 0.35$

- a EMC modified set with $\int \Delta G \simeq 3$ derived in the following way [6]:

$$\Delta G(x, Q^2) = \begin{cases} xG(x, Q^2)/.2 & \text{if } x < .2 \\ G(x, Q^2) & \text{if } x \geq .2 \end{cases}$$

So now, one can ask what is the best observable to test ΔG ? To answer to this question, it is important to keep in mind that to have some good tests of the theory, we have to know not only $\int \Delta G$ but the shape of ΔG with x and especially the small x behaviour. We consider the following observables.

3.1 P_t dependence for one jet: $\frac{d\Delta\sigma}{dydP_t}$ and $A_{LL} = \frac{d\Delta\sigma}{dydP_t} / \frac{d\sigma}{dydP_t}$

The polarized cross-section is given by:

$$\frac{d\Delta\sigma}{dydP_t} = 4P_t \sum_{i,j} \int_{x_0}^1 dx_a \frac{x_a x_b}{(2x_a - x_T \exp(y))} [\Delta F_i(x_a, Q^2) \\ \times \Delta F_j(x_b, Q^2) \frac{d\Delta\hat{\sigma}_{ij}}{d\hat{t}} + (x_a \leftrightarrow x_b)] \qquad (2)$$

with

$$x_T = \frac{2P_t}{\sqrt{S}}, \quad x_0 = \frac{x_T \exp(y)}{2 - x_T \exp(-y)}, \quad x_b = \frac{x_a x_T \exp(-y)}{2x_a - x_T \exp(y)}.$$

We present, in Fig.1, $A_{LL}|_{y=0}$ against P_t for $\sqrt{S}=$ 600 Gev, using the two sets of polarized distributions.

3.2 Rapidity dependence for one jet: $\frac{d\Delta\sigma}{dy}$ and $A_{LL}^y = \frac{d\Delta\sigma}{dy} / \frac{d\sigma}{dy}$

It is given by:

$$\frac{d\Delta\sigma}{dy} = \int_{P_{tmin}}^{\sqrt{S}/2} dP_t \frac{d\Delta\sigma}{dydP_t}. \qquad (3)$$

This observable is good for testing the small x behaviour of the partonic distributions. Indeed, as can be seen above, x_b is a function of x_a and y for fixed value of P_t. This function gets its minimum for $x_a = 1$ and $y = -\log(x_T)$. For instance, at $\sqrt{S} = 600$ Gev, $P_t = 10$ Gev the minimum value of x_b is 10^{-3} for $y = 3.4$. On the contrary, if we fix y to 0 the minimum value that x_b can reach is only 10^{-2}. We also make some comparison with up-to-date polarized distributions [8] in order to be sure that our parametrisation is not foolish. These results are shown in Table I.

3.3 Full differential cross-section for two jets: $\frac{d\Delta\sigma}{dy_1 dy_2 dP_t}$

The cross-section is given by:

$$\frac{d\Delta\sigma}{dy_1 dy_2 dP_t} = 2\pi P_t \tau \sum_{ij} [\Delta F_i(x_a, Q^2) \Delta F_j(x_b, Q^2) \frac{d\Delta\hat{\sigma}_{ij}}{d\hat{t}} + (x_a \leftrightarrow x_b)] \qquad (4)$$

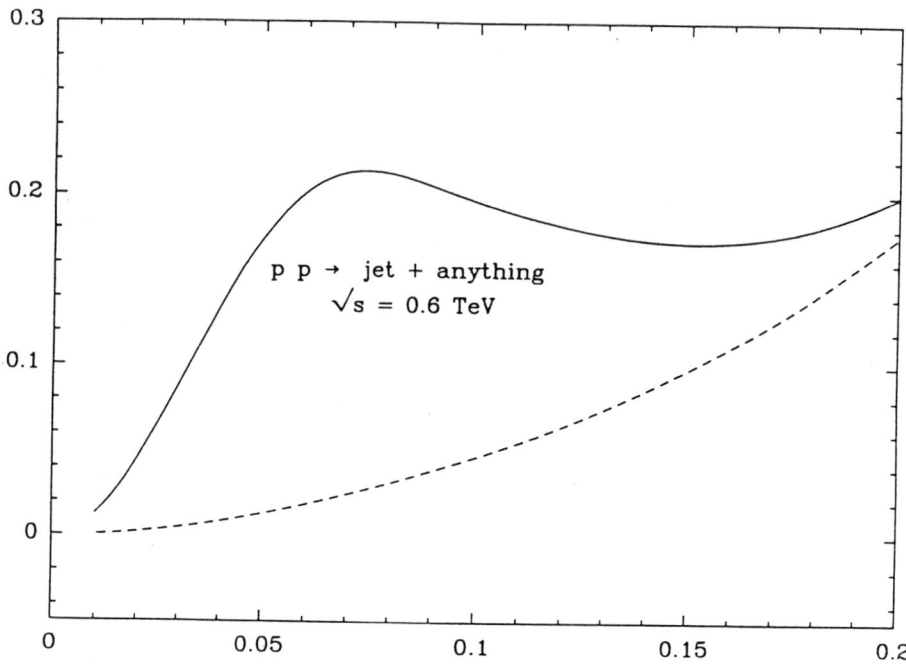

Figure 1: $A_{LL}|_{y=0}$ in % against P_t in Tev. Full curve: the new set of polarized distribution. Dashed curve: the standard set.

ΔF	$y=0$	$y=1$	$y=2$	$y=3$	$y=4$
set1	$0.79 10^6$	$0.60 10^6$	$0.21 10^6$	$0.12 10^5$	0.33
set2	$0.38 10^5$	$0.30 10^5$	$0.13 10^5$	$0.17 10^4$	0.06
set3	$0.10 10^7$	$0.82 10^6$	$0.33 10^6$	$0.30 10^5$	0.26

Table I: $d\Delta\sigma/dy$ in pb for $P_{tmin} = 10$ Gev, $\sqrt{S} = 600$ Gev. The set1 corresponds to the new set, the set2 to the standard set and the set3 to the set derived in ref.[8].

$y_2 \backslash y_1$	-2	-1	0	1	2
-2	**116.0**	**421.8**	**165.3**	**19.7**	**.3**
	36.1	100.3	29.3	2.9	.1
-1	**1141.5**	**7773.0**	**6585.5**	**1034.8**	**19.7**
	272.6	906.5	491.0	72.8	2.9
0	**1203.0**	**17824.5**	**23987.7**	**6585.5**	**165.3**
	216.5	1335.5	1672.8	491.0	29.3
1	**376.5**	**7529.5**	**17824.5**	**7737.0**	**421.8**
	58.9	537.7	1335.5	906.5	100.3
2	**15.8**	**376.5**	**1203.0**	**1141.5**	**116.0**
	5.2	58.9	216.5	272.6	36.1

Table II: $d\Delta\sigma/dy_1 dy_2 dP_t|_{P_t=30}$ in pb/Gev for $\sqrt{S} = 600$ Gev. The bold numbers are those obtained with the new set, the normal ones are obtained using the standard set.

with

$$x_a = \sqrt{\tau} \exp\left[\frac{1}{2}(y_1 + y_2)\right], \quad x_b = \sqrt{\tau} \exp\left[-\frac{1}{2}(y_1 + y_2)\right]$$

and

$$\tau = \frac{4P_t^2}{S} \cosh^2\left[\frac{1}{2}(y_1 - y_2)\right].$$

This observable is good for the shape of the polarized distributions since there is no integration on x. We have computed, in Table II, this cross-section at $P_t = 30$ Gev and by changing y_1 and y_2 between -2 and 2.

4 Mass spectroscopy

The structure of the W and Z resonance peaks can be studied in the di-jet invariant mass distribution [9]. The UA2 experiment performed this study in the unpolarized ($p\bar{p}$) collisions [10]; the QCD background is 10 times the W peak (which is higher than the Z one). In pp collisions, it is worse because the $q\bar{q}$ annihilation which gives the main contribution to the peak of the W and Z is reduced by a factor valence over sea. In this case the QCD background is 100 times the W peak.

A priori, polarized beams can improve the situation, because of the axial coupling of the W and Z to the fermions. We consider two cases.

- Parity conserving combination. It is not interested in pp collisions because the QCD background has not been sufficently reduced.

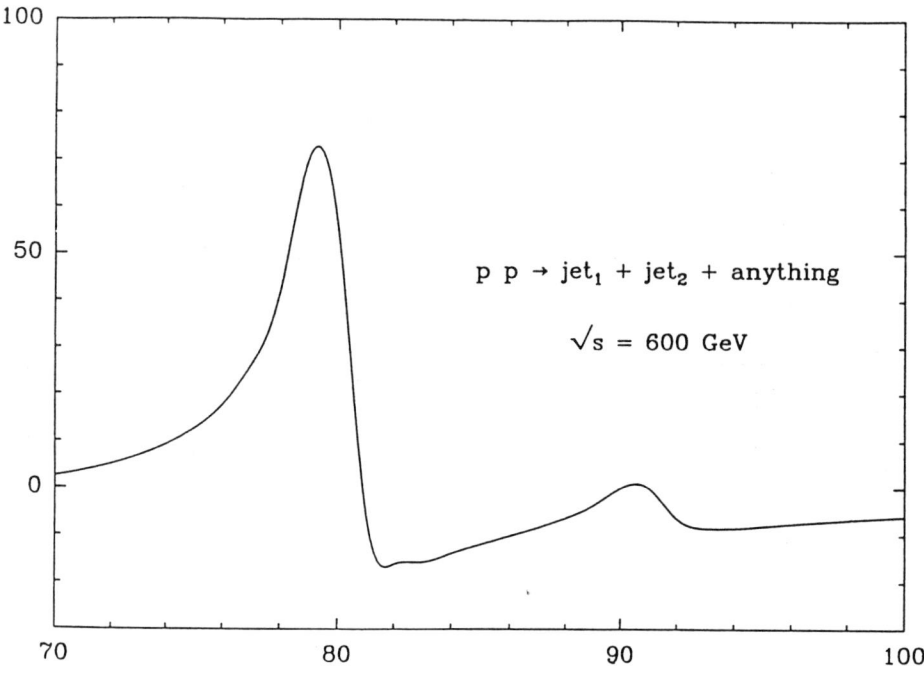

Figure 2: $d\Delta_p\sigma/dM$ in pb/Gev against M in Gev. We use the standard set of polarized distribution.

- Parity violating combination (or only one polarized beam). This case is interesting because the QCD background is stricly 0 due to the parity conservation of QCD. We see only the electroweak signal (see Fig.2). It can be very interesting at LHC or SSC energies to discover new gauge bosons.

5 conclusion

The physic with jets is interesting to test the parton dynamic and even to discover new particles but an unknown part remains: the experimental abilities.

References

[1] F. Aversa, P. Chiappetta, M. Greco, J.P. Guillet, Phys. Rev. Lett. **65** (1990), 401.

[2] S. Ellis, Z. Kunszt, and D. Soper, Phys. Rev. Lett. **64** (1990), 2121.

[3] R. Ellis and J. Sexton, Nucl. Phys. **B269** (1986), 475.

[4] J. Ellis and M. Karliner, Phys. Lett. **B213** (1988) 73.

[5] See A.V. Efremov in these proceedings and references therein.

[6] E.L. Berger and J.W. Qiu, Phys. Rev **40** 778.

[7] C. Bourrely, F.M. Renard, J. Soffer and P. Taxil, Phys. Rep. 177(1989) 320.

[8] P. Chiappetta and G. Nardulli, Preprint BARI-TH/90-71

[9] G. Ranft and J. Ranft, Nucl. Phys. **B165**(1980) 395

[10] J. Alitti et al., Preprint CERN-PPE/90-105

DIRECT PHOTON PRODUCTION IN POLARIZED PP COLLISIONS

Edmond L. Berger
High Energy Physics Division, Argonne National Lab, Argonne, IL 60439

Jianwei Qiu*
Institute for Theoretical Physics, SUNY, Stony Brook, New York 11794-3840

ABSTRACT

We argue that inclusive direct photon production at large transverse momentum in proton-proton interactions with longitudinally polarized beam and target is an incisive probe of the polarized gluon distribution in a proton. We provide predictions of cross sections for a range of reasonable choices of the polarized gluon distribution. At current fixed target energies, the cross sections are small but measurable. At collider energies, such as the SSC energy, isolated direct photon production should be considered in order to minimize the nonperturbative contribution due to fragmentation.

I. INTRODUCTION

Recent data on the polarization asymmetry in deeply inelastic scattering (DIS) of polarized muons on polarized protons have renewed interest in the relationship between the spin of a proton and the polarization of its constituents (quarks and gluons).[1,2] This asymmetry measures the spin-dependent structure function $g_1^p(x, Q^2)$ of a proton.[1] Here x is the usual Bjorken scaling variable, and Q^2 is the square of the four-momentum transfer. The measurement of $g_1^p(x, Q^2)$ and its first moment, combined with the Bjorken sum rule and the F/D ratio obtained from semileptonic hyperon decays, permits a determination of the first moment of the SU(3)-singlet combination of the spin-dependent quark distributions, $\Delta\Sigma$. The result is that $\Delta\Sigma$ is consistent with zero.[1] This result has led to a good deal of discussion in the literature.[3] At the core of the discussion is how to interpret (or to define) the "right" spin-dependent quark distributions, $\Delta q(x, Q^2)$, and whether there is a large gluonic contribution to the first moment of $g_1^p(x, Q^2)$. Of particular interest to us is the spin-dependent gluon distribution in a polarized proton and its independent experimental measurement.[4]

Direct access to the spin-dependent gluon distribution should be provided by processes in which the gluons contribute *dominantly* to the cross sections. In the case of unpolarized proton-proton scattering, direct photon production at large transverse momentum, p_T, is known to be dominated by the "Compton" subprocess $qg \to \gamma q$ at fixed target energies.[5] Here q and g stand for quark and gluon constituents. Data, therefore, allow a direct determination of the (spin-averaged) gluon distribution $G(x, Q^2)$, provided the quark distributions are well-determined in other processes (*e.g.*, in DIS). When both the beam and target protons are polarized, the cross section due to the "Compton" subprocess can be expressed directly in terms of the polarized gluon distribution $\Delta G(x, Q^2)$.[6] Based on the DIS data and a range of reasonable choices of spin-dependent gluon distributions, we shall argue that the inclusive direct photon production in polarized pp collisions is also dominated by the "Compton" subprocess. As a result, such experiments will provide direct access to the spin-dependent gluon distribution in a polarized proton.

---

*SSC Fellow

In Sec. II we explain why the "Compton" subprocess dominates the cross section for unpolarized hadronic prompt photon production. We then argue that the same conclusion can be reached for the polarized case. In Sec. III we provide specific predictions for the cross sections and polarization asymmetries expected for direct photon production in polarized pp collisions at fixed target energies. We discuss our expectations for polarized direct photon production at collider energies in Sec. IV. As shown explicitly in Sec. IV, at leading order, our predictions are not very sensitive to what interpretation or definition one adopts for the spin-dependent quark distributions. Instead, our prediction is directly related to the experimentally measured structure function $g_1^p(x, Q^2)$ as a whole. Conclusions are also found in Sec. IV.

II. PROMPT PHOTONS AT LARGE p_T

Measurement of inclusive direct photon production at large p_T in pp collisions is known to be a clean probe of the (spin-averaged) gluon distribution in a proton at fixed target energies. However, at collider energies, *isolated* (instead of inclusive) direct photon production must be considered in order to minimize the large contribution from nonperturbative photon fragmentation.[7] Many estimates are available of cross sections for direct photon production in *unpolarized pp* collisions.[5] Data are generally in accord with the calculations. To understand how good direct photon production in *polarized pp* collisions may be in determining the spin-dependent gluon distribution, it is important to examine why the process is good for determining the gluon distribution in the case of unpolarized pp collisions.

At large p_T, a photon can be produced through a pointlike, perturbatively calculable hard scattering term, as well as through a nonperturbative fragmentation process (generalization of the bremsstrahlung process). Because of the nonperturbative nature of the fragmentation process, perturbative QCD alone cannot predict the actual size of the inclusive prompt photon cross section. In principle, whether in the case of unpolarized or polarized interactions, we can expect direct photon production to give good information on parton (gluon or quark) distributions only if the nonperturbative fragmentation process can be controlled.

With approximate leading logarithmic photon fragmentation functions, we estimate that the fragmentation contribution in the unpolarized case is only about a 10% effect at current fixed target energies. In contrast, it amounts to a 50% (or even larger) effect at collider energies.[7] Because of the poorly known photon fragmentation function, we do not expect that inclusive direct photon production at high energy can yield clean information on gluon distributions. However, at fixed target energies, inclusive prompt photon is a relatively good probe of short-distance physics.

At leading order, short-distance prompt photon production is governed by the tree "Compton" $qg \to \gamma q$ and annihilation $q\bar{q} \to \gamma g$ subprocesses. The "Compton" (or annihilation) contribution to the cross section can be written in general as

$$\sigma_{\text{Com}} = (q \otimes G) \otimes \hat{\sigma}_{\text{Com}} \quad \text{"Compton"}$$
$$\sigma_{\text{anni}} = (q \otimes \bar{q}) \otimes \hat{\sigma}_{\text{anni}} \quad \text{Annihilation,} \qquad (1)$$

where $\hat{\sigma}$ is a parton level cross section – the short-distance hard scattering term, q ($\bar{q}$) and g are quark (antiquark) and gluon distributions, and $\otimes$ means a convolution between the distributions and the hard scattering term. The expression in parentheses () can be thought of as the parton flux. The statement that inclusive direct photon production is a good probe of the gluon distribution for the unpolarized case is based on the fact that the net "Compton" contribution in pp interactions is *numerically* much larger than the annihilation contribution. At the parton level, the "Compton" contri-

Direct Photon Production in Polarized pp Collisions

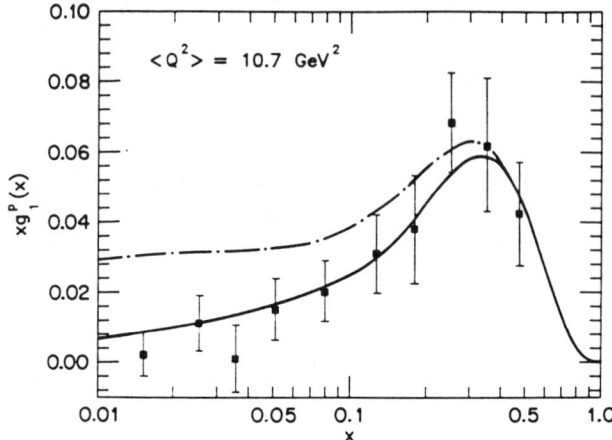

Figure 1: The dot-dashed line is the contribution to $xg_1^p(x,Q^2)$ from the polarized valence quark distributions at $Q^2 = 10.7$ GeV^2. The solid line is our parametrization of $xg_1^p(x,Q^2)$ including effective sea distributions (Ref. 4). The data are from EMC (Ref. 1).

bution is actually *smaller* than the annihilation contribution, $\hat{\sigma}_{\text{Com}} < \hat{\sigma}_{\text{anni}}$. The reason the net "Compton" contribution is larger than the annihilation contribution is that in a proton (spin-averaged) the gluon density is much larger than the antiquark density. Equivalently, the difference in the parton flux controls the answer. For example, if $x_T = 2p_T/\sqrt{s}$ and $x_a \sim x_b \sim x_T \sim 0.1$, $\hat{\sigma}_{\text{Com}} \sim (1/2)\hat{\sigma}_{\text{anni}}$; but $G(x) \sim 20 \, \bar{q}(x)$ for $x \sim 0.1$. Here x_a and x_b are momentum fractions carried by initial partons participating in the hard scattering. Because the observed cross section is dominated by the region where $x_a \sim x_b \sim x_T$, the above simple estimate suggests that $\sigma_{\text{Com}} \sim 10 \, \sigma_{\text{anni}}$, about the same as is obtained in a more precise numerical calculation.

In the case of polarized interactions, the hard part for the annihilation subprocess is the same as that for the unpolarized case up to a sign difference. But, the hard part for the "Compton" subprocess is even smaller than that in the unpolarized case. As a result, $\Delta\hat{\sigma}_{\text{Com}} \sim (1/4)|\Delta\hat{\sigma}_{\text{anni}}|$ for $x_a \sim x_b \sim x_T \sim 0.1$. Therefore, inclusive direct photon production in polarized pp collisions can yield useful information on ΔG iff ΔG is much bigger than $\Delta \bar{q}$, which is not a priori obvious. The spin-averaged gluon distribution is much larger than the antiquark distribution because the evolution kernel for gluons has a $1/x$ pole. In contrast, the evolution kernel for the spin-dependent gluon distribution does not have the $1/x$ pole. Consequently, ΔG is much smaller than G when x is small. The question is now whether $\Delta \bar{q}$ is even smaller.

To estimate the size of $\Delta \bar{q}$, we assume that the Carlitz and Kaur[8] model is approximately valid for the spin-dependent valence quark distributions. We justify this assumption by noting that the valence quark part of the structure function $g_1^p(x,Q^2)|_{\text{Val}}$, the dot-dashed curve in Fig. 1. fits the data in the large x region. Then, we use the difference between the data and $g_1^p(x,Q^2)|_{\text{Val}}$ in the small x region to parametrize the effective spin-dependent sea and anti-quark distributions. From Fig. 1, for example, we can estimate $\Delta\bar{q}(x)$ for $x \sim 0.1$ as follows:

$$xg_1^p(x,Q^2)|_{\text{data}} - xg_1^p(x,Q^2)|_{\text{Val}} \approx 2\left[2\left(\frac{1}{3}\right)^2 + \left(\frac{2}{3}\right)^2\right] x\Delta\bar{q}(x,Q^2) \tag{2}$$

where $xg_1^p(x,Q^2)|_{\text{data}}$ means the solid curve, and an SU(3) symmetric sea is assumed. Here the scale Q^2 is about 10.7 GeV^2. For $x \sim 0.1$, $xg_1^p(x,Q^2)|_{\text{data}} - xg_1^p(x,Q^2)|_{\text{Val}} \approx 0.01$, and consequently, $\Delta\bar{q}(x,Q^2) \sim -0.075$ when $x \sim 0.1$.

Because $\Delta G(x,Q^2)$ should be much smaller than $G(x,Q^2)$ in the small x region, a conservative assumption is that $\Delta G(x,Q^2) \approx xG(x,Q^2)$. Taking $xG(x,Q^2) \sim 3(1-x)^5$, we have $\Delta G(x,Q^2) \sim 1.8$ for $x \sim 0.1$. Comparing with $|\Delta\bar{q}(x,Q^2)| \sim 0.075$, we observe that $\Delta G(x,Q^2)$ is much larger than $\Delta\bar{q}(x,Q^2)$ in the region where $x \sim 0.1$. We conclude that inclusive direct photon production in polarized pp collisions is a good probe of the spin-dependent gluon distribution, $\Delta G(x,Q^2)$, of a polarized proton.

III. NUMERICAL RESULTS

To obtain quantitative predictions for the cross sections for inclusive direct photon production in polarized pp collisions, we need parametrizations of spin-dependent quark and gluon distributions. The general procedure to obtain the spin-dependent parton distributions is first to parametrize the distributions at an arbitrary reference value $Q^2 = Q_0^2$, and then to use the spin-dependent Altarelli-Parisi evolution equations[9] to obtain the distributions at any other scale Q^2. To parametrize the spin-dependent input parton distributions, we adopt the Carlitz and Kaur model for spin-dependent valence quark distributions. With the assumption of an SU(3) symmetric sea, we parametrize the spin-dependent sea quark and antiquark distributions by fitting the difference between the data and the prediction from the Carlitz and Kaur model. For the spin-dependent gluon distribution $\Delta G(x,Q^2) = G^{(+)}(x,Q^2) - G^{(-)}(x,Q^2)$, the difference between the numbers of gluons with positive and negative helicity, we select a simple one-parameter function[4]

$$\Delta G(x,Q_0^2) = \begin{cases} G(x,Q_0^2), & x_c \leq x \leq 1, \\ \frac{x}{x_c}G(x,Q_0^2), & 0 \leq x \leq x_c. \end{cases} \tag{3}$$

The parameter x_c with $0 < x_c \leq 1$ determines the net helicity carried by gluons. When the value of x_c is adjusted, this simple parametrization covers a spectrum of reasonable choices of the polarized gluon distribution. The behavior of our polarized parton distributions is displayed numerically in Ref. 4.

In leading-order QCD perturbation theory, the cross section for prompt photon production in *polarized pp* reactions, $\vec{p}_A + \vec{p}_B \to \gamma + X$, is given by

$$E_\gamma \frac{d\Delta\sigma_{AB}}{d^3p_\gamma}(s,x_F,p_T) = \sum_{a,b} \int dx_a dx_b\, \Delta P_A^a(x_a,Q^2)\, \Delta P_B^b(x_b,Q^2)$$

$$\times E_\gamma \frac{d\Delta\hat{\sigma}_{ab}}{d^3p_\gamma}(\hat{s},x_F,p_T), \tag{4}$$

where a and b run over all possible quark flavors and the gluon; $\Delta\sigma_{pp} \equiv \frac{1}{2}[\sigma_{pp}(++) - \sigma_{pp}(+-)]$; and $\sigma_{pp}(++)$ denotes the cross section when the two incident protons have their spins oriented along the directions of motion of the protons in the overall pp center-of-mass system. Similarly, for the unpolarized case,

$$E_\gamma \frac{d\sigma_{AB}}{d^3p_\gamma}(s,x_F,p_T) = \sum_{a,b} \int dx_a dx_b\, P_A^a(x_a,Q^2)\, P_B^b(x_b,Q^2)\, E_\gamma \frac{d\hat{\sigma}_{ab}}{d^3p_\gamma}(\hat{s},x_F,p_T). \tag{5}$$

Direct Photon Production in Polarized pp Collisions

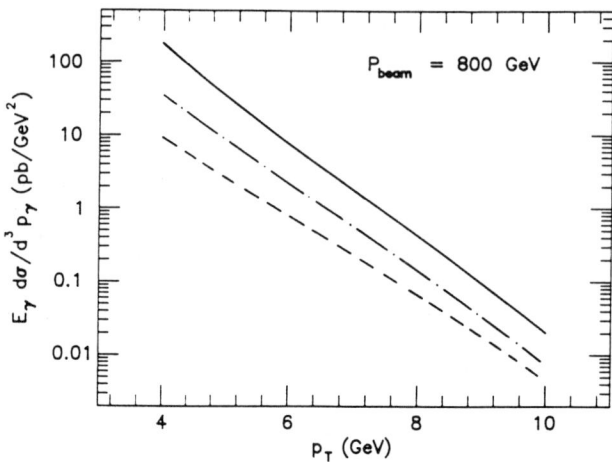

Figure 2: Predicted cross sections for prompt photons as a function of p_T for $x_F = 0$ in pp interactions at $p_{\text{beam}} = 800$ GeV/c. The solid line shows the rate in unpolarized pp interaction, whereas the dashed and dot-dashed curves show a range of rates in the polarized case.

In Eq. (4), $\Delta P_A^a(x_a, Q^2)$ and $\Delta P_B^b(x_a, Q^2)$ are spin-dependent parton distributions, and in Eq. (5), $P_A^a(x_a, Q^2)$ and $P_B^b(x_a, Q^2)$ are spin-averaged parton distributions. The short-distance hard scattering terms $\Delta \hat{\sigma}$ and $\hat{\sigma}$ include the parton level "Compton" and annihilation subprocesses. Their analytical expressions can be found in Ref. 4.

Having the parton distributions and expressions for the cross sections, we can calculate rates for direct photon production at different energies.[4] In Fig. 2 we plot the production rates as a function of the photon's transverse momentum p_T for incoming beam energy equal to 800 GeV^2. The solid line represents the rate in unpolarized pp collisions. The dashed line represents the rate in polarized pp interactions with parameter $x_c = 1.0$ for the polarized input gluon distribution, and dot-dashed line for $x_c = 0.2$. These two lines in the polarized case represent roughly a reasonable range expected for the polarized cross section at fixed target energies. They correspond to values of the integral (integral of ΔG over all x) of 0.5 and 2.5 at $Q \sim 2$ GeV. The predicted energy dependence of the cross sections is shown in Fig. 3. Notice in Fig. 3 that when energy increases, the unpolarized cross section increases faster than the polarized one. At higher energies, smaller values of x contribute, and at small x, the unpolarized gluon distribution is much larger than its polarized counterpart. From Fig. 2, we notice that the cross section is small for direct photon production with transverse momentum $p_T > 4$ GeV, but it is still measurable.[4]

IV. DISCUSSION AND CONCLUSIONS

We have argued that at fixed target energies, inclusive direct photon production in polarized pp collisions is a good probe of the spin-dependent gluon distribution in a proton. At collider energies (e.g., SSC energy), the inclusive process is no longer ideal for measuring the gluon distribution because of the large nonperturbative contribution due to fragmentation processes. However, our experience with the treatment of the

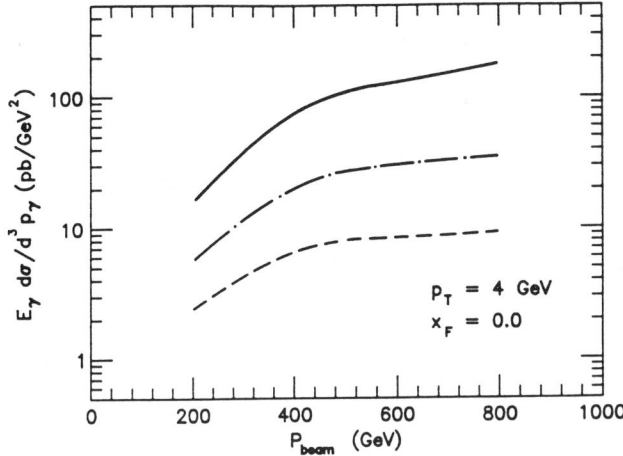

Figure 3: Predicted cross sections for prompt photons as a function of beam energy p_{beam} for $x_F = 0$ and $p_T = 4$ GeV. The solid, dashed and dot-dashed lines are defined as in Fig. 2.

unpolarized case [7] at collider energies leads us to believe that isolated direct photon production in polarized pp interactions should be an appropriate process from which to extract the spin-dependent gluon distribution, as long as the isolation cut is imposed properly in both experimental measurements and theoretical calculations.

Our numerical predictions for the rate of direct photon production at large p_T depend on the polarized parton distributions. Experiments measure only a combination of the distributions, for example, the structure function $g_1^p(x, Q^2)$. One question of interest is how sensitively our predictions depend on the different parametrizations (or separations) of the valence and sea quark distributions. Our conclusion is that at leading order, the cross section is actually insensitive to the different choices of parametrizations of parton distributions, as long as these distributions fit the DIS data. This result is due to dominance of the "Compton" subprocess. To make our point more transparent, we rewrite the cross sections given in Eqs. (4) and (5) approximately as

$$E_\gamma \frac{d\Delta\sigma_{AB}}{d^3p_\gamma}(s, x_F, p_T) = \int dx_a dx_b \Big[2g_1^p(x_a, Q^2) \Delta G(x_b, Q^2) E_\gamma \frac{d\Delta\hat{\sigma}_{qg}}{d^3p_\gamma}(\hat{s}, x_F, p_T) + (x_a \leftrightarrow x_b) \Big] + \text{"Annihilation term"} \quad (6)$$

and

$$E_\gamma \frac{d\sigma_{AB}}{d^3p_\gamma}(s, x_F, p_T) = \int dx_a dx_b \Big[\frac{F_2(x_a, Q^2)}{x_a} G(x_b, Q^2) E_\gamma \frac{d\hat{\sigma}_{qg}}{d^3p_\gamma}(\hat{s}, x_F, p_T) + (x_a \leftrightarrow x_b) \Big] + \text{"Annihilation term"} \quad (7)$$

where $g_1^p(x, Q^2)$ and $F_2(x, Q^2)$ are the two proton structure functions measured in polarized and unpolarized DIS. Neglecting the annihilation terms, one sees immediately that the cross sections for direct photon production are proportional to measured structure functions and are therefore not sensitive to the detailed parametrizations of the quark distributions of different flavors. More important is that Eqs. (6) and (7) show that large p_T direct photon production in pp interactions measures the gluon distribution *directly*, provided the proton structure functions are well measured in DIS.

The x range explored by fixed target experiments is limited. Based on fixed target data, it will be difficult to derive an accurate measurement of the *first moment* of the polarized gluon distribution, which we would like to know. Nevertheless the sign of the polarized gluon distribution as well as its magnitude in the large and intermediate x regions should be accessible experimentally. This information would be particularly interesting.

We conclude that inclusive direct photon production at large p_T in polarized pp collisions at fixed target energies is dominated by the gluon-induced "Compton" subprocess. Correspondingly, fixed target experiments with longitudinally polarized proton beams and targets would measure the sign and magnitude of the polarized gluon distribution in a proton in interesting regions of x. At fixed target energies, the cross section is small, but it should be measurable. The production rate will be greater at higher energies (such as the SSC energy), but at collider energies an isolation cut on the photon may be required to minimize the nonperturbative contribution due to fragmentation processes.

Acknowledgements

We have benefitted from discussions with J.C. Collins, G. Sterman, and W.K. Tung. The research was supported in part by the U.S. Department of Energy, Division of High Energy Physics, Contract W-31-109-ENG-38; and by the National Science Foundation under grant No. PHY-89-08495.

REFERENCES

1. European Muon Collaboration, J. Asham *et al*, Phys. Lett. 206B, 364 (1988).
2. M.J. Alguard *et al*, Phys. Rev. Lett. 37, 1258 (1976); 41, 70 (1978); G. Baum *et al*, *ibid*. 51, 1153 (1983).
3. For example, see articles in these proceedings by A. Efremov, by J. Manohar, and by G. Bodwin, respectively.
4. E.L. Berger and J. Qiu, Phys. Rev. D40, 778 (1989); D40, 3128 (1989).
5. E.L. Berger, E. Braaten, and R.D. Field, Nucl. Phys. B239, 52 (1984), and references therein; J.F. Owens, Rev. Mod. Phys. 59, 465 (1987).
6. N.S. Craigie, K. Hidaka, M. Jacob, and F.M. Renard, Phys. Rep. 99C, 69 (1983); C. Bourrely, F.M. Renard, J. Soffer and P. Taxil, Phys. Rep. 177, 319 (1989).
7. E.L. Berger and J. Qiu, Phys. Lett. 248B, 371 (1990); Argonne preprint ANL-HEP-PR-90-104.
8. R. Carlitz and J. Kaur, Phys. Rev. Lett. 38, 673 (1977).
9. G. Altarelli and G. Parisi, Nucl. Phys. B126, 298 (1977).

BEYOND THE STANDARD MODEL WITH POLARIZED BEAMS AT FUTURE COLLIDERS

P. Taxil

Centre de Physique Théorique, CNRS, Luminy, Case 907
F-13288 Marseille Cedex 9, France

ABSTRACT

The availability of high-intensity polarized beams at the future hadronic colliders would allow to reveal interesting spin effects as possible manifestations of the presence of New Physics beyond the Standard Model. Illustrative examples are given, which concern supersymmetry, compositeness and new gauge bosons.

INTRODUCTION

In this report we are forced by the subject to consider mainly the physics at future supercolliders : the SSC with a center of mass energy of 40 TeV and a luminosity of $\mathcal{L} \approx 10^{33}$ cm^2 s^{-1} [1], and the CERN project LHC with a c.m. energy of 16 TeV but with a higher luminosity $\mathcal{L} \approx 4 \; 10^{34}$ cm^2 s^{-1} [2]. Note that, as we will see, the regime where New Physics could manifest is perhaps not out of reach for a collider such as the RHIC project with polarized proton beams at a c.m. energy of 600 GeV. Indeed, the goal of these supercolliders is the exploration of the TeV scale which is a very important domain of energy for two complementary reasons :

- In the framework of the Standard Model (SM), a definite exploration of the mechanism of electroweak symmetry breaking will require collisions between fondamental particles (e$^+$e$^-$ or partons) at a c.m. energy of the order of 1 TeV. The Higgs boson has to be found with a mass below or not much above that scale which is a bound from perturbative unitarity arguments. If there is no Higgs boson below 1 Tev, it means that perturbative theory breaks down and that the "weak interactions" between W's and Z become strong. Then new phenomena must show up.

- On the other hand it is widely accepted that the SM must be incomplete : it has too many parameters and also it does not explain how the relatively low scale of the electroweak symmetry breaking is maintained in the presence of quantum corrections. For instance, contributions from quantum loops to the Higgs mass are quadratically divergent in the cutoff Λ, a scale beyond which the SM is no longer valid. This is the source of the famous hierarchy or naturalness problem [3] : if Λ is of the order of , say, the Grand Unification scale ($\approx 10^{15}$ GeV) it is quite difficult to substract this divergence in order to maintain a relatively light Higgs boson. Any way to solve this problem is a potential source of New Physics.

The most elegant way is perhaps supersymmetry (SUSY) where partners of ordinary particles enter into new loops, a phenomenon which stabilizes the Higgs mass. The cutoff Λ is then the scale of SUSY breaking, a scale which must not be larger than 1 TeV to help solving the naturalness problem, which in turn implies that SUSY partners would not be too heavy. Composite models provide an other way, Λ being now the compositeness scale of fermions and/or gauge bosons. Finally, extending the gauge group is also a way to go beyond the SM with the prediction of the presence of new gauge bosons and new interactions. We will discuss in turn some examples in the framework of these scenarios, showing the results of illustrative calculations performed with a "reasonable" set of polarized parton distributions [4]. Of course these distributions are an essential ingredient of any calculation : one can hope they will be better known in the future thanks to forthcoming deep inelastic scattering experiments with a polarized lepton beam on a polarized target [5]. In fact, the most important point is that the studies of SM processes themselves, at polarized supercolliders, would allow a good calibration of spin-dependent as well as spin-independent distributions [4,6].

SUPERSYMMETRY

SUSY partners of ordinary particles differ from the latter essentially by the value of their spin (also by their masses since SUSY is broken), hence their production should present characteristic spin effects. Consider first SUSY partners of quarks and gluons : squarks and gluinos. They would be produced via QCD-like subprocesses leading to jets plus missing energy events. Note that the decay pattern of these objects could be quite complicate if e.g. gluinos are heavier than some SUSY partners of the electroweak gauge bosons, leading to events whose interpretation could be difficult. The production of these sparticles is copious. For example, even for 1 TeV gluinos, 10^4 gluinos/year are expected at LHC with an integrated luminosity of 10^4 pb^{-1} only (the present limits on masses are around 75-100 Gev for squarks and gluinos [7]).

Concerning spin effects, since parity is conserved in these strong processes, only double spin asymmetries $A_{LL} = (d\sigma^{++} - d\sigma^{+-})/(d\sigma^{++} + d\sigma^{+-})$ could be non-zero. Here the superscripts +(-) refer to the helicities of the colliding protons. Indeed, the subprocesses asymmetries for the production reactions are all $\hat{a}_{LL} = -100\%$ in contrast with the majority of standard QCD subprocesses (this is true in the massless limit but this pattern survives mass effects). As a consequence, the observable A_{LL} which is displayed in Fig. 1, is opposite in sign and larger in magnitude than the same quantity in the production of standard QCD jets [4,6].

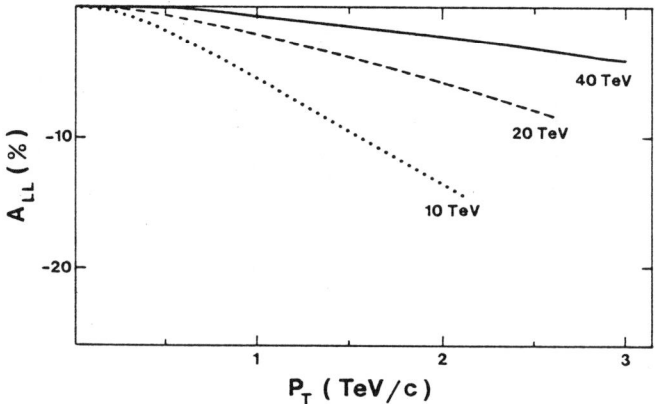

Fig.1. A_{LL} in the production of one strongly interacting sparticle at zero rapidity, versus p_T for three different energies.

In the SUSY electroweak sector, one can get now parity violating single spin asymmetries $A_L = (d\sigma^- - d\sigma^+)/(d\sigma^- + d\sigma^+)$. For instance, a measurement of A_L in scalar lepton pair production via a generalized Drell-Yan process could allow to separate the two different scalars $\tilde{l}_L, \tilde{l}_R$ associated to each helicity state of a lepton l, if these two sparticles are not degenerate in mass [4]. A_L in the production of pairs of neutral gauginos ($\tilde{\gamma}\tilde{Z}, \tilde{Z}\tilde{Z}$), going through $\tilde{q}_{L;R}$ exchange, is sensitive to the same kind of phenomenon. Moreover, processes involving charged Winos : $\bar{q}_i q_j \rightarrow \tilde{W}^\pm \tilde{\gamma}, \tilde{W}^\pm \tilde{Z}$ display "universal asymmetries" as in the case of standard W production [4,6].

A first conclusion for this section is that the best signature for the production of SUSY particles would be the presence of missing energy events associated to a specific polarization asymmetry.

Finally, in SUSY extension of the SM, the Higgs sector also is more elaborate. Even in the minimal case two Higgs doublets are requested (plus their fermionic partners, the Higgsinos) which implies a new phenomenology : more neutral scalars and also charged Higgses, these ones necessarily heavier than the W's. Note that such charged scalars are also present in any (non necessarily SUSY) two-doublet extension of the minimal version of the SM. A_L in H^+H^- production can be found in ref.[4]. It is interesting to remember that it is independant of the free parameters of the model (mixing angles ..). Moreover the same asymmetry is obtained in the production of a pair of technipions in technicolor theories. This leads us to an other scenario which is compositeness.

COMPOSITENESS

We will leave aside technicolor to concentrate first on composite quarks and leptons. There is no standard composite model on the market : in phenomenological studies one assumes that subconstituents could interact by means of a new "contact" interaction, normalized to a certain compositeness scale Λ. New amplitudes have to be taken into account, whose presence would be revealed first through their interference with standard QCD amplitudes. From current experiments, the limits on Λ are of the order of 1 TeV. The discovery limits of HERA in e-p collisions is $\Lambda \approx 5$ TeV and (unpolarized) LHC (SSC) will probe Λ up to 20 (30) TeV according to some recents analysis [8]. It has been emphasized that the chirality structure of this new interaction is arbitrary : in particular there is no reason to assume that it conserves parity ! This could be the origin of very dramatic single spin effects in various channels [4] : lepton pair production, single photon or jet production. In Fig. 2 we show for example the parity violating A_L one gets in single jet production according to QCD + a purely left-handed contact interaction.

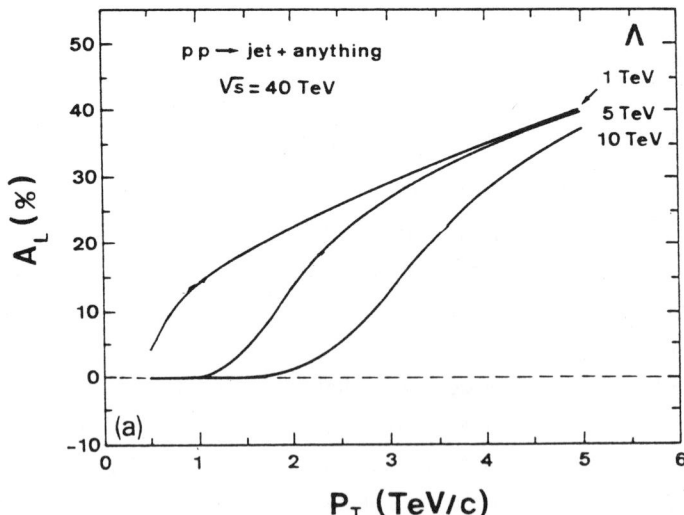

Fig. 2 A_L in pp $\to$ jet + X at y = 0, versus p_T, at SSC with different values of the pure left-handed contact interaction scale Λ.

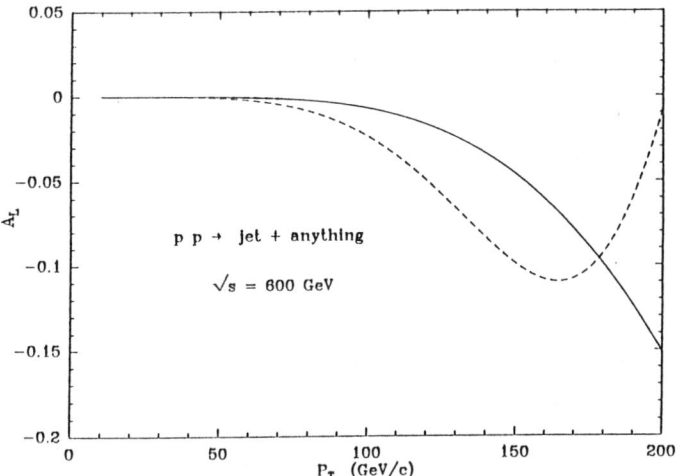

Fig.3. Same as Fig.2 but at RHIC energy, with $\Lambda = 1$ TeV (dashed) or 2 TeV (solid)

Recents calculations [9], whose results are displayed in Fig.3, show that even at RHIC ($\sqrt{s} = 600$ GeV) with one polarized proton beam one could expect to find a signal in A_L (which could be measured to a 1% accuracy [10]) provided Λ is not too large. In fact, in this case, as in the case of higher energies, polarization would increase notably the sensitivity to the presence of a new contact interaction. Indeed, pure cross section studies are limited to the search for an abnormal increase in the number of events at large p_T and this is obscured by theoretical uncertainties in the QCD predictions (at least of the order of 40%) due to unknown higher order corrections plus uncertainties in the partonic distributions. On the contrary, as soon as A_L is non zero in jet production, a new behavior is revealed whose origin has nothing to do with QCD ! Also, of course, a unique information could be obtained on the chirality nature of this new contact interaction.

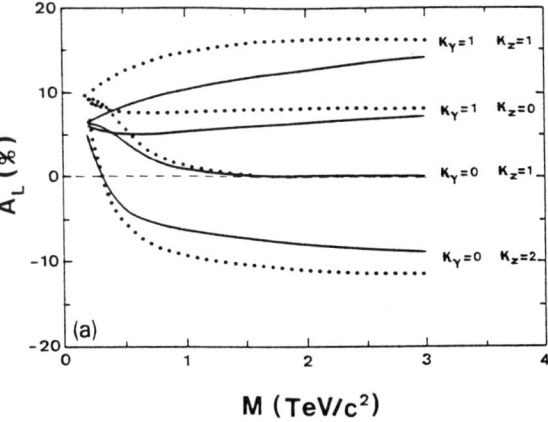

Fig. 4 A_L in W pair production at $\sqrt{s} = 10$ TeV (dotted curves) and 40 TeV (solid curves) versus M, the invariant mass pair, with various values of the anomalous magnetic moments ($\kappa_\gamma = \kappa_Z = 1$ corresponds to the case of the Standard Model).

In addition, compositeness could manifest itself in the gauge boson sector through several types of anomalous behaviors : new bosonic states could appear and also other contact terms could be at the origin of large spin effects in vector boson pair production. The consequence of anomalous magnetic couplings at the famous γW⁺W⁻ and ZW⁺W⁻ vertices have received much attention (see [4] and refs. therein). As shown in Fig. 4 A_L in pp → W⁺W⁻ is very much sensitive to the values of these couplings.

EXTENSIONS OF THE GAUGE GROUP

There are many proposals to incorporate the standard electroweak gauge group $SU(2)_L \otimes U(1)_Y$ in a larger framework. One possibility is provided by left-right (LR) symmetric models where parity is restored at high energy [11]. New vector bosons are present in these theories : massive right-handed W's whose production should display a typical A_L [4] and also a new neutral Z'. Independently, E6-type models [12], inspired or not by superstring theories, imply also the existence of two new neutral gauge bosons. We display in Fig.5 the results of a recent calculation [13] which shows that polarization could be of great help to find the origin of a new Z' since the single spin asymmetries have quite different behaviors in E6 and LR models.

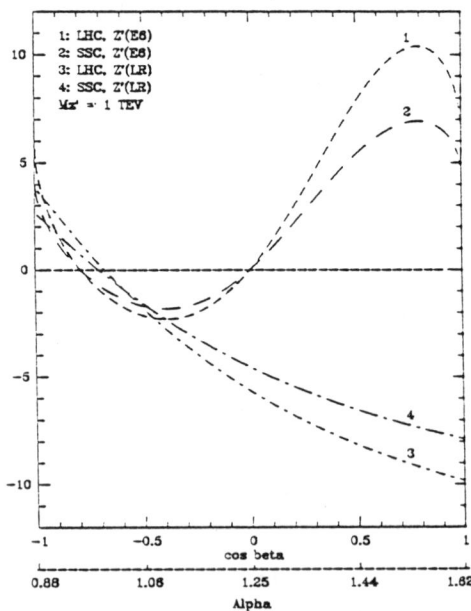

Fig. 5 A_L (in %) in the production of a 1 TeV Z' at LHC and SSC, in E6 and LR models, versus cosβ for E6 and α for LR models.

Here, cosβ is a mixing angle parameter between two extreme cases of Z' in E6 models and α = [($g_R \cos\theta_W / g_L \sin\theta_W)^2 - 1$]$^{1/2}$ is a typical parameter of $SU(2)_R SU(2)_L \otimes U(1)_Y$ models with gauge couplings g_R and g_L.

Note that such a study with polarized beams would be complementary and perhaps more accurate than the analysis of forward-backward asymmetries in massive $\mu^+\mu^-$ pair production in the unpolarized case [14].

CONCLUSIONS

In conclusion, it can be said that, in addition to the very exciting polarization program at RHIC, the availability of polarization at supercolliders would allow to study more carefully various possible manifestations of New Physics. In fact, in this very high energy domain where new phenomena, some of them completely unexpected, will certainly happen, polarized beams could appear as an essential tool for understanding what's going on, and then polarization only would allow for a full exploitation of the future machines.

REFERENCES

[1] See the proceedings of the various Snowmass meetings edited by the APS.
[2] See the proceedings of the 1990 Aachen meeting on the physics at the LHC, CERN report, to appear.
[3] See for example, H.E. Haber and G.L. Kane, Phys. Rep. 117, 75 (1985) and refs. therein
[4] C. Bourrely, F.M. Renard, J. Soffer and P. Taxil, Phys. Rep., 177, 319 (1989)
[5] V. Hughes, R. Milner, P. Souder, these proceedings
[6] J. Soffer, Marseille preprint, CPT-91/P.2505, these proceedings
[7] Particle Data Group, Phys. Lett. 239B (1990)
[8] P. Chiappetta and M. Perrottet, Marseille preprint, CPT-90/P.2440 (to appear in Phys. Lett. B), see also [1] and [2]
[9] C. Bourrely, J.Ph. Guillet and J. Soffer, Marseille preprint, CPT-90/P.2474
[10] G. Bunce, these proceedings
[11] See R.N. Mohapatra and G. Senjanovic, Phys. Rev. D23, 165 (1981) for a list of references following the original J. Pati and A. Salam proposal
[12] See J.L. Hewett and T.G. Rizzo, Phys. Rep. 183, 193 (1989) and refs. therein
[13] A. Fiandrino and P. Taxil, Marseille preprint, CPT-91/P.2504, in preparation
[14] P. Langacker, R.W. Robinett and J.L. Rosner, Phys. Rev.D30, 1470 (1984)
V. Barger et al. Phys. Rev.D35, 2893 (1987), F. Del Aguila, M. Quiros and
F. Zwirner, Nucl. Phys. B287, 419 (1987), see also various contributions to [2]

TRANSVERSE POLARIZATION IN DEEP INELASTIC COLLISIONS

X. ARTRU
Institut de Physique Nucléaire de Lyon
43 boulevard du 11 novembre 1918, F-69622 Villeurbanne Cedex, France

ABSTRACT

An introduction to transverse polarization phenomena in deep inelastic processes is given. A transversely polarized state of a massive or massless particle is defined as a coherent superposition of different helicity states; for a quark, it corresponds to transverse spin; for a gluon, to linear polarization. A simple toy model shows that quarks could have a sizeable transverse polarization in a transversely polarized baryon. A transverse polarization asymmetry corresponds to the exchange of a parton-antiparton pair of nonzero total helicity in the t-channel of the unitarity diagram. Conservation of t-channel helicity is used to select experiments which are sensitive to transverse polarization to leading order in α_s and $1/Q$. The transversely polarized quark and gluon densities have independent evolutions with Q^2.

INTRODUCTION

Spin of massless particles seems of quite different nature from spin of massive ones; for instance the spin of a photon, a graviton or a neutrino can not be oriented transversely to the direction of propagation. One might think this to be also the case of massive particles in the ultrarelativistic limit: The relativistic generalisation of the spin vector $\vec{s}$, the Pauli-Lubanski four-vector s^μ, has its longitudinal components dilated by the Lorentz factor $\gamma = E/m$, whereas the transverse components stay of order 1. In the domain of deep inelastic collisions, where parton masses can be neglected, this has led many theorists to think that only helicity is a "good" polarisation observable. This attitude has been reinforced by the fact that, in Deep Inelastic Lepton Scattering (DILS), transverse polarization asymmetries appear to be suppressed by a factor $1/Q$; also, they involve the structure function $g_2(x)$ which contains nonleading twist contributions.

On the experimental side, only the "longitudinally polarized structure function" $g_1(x)$ of the proton has been measured up to now. The disagreement with expectation of the naive, nonrelativistic quark model[1] has triggered a lot of theoretical works but all these are about longitudinal spin. It should be recalled that the "spin crisis" is a "longitudinal" one.

The orientation of the Pauli-Lubanski vector is misleading: Take a spin $\frac{1}{2}$ particle at rest in the spin state

$$|\vec{s}> = 0.8|+> +0.6|->,$$

for which $<s_x> = +0.48$, $<s_z> = +0.14$, and boost it at high velocity in the $\hat{z}$ direction. The particle will remain in the helicity state $0.8|+> +0.6|->$ despite the fact that the space components of s^μ will become almost parallel to $\vec{p}$ (Fig.1), naively suggesting $<s_x> \to 0$, $<s_z> \to +\frac{1}{2}$.

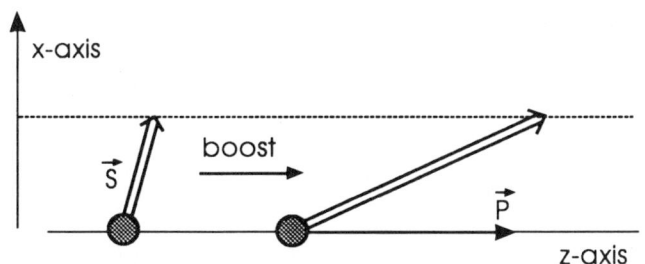

Fig.1. Boost of the Pauli-Lubanski vector.

The linear combinations of helicity states

$$|\hat{x}> = 2^{-1/2}(|+> +|->) \, , \, |-\hat{x}> = 2^{-1/2}(|+> -|->) \qquad (1)$$

have to be interpreted as transversely spinning states, even for a massless fermion (in the later case, you may take it as a definition). The analogous states for spin-one particles

$$|\hat{x}> = 2^{-1/2}(|+> +|->) \, , \, |\hat{y}> = 2^{-1/2}(|+> -|->) \qquad (2)$$

are states of transverse linear polarization. Note that the spin-1 analogue of $|-\hat{x}>$ is $|\hat{y}>$.

Contrarily to the spin $\frac{1}{2}$ case, the states (2) are not transversely spinning states (in fact $<S_x> = <S_y> = 0$). But the formalism of transverse spin for quarks follows so closely that of linear polarization for gluons that I shall here call both polarizations "transverse", as opposed to "longitudinal spin polarization" (although, strictly speaking, "transverse" means "of nonzero helicity" for spin 1). Here I shall review these two kinds of phenomena in the deep inelastic regime, thus excluding soft collisions or hard elastic ones (for more details and references, see Ref.[2]).

TRANSVERSELY POLARIZED PARTON DISTRIBUTIONS

a) transversely polarized quarks. Take a proton in the $|+\hat{x}>$ state. We define

$$\Delta_1 q(x) \equiv q_{\hat{x}}(x) - q_{-\hat{x}}(x) \, , \qquad (3)$$

where x is the fraction of hadron momentum carried by the quark, $q_{\hat{n}}(x)$ the distribution of quarks polarized in the $\hat{n}$ direction. The nonrelativistic quark model predicts

$$\Delta_1 q(x) = \Delta q(x) \, , \, \int_0^1 \Delta q(x) dx = 1 \, ,$$

where $\Delta q(x) \equiv q_+(x) - q_-(x)$ is the usual quark helicity distribution, but this model is too naive; the second relation is badly violated. Will there be also a transverse spin crisis?

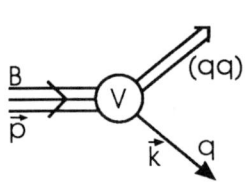

Fig.2. The B − q − (qq) vertex.

A simple covariant toy model.[2,3]
Current quark have small masses and one should use a relativistic model. Let us, for instance, consider the baryon as a bound state of a quark and a scalar elementary diquark (Fig.2), with a baryon-quark-diquark vertex of the form:

$$<\hat{s}_q|V|\hat{s}_B> = g(k^2)\ \bar{u}(\vec{k},\hat{s}_q)\ u(\vec{p},\hat{s}_B)\ . \tag{4}$$

The polarised quark distibutions can be obtained by a generalization of the Weisszäker-Williams method[4,5]:

$$dN_q = \frac{1}{16\pi^3} \times \frac{x\,dx}{1-x}\,d^2\vec{k}_T\,\left|\frac{<\hat{s}_q|V|\hat{s}_B>}{k^2 - m_q^2}\right|^2\,, \tag{5}$$

where $k^2 = x\,m_B^2 - (k_T^2 + x\,m_{qq}^2)/(1-x)$ is the the quark virtuality. This model predicts

$$\Delta_1 q(x) = (x\,m_B + m_q)^2\ f_0(x) = q_+(x)\,,$$
$$q_-(x) = f_2(x)\,, \tag{6}$$

with $$f_n(x) \equiv \frac{1}{16\pi^2\,(1-x)} \int_0^\infty k_T^n\,dk_T^2\,\left[\frac{g(k^2)}{k^2 - m_q^2}\right]^2\,. \tag{7}$$

Thus, in this model,
- $\Delta_1 q(x)$ is different from zero even if $m_q = 0$
- $\Delta_1 q(x) \neq \Delta q(x)$. This shows that $\Delta_1 q$ is not a redundant information on hadron structure.

b) linearly polarized gluons. Take a meson of spin $S \neq 0$ and quark orbital angular momentum $l \neq 0$, and polarize it linearly along $\hat{x}$. The quark-antiquark axis will be preferentially oriented parallel to $\hat{x}$. So will be the static chromo-electric field (Fig.3). Boost the whole system in the $\hat{z}$ direction. According to the Weisszäcker-Williams picture, the static gluon field will become a cloud of quasi-real gluons. These gluons will be partially polarized along $\hat{x}$.

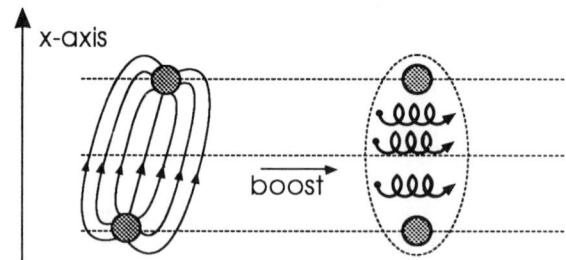

Fig.3. Boost of a linearly polarized meson, in the flux-tube model.

We define the linear polarization distribution as
$$\Delta_2 G(x) = G_{\hat{x}}(x) - G_{\hat{y}}(x) , \qquad (8)$$
where $G_{\hat{n}}(x)$ is the distribution of gluons linearly polarized along the transverse direction $\hat{n}$. A target or a beam of such mesons is unrealistic but a linearly polarized beam of photons could do the job, according to the Vector meson Dominance Model (VDM). We only require that the vector meson has a sizeable D-wave component. It has also be proposed to look for linearly polarized gluons in linearly polarized nuclei.[6] This would reveal an exotic component of the nucleus wave function.

OBSERVABLE EFFECTS AT LEADING TWIST
A) INITIALLY POLARIZED QUARKS

Any transverse polarization asymmetry results from an interference between different helicity amplitudes. Consider, for instance, the subprocess *quark a+ quark b* $\to f$ (Fig.4) and calculate
$$\Delta_1 \hat{\sigma} \equiv \hat{\sigma}(\hat{s}_a = +\hat{x}, \hat{s}_b) - \hat{\sigma}(\hat{s}_a = -\hat{x}, \hat{s}_b) . \qquad (9)$$
($\hat{s}_b$ is not specified for the moment). Using (1), we find
$$\Delta_1 \hat{\sigma} = <f|T|+,\hat{s}_b><f|T|-,\hat{s}_b>^* +\text{complex conjugate}$$
$$= <-,\hat{s}_b|T^\dagger|f><f|T|+,\hat{s}_b> +\text{complex conjugate} . \qquad (10)$$

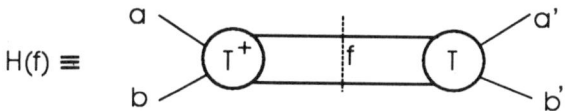

Fig.4. Unitarity Diagram for the subprocess.

$\Delta_1 \hat{\sigma}$ is made of *helicity-flip* (for quark a) terms of the partial discontinuity
$$H(f) \equiv T^\dagger |f><f|T \qquad (11)$$
of the forward $a + b \to a + b$ scattering amplitude. In the *t-channel* of the unitarity diagram, the $a\bar{a}$ pair has total helicity ± 1. Similarly, $\Delta_1 q$ is a term of helicity-flip 1 in the imaginary part $<\lambda'_A, \lambda_a|\bar{\Gamma}|\lambda_A, \lambda'_a>$ of the forward quark-hadron scattering amplitude (upper and lower blobs of Fig.5):
$$\Delta_1 q(x) = <-1/2, +1/2|\Gamma|+1/2, -1/2> . \qquad (12)$$

Selection rules from helicity conservation. *t*-channel total helicity is conserved from top to bottom of the unitarity diagram of Fig.5 because (i) In the upper and lower blobs J_z is conserved, (ii) the hard subprocess conserves total quark helicity to leading order in $1/Q$. Therefore, both A and B must be transversely polarized.

If the *t*-channel carries the helicity ± 1, then J_z is not conserved in the *s*-channel: $J_z - J'_z = \pm 2$. On the other hand, if we integrate H over the azimuth of the final state, we get $J_z = J'_z$, due to cylindrical symmetry. These conflicting results mean that any transverse polarization asymmetry is washed out by azimuthal integration. In fact, the transversely polarized qq or $q\bar{q}$ cross section is

180 Deep Inelastic Collisions

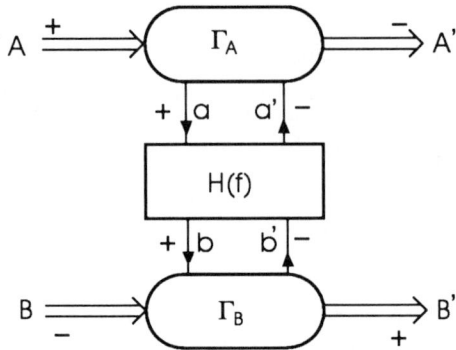

Fig.5. Whole unitarity diagram for the hard hadron-hadron collision. Helicity signs contributing to transverse the asymmetry are displayed

$$\frac{d\hat{\sigma}}{d\Omega} = \left(\frac{d\hat{\sigma}(\hat{s},\hat{\theta})}{d\Omega}\right)_{unpol.} \times [1 - P_a P_b \, \hat{A}_{NN}(\hat{\theta}) \, \cos(2\phi - \phi_a - \phi_b)] , \quad (13)$$

where $\vec{P}_a$ is the polarization vector of a, ϕ_a its azimuth (*idem* for b), and $\hat{A}_{NN}$ the normal-normal spin correlation parameter. For instance, in Drell-Yan mechanism,

$$\hat{A}_{NN}(\hat{\theta}) = -\frac{\sin^2 \hat{\theta}}{1 + \cos^2 \hat{\theta}} . \quad (14)$$

Convoluting (13) with spin-dependent quark distributions, we get

$$d\sigma(\uparrow A + \uparrow B \to f + X) = \int dx_a \, dx_b \, d\hat{\sigma}(a + b \to f)_{unpol.}$$

$$\times [\, a(x_a) \, b(x_b) - P_A \, P_B \, \Delta_1 a(x_a) \, \Delta_1 b(x_b) \, \hat{A}_{NN}(\hat{\theta}) \, \cos(2\phi - \phi_A - \phi_B)] . \quad (15)$$

(15) was first written by Ralston and Soper for muon pair production.[7]

Helicity conservation works for each quark line separately ; therefore a and a' must not be on the same quark line (*idem* for b and b'). The hard subprocesses must be either $q\bar{q}$ annihilation (Fig.6a) or the crossed term in the scattering of identical quarks (Fig.6b). The interference term in $q\bar{q}$ Bhabha scattering (Fig.6c) also works. In the annihilation class, besides Drell-Yan process, we have $q\bar{q} \to \gamma\gamma$, $\gamma + gluon$ or $gluon + gluon$; $q\bar{q} \to c\bar{c}$ or $b\bar{b}$. In qq scattering, unfortunately, $\hat{A}_{NN}$ is only $-1/11$ at $90°$).

By the same argument, no transverse asymmetry is expected in *ordinary* DILS (Fig.7), to leading order in $1/Q$ (it does not help to polarize the electron), as well as in quark-gluon scattering.

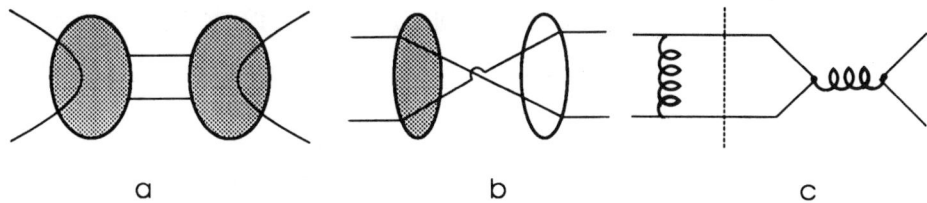

Fig.6. Unitarity diagrams allowing non-zero t-channel helicity.

B) FINAL POLARIZED QUARKS

Consider the *semi-inclusive* DILS experiment
$$e^- + \uparrow p \to e^- + \uparrow \Lambda + X \,, \tag{16}$$
Its unitarity diagram (Fig.7b) is obtained from Fig.7a by opening the horizontal quark line and inserting a blob which represents the fragmentation function of the quark into Λ. The helicity argument which impeded transverse polarization effects in ordinary DILS is overcome. The transverse polarization of the proton (t-channel helicity ± 1) is (partially) transmitted to the incoming quark, then to the scattered quark and finally to the Λ, which serves as a quark "polarimeter". The magnitude of Λ polarization will be
$$P_\Lambda = P_p \times \frac{\Delta_1 q(x)}{q(x)} \times \hat{D}_{NN}(\hat{\theta}) \times \frac{\Delta_1 f(z)}{f(z)} \,, \tag{17}$$
where $\Delta_1 f(z)$ is the transverse spin asymmetry of the fragmentation function and $\hat{D}_{NN}(\hat{\theta})$ the so-called "depolarization parameter" of the subprocess ($e^- + q \to e^- + q$):
$$\hat{D}_{NN}(\hat{\theta}) = \frac{4(1+\cos\hat{\theta})}{4 + (1+\cos\hat{\theta})^2} \,. \tag{18}$$
$\vec{P}_\Lambda$ will have the same angle with the scattering plane as $\vec{P}_p$. This experiment is easier to do than the polarized Drell-Yan one.

Finally, $\Delta_1 f(z) \times \Delta_1 f(z')$ could be independently measured in
$$e^+ e^- \to \uparrow \Lambda + \uparrow \bar{\Lambda}(\text{back}-\text{to}-\text{back}) + X \,. \tag{19}$$

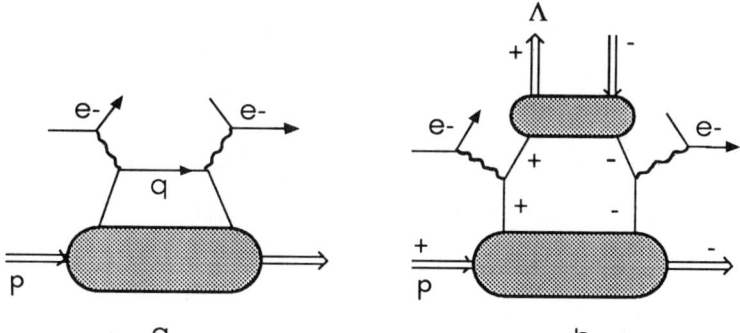

Fig.7. Unitarity diagram for deep inelastic electron scattering; a) totally inclusive b) semi-inclusive Λ production.

C) GLUON LINEAR POLARIZATION EFFECTS

The formalism for transverse quark spin can be translated in a straightforward way to the case of linear gluon polarization: (10) applies almost without modification to

$$\Delta_2 \hat{\sigma} \equiv \hat{\sigma}_{\hat{x}\hat{n}} - \hat{\sigma}_{\hat{y}\hat{n}} \,, \tag{9'}$$

which has helicity-flip 2. Similarly,

$$\Delta_2 G(x) = < -1, +1 | \Gamma | +1, -1 > \quad \text{for gluons.} \tag{12'}$$

This imply a parent hadron of spin $S \geq 1$. In the subprocess, total gluonic helicity is not always conserved, but, in the $2 \to 2$ case and to lowest order in α_s, it is conserved. The second line of (15) is replaced by

$$\times [G(x_a)G(x_b) + P_A P_B \Delta_2 G(x_a)\Delta_2 G(x_b)\hat{A}_{NN}(\hat{\theta})\cos(4\phi - 2\phi_A - 2\phi_B)] \,. \tag{15'}$$

Collisions of two polarized initial gluons are problematic (one has to collide two polarized photon beams...). Instead, we can obtain a final linearly polarized gluon from QCD "Compton" scattering of a linearly polarized photon (Fig.8). Here a vector meson serves as gluon polarimeter. This experiment measures $\Delta_2 f(z)$, which is *a priori* as interesting as $\Delta_2 G(x)$. $\Delta_2 f(z) \times \Delta_2 f(z')$ could also be measured in

$$p\bar{p} \to \uparrow V + \uparrow V'(\text{at opposite large} p_\perp) + X \,, \tag{19'}$$

via the $q\bar{q} \to gluon + gluon$ subprocess.

In *disconnected double gluon scattering* (see, for instance Fig.9), the polarizations of two incoming gluons from the *same* hadron may be correlated[8] (e.g., in the model of Fig.3, their linear polarizations are nearly parallel). It should result in an azimuthal correlation between the c-jet and the b-jet of the form

$$d\sigma/(d\phi_c d\phi_b) = A + B \cos[2(\phi_c - \phi_b)] \,. \tag{20}$$

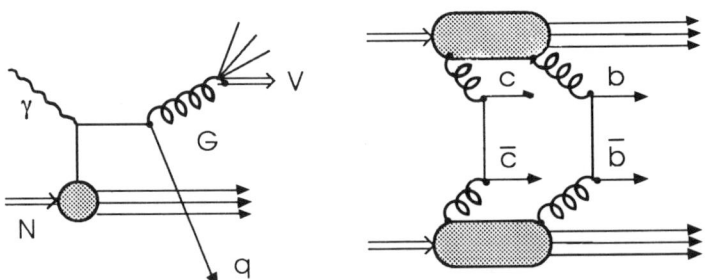

Fig.8. Q.C.D. Compton process.   Fig.9. Double gluon scattering.

QCD EVOLUTIONS OF $\Delta_1 q(x,Q^2)$ AND $\Delta_2 G(x,Q^2)$

Transversely polarized parton densities evolve via ladder diagrams carrying total helicity ± 1 (for quarks) and ± 2 (for gluons) in the t-channel. There is no mixing between quark and gluon densities. The kernels may be found in Refs.[9,2] for $\Delta_1 q(x,Q^2)$ and in Refs.[2,10] for $\Delta_2 G(x,Q^2)$. All integer moments of these distributions decrease with Q^2 (contrarily to case of $\Delta q(x,Q^2)$ and $\Delta G(x,Q^2)$ for $n=1$). There is apparently no conserved quantity.

SUMMARY AND OUTLOOKS

The transversely polarized parton distributions, $\Delta_1 q(x)$ for a quark in a nucleon and $\Delta_2 G(x)$ for a gluon in a spin ≥ 1 hadron, are well-defined even for massless partons (more properly, $\Delta_2 G$ corresponds to linear polarization). $\Delta_1 q$ and $\Delta_2 G$ are helicity-flip terms of the discontinuity of the hadron-parton forward scattering amplitude. They could be large and constitute a piece of information about hadron structure which is not redundant with the longitudinal spin distributions $\Delta q(x)$ and $\Delta G(x)$. Similarly for the transversely polarized fragmentation functions $\Delta_1 f(z)$ and $\Delta_2 f(z)$.

Transverse spin asymmetries (for quarks) and linear polarization asymmetries (for gluons) are of different nature and do not mix. There exists a large class of deep inelastic processes where they are observable at leading order in α_s and $1/Q$. They involve two transversely polarized (initial or final) hadrons. For *quarks*, we have in particular

$$\uparrow p + \uparrow \bar{p} \to \mu^+\mu^-, \gamma + gluon\, jet,\, 2\, gluon\, jets,\, one\, pair\, of\, heavy\, quark\, jets;$$

$\uparrow p + \uparrow p \to \pi^+ \pi^+$ at large and opposite $p_\perp$, via the scattering of two *up* quarks. Totally inclusive Deep Inelastic Lepton Scattering does not belong to this class, but the semi-inclusive reaction $e^- + \uparrow p \to e^- + \uparrow \Lambda + X$ does.

For *gluons* things are experimentally more difficult. Initial linearly polarized gluons can be found in the hadronic component of photons. In double disconnected gluon scattering, possible correlations exist between the linear polarizations of two gluons of the same incoming hadron; they should result in azimuthal correlations between the two pairs of jets. Final linearly polarized gluons should be produced via $\gamma + q \to gluon + q$ or (correlated polarizations) $q + \bar{q} \to 2\, gluons$. Fragmentation into a vector meson should serve as final gluon polarimeter.

On the theoretical side, much work remains to be done, e.g., are there transversely polarized sum rules? what is the connection between $\Delta_1 q(x)$ and $g_1(x) + g_2(x)$? As for the experimental side, *all* is to be done.

Acknowledgment. I am very much indebted to Pr. Mustapha Mekhfi who introduced me to the subject of spin correlations. A large part of this work has been done in collaboration with him.

REFERENCES

1. J. Ashman *et al* (EMC collaboration), Phys. Lett. B **206**, 364 (1988)
2. X. Artru and M. Mekhfi, Z. Phys. C, **45**, 669 (1990).
3. See also H. Meyer and P.J. Mulders, NIKHEF-P-12 (July 1990)
4. P. Kessler, Nuovo Cimento **17**, 809 (1960).
5. G. Altarelli and G. Parisi, Nucl. Phys. **B126**, 298 (1977).
6. R.L. Jaffe and A. Manohar, Phys. Lett. B **223**, 218 (1989)
7. J.P. Ralston and D.E. Soper, Nucl. Phys. **B152**, 109 (1979)
8. X. Artru and M. Mekhfi, Orsay preprint LPTHE 89/11 (1989)
9. M. Mekhfi and X. Artru, XIXth International Symposium on Multiparticle Dynamics (Arles, France, 1988), D. Schiff, J. Tran Thanh Van (eds) p. 111; Editions Frontières.
10. E. Sather and C. Schmidt, HUTP-90/A023 (May 1990)

LEPTON PAIR PRODUCTION IN POLARIZED COLLIDER EXPERIMENTS AS A METHOD TO MEASURE THE TRANSVERSE POLARIZATION OF QUARKS IN THE PROTON

J.L. Cortes[1], B. Pire[2] and J.P. Ralston[2]

1. Dep. de Fisica Teorica, Univ. de Zaragoza, 50009 Zaragoza (Spain)
2. Centre de Physique Théorique, Ecole Polytechnique
91128 Palaiseau Cedex(France)

ABSTRACT

The transverse spin distribution of quarks in protons contributes at leading twist to the angular distribution of lepton pairs produced in doubly polarized proton-proton collisions. It is the last classic parton model distribution left to be measured and it is not as subtle as often assumed.

TRANSVERSE SPIN IS WELL DEFINED

The SLAC and EMC experiments which have measured the longitudinal spin structure of quarks in the proton have been surprising and interesting. Perhaps even more surprising, to this day next to nothing is known about the distribution of transversely polarized quarks in a transversely polarized proton. Yet transverse polarization, due to its chiral properties, is probably more interesting than longitudinal polarization. The state of ignorance would be understandable if the concept of quarks spinning "sideways" in the high energy limit were subtle, or defined in terms of some power-supressed process. The truth is neither : quarks spinning sideways are just as well defined in the QCD parton model as quarks spinning longitudinally. Moreover, transversely polarized quarks can be measured with straightforward, leading order observables in some experiments. In this paper we emphasise using the Drell-Yan (DY) lepton pair production process with polarized beam and target to measure quark transverse polarization. We understand that the technology to do the doubly-polarized experiment now exists. We will show that the data is not so difficult to gather or analyze to directly measure the quark and anti-quark transverse spin distributions. Theoretically, every link in the logical chain relating parton distributions to QCD matrix elements can be completed for transverse spin, as for longitudinal spin, and only data is lacking.

One of the problems with transverse polarization of quarks is historical. Feynman[1] used the concept of transverse spin in his parton model of the deep inelastic scattering (DIS) structure function g_2. However, in this case transversely polarized parton distributions contribute a term of order $1/\sqrt{Q^2}$, compared to the same type of calculation for g_1, where Q^2 is the squared magnitude of the virtual photon momentum. Such "higher twist" contributions are tricky in the parton model, and one finds that there are unfortunately a few other contributions not corresponding to transversely polarized parton densities. Subsequent attempts to *define* transverse spin using g_2 seemed to increase the confusion, and the whole subject of transverse spin was sometimes dismissed as "higher twist".

MISCONCEPTIONS

There is a false argument springing from this history which we should review; its resolution clarifies what we have to say. "Consider" (goes the argument) "the boost of a longitudinal spin vector to the infinite momentum frame. The longitudinal spin increases by the same Lorentz factor as the longitudinal momentum k and so longitudinal spin effects are large. The transverse spin, being invariant under the boost, is smaller by a factor of M/k and so are transverse spin effects".

If the spin were measured this argument would be right. However, we actually measure the density matrix of quarks spinning in a proton. In the covariant density matrix, there are several terms proportional to the large momentum k, including the transverse spin density, and transverse spin effects are not kinematically suppressed.

DEFINITIONS

The parton momentum and spin densities $P_{\alpha\beta}$ can be defined[3] in terms of proton matrix elements of the quark field operator $\psi_\alpha(x)$. The definition is (with usual light cone coordinates $P^\pm = (P^0 \pm P^3)/\sqrt{2}$):

$$P_{\alpha\beta}(x, Q^2) = \frac{xP^+}{2\pi} \int dy^- e^{ixP^+y^-} <Ps|\overline{\psi}_\beta(0) \psi_\alpha(0,0,y^-)|Ps> \qquad (1)$$

Our definition of the proton spin is $s^\mu = \frac{\lambda}{m_p}(p^\mu - m^2 n^\mu) + s_T^\mu$, $s^2 = -1$. We let n^μ be a light-like gauge-fixing vector satisfying $P \cdot n = 1$. In the expansion of $P_{\alpha\beta}$ of terms allowed by parity, time reversal, and hermiticity, there are 8 possible terms[3]. In the parton model for leading order calculations we are indeed only concerned with the "big" terms in the infinite momentum limit. These are :

$$P_{\alpha\beta}(x, Q^2) = \mathcal{P}(x, Q^2) \slashed{k}_{\alpha\beta}/2 - \lambda h_L(x, Q^2)(\slashed{k}\gamma_5)_{\alpha\beta}/2 - h_T(x,Q^2)(\gamma_5 \slashed{k}\slashed{s}_T)_{\alpha\beta}/2 \quad (2)$$

In the last term above we identify $h_T(x,Q^2)$ as the transversly polarized quark distribution[3].

The Q^2 evolution of each term above is known, and of the usual logarithmic type. It is interesting that h_T evolves autonomously[2], i.e. its Altarelli-Parisi equation does not require knowledge of $\mathcal{P}$, h_L or "higher-twist" terms. This is consistent with h_T being an independent leading order distribution. Henceforth we drop the Q^2 dependence, which will become implicit, to simplify our notation.

That $h_T(x)$ measures quarks spinning sideways is easy to see by comparing (2) to the density matrix of a free, polarized quark :

$$u_\alpha(k,s)\bar{u}_\beta(k,s) = \frac{\slashed{k}+m}{2}\frac{(1+\gamma_5 \slashed{s})}{2} \to (\slashed{k} - \lambda\slashed{k}\gamma_5 - \gamma_5\slashed{k}\slashed{s}_T)_{\alpha\beta} \quad (3)$$

On the right-hand side we took the high energy limit, which is straightforward for the transverse piece. The slightly tricky part is the longitudinal (helicity) piece, because there is a term of order m/m surviving as m/k $\to$ 0. The result (3) reflects the fact that a high-energy quark still lives in a (1/2, 1/2) representation of the Lorentz group. Superposition of ± helicity state is perfectly legitimate ; it should not be confused with the case of a massless "neutrino" (0, 1/2) representation with a unique helicity. Transverse spin represents a fundamental quark degree of freedom.

MEASURING TRANSVERSE SPIN

The fundamental difference between the h_T and other parton densities is the transformation properties under chiral symmetry. Note that the $h_T(x)$ term comes with 2 gamma matrices ; it is "even" and commutes with γ_5. The other terms have 1 gamma matrix, are "odd" and anti-commute with γ_5. This symmetry imposes some selection rules. Specifically, probes that are good at measuring "odd" terms are not good at measuring "even" term.

Elsewhere we discuss this in detail[4]; here there is space only for a brief summary. In DIS the chirally invariant electromagnetic current operator is good only for measuring "odd" terms, i.e. the unpolarized and longitudinally polarized terms. The bad "higher twist" measurement of $h_T(x)$ in DIS occurs because the transverse spin term is "even". We need another probe, of different chirality, to measure $h_T(x)$. The DY process is ideal to measure transverse spin, because the anti-quark serves as a probe of the quark, in conjunction with the current. This is easily seen by bringing the anti-quark field into the DY matrix-element:

$$< qs\,\bar{q}\,\bar{s}|J^\mu(x)\,J^\nu(0)|qs\,\bar{q}\,\bar{s}> \rightarrow <qs|\psi_\alpha(y)\,J^\mu(x)\,J^\nu(0)\,\overline{\psi}_\beta(z)|qs>,$$

in obvious notation. The combination $\psi_\alpha(y)...\overline{\psi}_\beta(z)$ has <u>mixed</u> chirality, showing that the antiquark can measure the quark.

A simple calculation confirms this. Consider DY dilepton production with Q^μ the lepton pair momentum. Let the pair rapidity be y. The angular distribution involves a polar angle θ, and an azimuthal angle ϕ around the "beam" (Collins-Soper) axis. Then the production from protons A and B with transverse spin vectors oriented along angles ϕ_A and ϕ_B is[3]

$$\frac{d\sigma}{dQ^2 dy d\Omega} = \frac{\alpha^2 e_a^2}{12Q^2 s}\{[\mathcal{P}_{a/A}\,\mathcal{P}_{\bar{a}/B} - h_L^{a/A}(x_A)\,h_L^{\bar{a}/B}(x_B)\,\lambda_A\,\lambda_B](1+\cos^2\theta)$$
$$+ h_T^{a/A}(x_A)\,h_T^{\bar{a}/B}(x_B)\,s_T^A\,s_T^B\,\sin^2\theta\,\cos(2\phi-\phi_A-\phi_B)\} \quad (4)$$

The sum over flavors a, symmetrized by A↔B, is implicit. Statistics are maximized by integrating over Q_T, so we have assumed this case.

Eq. (4) can be converted into an asymmetry. Let both protons be polarized in the "$\hat{x}$" direction, transverse to the beam. Divide the lepton pair phase space (looking down the beam) into 4 quadrants with lines running at ± 45° with respect to $\hat{x}$. Add the top and bottom quadrants and subtract the left and right quadrants to define $\Delta\sigma$. The transverse spin terms are thus isolated:

$$d\Delta\sigma/dQ^2 dy = \left[\int_{\pi/4}^{3\pi/4} d\phi + \int_{5\pi/4}^{7\pi/4} d\phi - \int_{3\pi/4}^{5\pi/4} - \int_{-\pi/4}^{\pi/4} d\phi \right] d\sigma/dQ^2 dy d\phi \ ;$$

$$d\Delta\sigma/dQ^2 \, dy = \frac{4}{9} \frac{\alpha^2 e_a^2}{Q^2 s} h_T^{a/A}(x_A) \, h_T^{\bar{a}/B}(x_B) \, s_T^A \, s_T^B \qquad (5)$$

Using the asymmetry, the measurement of h_T is no more difficult than conventional DY measurements. The factor of $s_T^A s_T^B$ is the price in reduced data collection paid for polarization : with our normalization, if beam and target are 50% polarized, then $s_T^A s_T^B = 0.25$.

One experiment with identical pp beams can measure both the quark and anti-quark transverse polarization. Recall that the pair rapidity measures x_A/x_B and the pair mass fixes $x_A x_B$; both are determined. The measurement is almost "model independent" in the sense that little further processing is needed provided the usual DY and parton model physics applies. An important simplifying fact is that $e_a^2 = 4/9$ for u-quarks and 1/9 for d- and s- quarks, so the measurement is practically specific to u-quarks. Gluons do not contribute at this order. Moreover, chirality implies no leading anomalous gluon contribution for transverse spin. One may ask whether one can trust the DY model for this new observable. Theoretically we see no subtle issues. Experimentally, there turn out to be numerous self-consistency checks in the angular distribution which are signals one has observed on-shell $q\bar{q}$ annihilation. In fact[3,5], eight of eleven possible terms in the angular distribution vanish identically in the DY model, leaving only the 3 terms shown in (4). This is why the asymmetry method can be used without an elaborate (and high statistics) full angular distribution measurement.

APPLICATIONS

Besides its intrinsic interest, measurement of quark transverse spin triggers a whole chain of parton model applications. The situation is similar to the DIS based phenomenology of a decade ago, except that DY is the "clean" defining experiment. For example, it turns out that an independent measurement of $h_T(x)$ is actually necessary to interpret the DIS measurements of $g_2(x,Q^2)$ which are attracting much interest. Let us outline how this works ; details are given elsewhere[4]. The operator product expansion (OPE) analysis of g_2 is well understood and consists of 3 independent families of matrix elements[6] : (1) the "Wandzura-Wilczek" part, twist-2 in disguise but contributing at twist 3 nonetheless. Assuming g_1 (x,Q^2) is measured, this piece can be subtracted away ; (2) a "gluon insertion" piece, of twist-3 and considerable theoretical interest, and (3) a "mass dependent piece" of twist-3 and subtractable if we know h_T (x,Q^2) from an independent measurement. Unfortunately there is *no model independent separation* of the two-families of twist-3 type without a separate measurement such as the DY process would provide. This is good news and bad news : bad news that a measurement of g_2 (x,Q^2) cannot (despite some claims) be identified with unique OPE local matrix elements ; good news that DY can save the day.

Another application, recently proposed by Artru and Mekhfi[7] is to semi-inclusive electroproduction of transversely polarized baryons such as Λ's : $e + p\uparrow \to \Lambda\uparrow + X$. The Λ is self-analyzing, and a corellation of the Λ spin and an initial proton polarization involves h_T (x,Q^2). The analysis is straightforward, leading order parton model but depend on the fragmentation function of quarks to Λ's. The process seems to be a good way to *measure* the polarization dependence of the fragmentation process once h_T (x) is known from other experiments.

Finally, transverse polarization can be useful for "new physics" applications[8], in providing a unique quark beam with characteristics not available from longitudinal polarization alone, as well as well as for the full spectrum (inclusive γ, large-p_T jet, etc) of high energy experiments. No doubt many of these would reveal surprises.

CONCLUSIONS

The early measurements of quark distributions led to a theoretical and experimental gold mine of information. Longitudinal and transverse spin distribution measurements offer the same impressive potential : many "spin-off" experiments become meaningful once one measurement is decisively made. The distribution of transversely polarized quarks is the last classic parton model quantity left to be measured and it is probably more interesting than longitudinal spin. Disregarding historical accidents, quark transverse spin is not subtle : we just don't know anything about it, and we need to know.

Acknowledgments : This work was supported on part under DOE Grant No. 85ER40214.A006 and NSE-CNRS International Cooperative Programs Grant Int 8914626.

REFERENCES

1. R.P. Feynman, Proton Hadron Interactions, Benjamin New York (1972).
2. A. Bukhvostov, E.A. Kuraev and L.N. Lipatov, JETP Lett. 37, 483 (1983) ; P.G. Ratcliffe, Nuc. Phys. B264, 493 (1986).
3. J.P. Ralston and D. Soper, Nucl. Phys. B152, 109 (1979).
4. J.L. Cortes, B. Pire and J.P. Ralston, Ecole Polytechnique preprint A017.1090.
5. J. Donohue and S. Gottlieb, Phys. Rev. D23, 2577 (1981) ; ibid 2581 (1981).
6. For reviews, see R.L. Jaffe and X. Ji, MIT preprint CT 1848 and X. Ji, these proceedings.
7. X. Artru and M. Mekhfi, Z. Phys. C45, 669 (1990) and these proceedings.
8. G. Ladinsky and J.C. Collins, these proceedings.

Single Spin Asymmetries in Muon Pair Production

Robert D. Carlitz and Raymond S. Willey
*Department of Physics and Astronomy, University of Pittsburgh,
Pittsburgh, PA 15260*

Theoretical analyses of polarized leptoproduction data suggest that the polarized gluon structure function might be large, but there has been no independent measurement of this quantity. Measurements of single spin asymmetries in the production of muon pairs from the scattering of two protons, one of which is longitudinally polarized, can be interpreted in terms of polarized gluon and polarized quark structure functions. Here we compute the asymmetries for the parton subprocesses that contribute to the measured muon pair production.

This talk discusses some recent calculations of the parton processes

$$q + \bar{q} \to \mu^+ \mu^- + G \tag{1}$$
$$G + q \to \mu^+ \mu^- + q \tag{2}$$

with one of the incident partons polarized along its direction of motion. The motivation for these calculations comes from the EMC spin experiments[1], which extended the previous SLAC-Yale experiments[2] for polarized leptoproduction down to Feynman x values as low as 0.015. The spin distribution functions extracted from the recent experimental data are inconsistent with naive models[3,4] of the proton's spin structure, which would have

$$\Delta u(x), \Delta d(x) \neq 0, \tag{3}$$
$$\Delta s(x) \simeq \Delta G(x) \simeq 0. \tag{4}$$

One interpretation[5,6] of the EMC data suggests that there might be a large gluon spin component. The gluon spin can influence deep inelastic scattering through the axial anomaly[7], which relates the measured matrix element of the axial current, $\Delta q'$, to the first moments of the quark and gluon spin distributions:

$$\Delta q' = \Delta q - \frac{\alpha_S}{2\pi} \Delta G. \tag{5}$$

Since the contribution of the gluon spin distribution to $\Delta q'$ enters with a coefficient of $\alpha_S/2\pi$, the gluon spin component of the proton can significantly influence the EMC data only if it is very large. The possibility that ΔG might be of order 5 to 10 (*i.e.* 500 - 1000%!) invites one to attempt to isolate ΔG in experiments that might be directly sensitive to it.

There are a number of consequences which follow from the fact that ΔG might be large in the EMC range

$$< Q^2 > \sim 12 \; (\text{GeV})^2. \tag{6}$$

The expectation of QCD is that $\Delta G(Q^2)$ should grow even larger at larger values of Q^2. Indeed, from the renormalization group[8] one finds that as $Q^2 \to \infty$,

$$\alpha_S(Q^2)\Delta G(Q^2) \to \text{constant} \tag{7}$$

i.e.,

$$\Delta G(Q^2) \to \infty. \tag{8}$$

Attempts to fit the EMC data by using a large gluon spin component typically require large gluon polarizations[9] at small x. Of course the gluon polarization must vanish at $x = 0$, but the possibility exists that highly polarized gluons might exist in the small x region.

High energy hadron colliders provide experimentalists with large numbers of small-x gluons. If these partons are polarized, then new machines could provide us with intense beams of highly polarized gluons. Obviously this possibility would be of considerable experimental interest. It makes sense to explore this possibility – insofar as that can be done – with existing accelerators. The results from these machines could help in the process of designing new machines and could help determine the advisability of adding polarization options to any new hadron collider.

The present work suggests one way of approaching this question experimentally. Rather than demanding the experimental complexity of a polarized beam and a polarized target, we consider experiments with a polarized beam *or* a polarized target. This adds somewhat to the theoretical complexity of analyzing these processes, but greatly simplifies the experimental situation. In order to find any longitudinal spin asymmetry in such a process it is necessary to measure at least 2 independent transverse momenta in the final state. If only one transverse momentum were measured, then the only axial vector that could be constructed from the experimentally-determined quantities would be the normal to the plane defined by this vector and the momenta of the initial particles. Correlations of the initial spin direction with this vector would correspond to transverse spin asymmetries rather than the longitudinal spin effects in which we are interested.

The second complicating feature in the analysis of single spin asymmetries is that they require the production amplitude to have an imaginary part. This implies that all single spin asymmetries are intrinsically of order α_S. The question of whether or not these effects are large at moderate values of Q^2 thus depends in detail upon the coefficient of α_S in the calculated asymmetry. There are a number of processes that one might consider in this regard. Among them are:

- muon pair production,
- direct γ + jet,
- 2 jets.

In the present talk we will discuss the muon pair production process

$$p^\uparrow + p \to \mu^+ \mu^- + \dots \qquad (9)$$

At the parton level this requires one to compute amplitudes for the processes (1) and (2) to order α_S. Previously Ralston and Pire[10] have examined the quark-antiquark contribution, Eq. (1). They claimed that this contribution should be small, owing to a color factor $1/2N$ in the dominant terms that they computed. They noted that the gluon-quark contribution, Eq. (2), need not be small at the parton level, but they assumed (consistent with the general outlook of that era) that gluons should not carry an appreciable fraction of the proton's spin. If, indeed, the quark-antiquark contribution to muon pair production is suppressed, then this process seems like a particularly good one in which to study the effects of gluon polarization.

Our new calculations fail to reproduce the color suppression factor of Ralston and Pire. We find comparable asymmetries at the parton level for either of the processes (1) and (2). We define

$$Q_\perp^2 = \tau Q^2 \qquad (10)$$

and consider the kinematic region

$$Q^2 \ll s. \qquad (11)$$

We let $\mathcal{A}(\tau)$ denote the asymmetry

$$\frac{d\sigma_+ - d\sigma_-}{d\sigma_+ + d\sigma_-} \qquad (12)$$

maximized over lepton orientations. The signs ± refer to the helicities of the polarized initial parton. For $\tau \simeq 0.65$, $\mathcal{A}(\tau)$ attains a maximum value of order

$$\mathcal{A} \simeq 0.6 \left(\frac{N^2 - 1}{2N} \right) \alpha_S. \tag{13}$$

In the figure below we display the asymmetry $\mathcal{A}(\tau)$ as a function of τ. This asymmetry applies to any of the contributing parton processes.

Some further work remains to be done on this problem. Since the current calculation disagrees with that portion which exists in the literature, we are recomputing the quark-antiquark process using the spinor helicity formalism[11]. This formalism actually offers a more convenient way of dealing with the complexities of the particle spins, and it organizes the calculation in a completely different manner from the invariant amplitude approach that we initially have employed. After verifying our calculations beyond any reasonable doubt, we will compute convolutions with model parton distributions to find the size of expected experimental effects. These calculations will indicate the likely ability of experimentalists to separate ΔG from Δq in the proposed measurements.

There is no need to restrict ones attention to muon pair production alone. Once one has a streamlined approach to computations of the type involved in this talk one can investigate some of the other processes mentioned above – such as multiple jet production. We anticipate that there should be sizable asymmetries for a number of experimentally accessible processes.

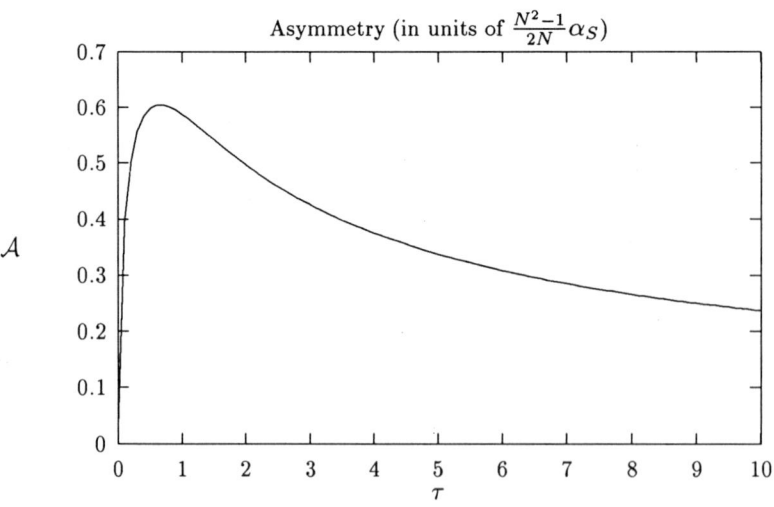

References

[1] J. Ashman *et. al.*, Phys. Lett. **B206**, 364 (1988).

[2] M. J. Alguard *et. al.*, Phys. Rev. Lett. **37**, 1261 (1976).

[3] J. Ellis and R. Jaffe, Phys. Rev. **D9**, 1444 (1974).

[4] R. D. Carlitz and J. Kaur, Phys. Rev. Lett. **38**, 673 (1977).

[5] G. Altarelli and G. G. Ross, Phys. Lett. **B212**, 391 (1988).

[6] R. D. Carlitz, J. C. Collins and A. H. Mueller, Phys. Lett. **B214**, 229 (1988).

[7] A. V. Efremov and O. V. Teryaev, JINR Report E2-88-287 (1988).

[8] M. Einhorn and J. Soffer, Nucl. Phys. **B274**, 714 (1986).

[9] E. Berger and J. Qiu, Phys. Rev. **D40**, 778, (1989).

[10] J. P. Ralston and B. Pire, Phys. Rev. **D28**, 260 (1983).

[11] Z. Xu, D.-H. Zhang and L. Chang, Nucl. Phys. **B291**, 392 (1987).

PREDICTIONS FOR JETS IN POLARIZED pp COLLISIONS FROM EMC EXPERIMENT

G. Nardulli

Dipartimento di Fisica dell'Universita', Bari
I.N.F.N., Sezione di Bari, Italy

Abstract

By using a model for polarized parton densities that fits EMC data on polarized deep inelastic scattering, we make predictions for double helicity asymmetries in 1 and 2 jet inclusive production from polarized proton proton scattering at RHIC energies.

The recent results on polarized deep inelastic scattering from EMC Collaboration [1], together with previous data from SLAC [2], allow to make predictions for jet production by polarized proton proton diffusion on a sounder theoretical base. As a matter of fact, EMC data shed light on a crucial ingredient which is needed in order to compute spin effects in jet production, namely the polarized parton densities ($f = q, \bar{q}$ or g):

$$\delta f(x, Q^2) = f_+(x, Q^2) - f_-(x, Q^2) \tag{1}$$

where f_+ and f_- are the probabilities of having the parton f with fractional momentum x and spin parallel or antiparallel to the proton spin, respectively. In the same notations the unpolarized parton densities are given by $f(x, Q^2) = f_+(x, Q^2) + f_-(x, Q^2)$; for $f(x, Q^2)$ we use the parametrization of Diemoz et al.[3].

In a recent paper [4] a model has been proposed for polarized parton densities; let us summarize its main aspects. The general formula

$$\delta f(x, Q_0^2) = C \frac{x^{-\alpha}(1-x)^\beta}{B(1-\alpha, 1+\beta)} \tag{2}$$

is assumed to be valid at $Q_0^2 \simeq 1 GeV^2$. The values of the parameters α, β and C are obtained as follows. For $x \to 0$ Regge behaviour is assumed, leading to $\alpha = -0.14$ for quarks and $\alpha = -0.60$ for gluons. For $x \to 1$, spectator quark counting rules and positivity constraints:

$$|\delta f| \leq f \tag{3}$$

lead to the values: $\beta_{u_v} = 3$, $\beta_{d_v} = 4$, $\beta_{\bar{s}} = 9$, $\beta_g = 6$. Finally the normalization constants C are obtained by using inputs from hyperon β decays, the EMC experiment and the $U_1(A)$ Goldberger Treiman relation[5] as follows. If we define

$$\Delta f(Q^2) = \int_0^1 dx \, \delta f(x, Q^2) , \qquad (4)$$

then

$$\Delta u = \Delta u_v + 2A\Delta \bar{s} = F + \frac{D + \Delta\Sigma}{3}$$

$$\Delta d = \Delta d_v + 2A\Delta \bar{s} = \frac{-2D + \Delta\Sigma}{3}$$

$$\Delta s = 2\Delta \bar{s} = -F + \frac{D + \Delta\Sigma}{3} \qquad (5)$$

$$\Delta\Sigma - N_f \Delta\Gamma = 0.14 \pm 0.18$$

$$\Delta\Sigma = \frac{\sqrt{N_f}}{2m_p} f_{\eta'} \, g_{\eta'NN}$$

A few remarks are in order. F and D are the SU(3) couplings of the axial vector current between baryon octet states (F=0.477, D=0.755); A=2.5 is given by Diemoz et al.[3] $\Delta\Sigma$ and $\Delta\Gamma$ are given by: $\Delta\Sigma = \Delta u + \Delta d + \Delta S$, $\Delta\Gamma(Q^2) = \frac{\alpha_s(Q^2)}{2\pi} \Delta g(Q^2)$ and it should be observed that $\Delta\Gamma$ remains constant when $Q^2 \to \infty$[6]. Finally the last equation in (5) is obtained by an extension to the U(1) sector of the Goldberger Treiman relation [5]; numerically we take $f_{\eta'} \simeq 130 MeV$ and $g_{\eta'NN} = 4.2 \pm 1.5$, which is the maximum value allowed by the positivity constraint (3) and is not too far from the SU(6) result $g_{\eta'NN} = 6.5$. From the polarized parton densities we can compute the polarized proton structure function which is reported in Fig.1 together with data from EMC and SLAC Collaborations.

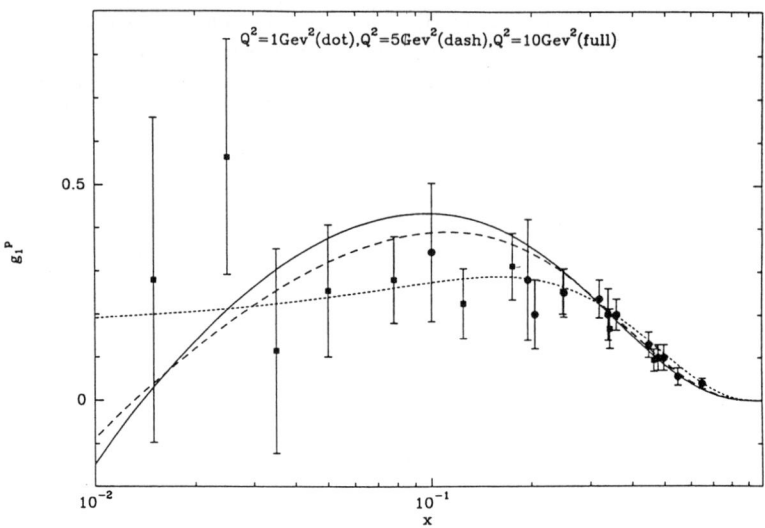

Fig.1 Proton polarized structure function $g_1^p(x)$ as a function of x. Theoretical curves are for $Q^2 = 1$, 5 and 10 GeV^2; data points are from EMC[1] (boxes) and SLAC[2] (circles).

We now turn to jet production. By using polarized parton densities, partonic cross sections and the factorization hypothesis one is able to make predictions for the double helicity asymmetries

$$A_{LL} = \frac{d\sigma_{++} - d\sigma_{+-}}{d\sigma_{++} + d\sigma_{+-}} \qquad (6)$$

where $d\sigma_{hh'}$ are differential cross section for the process:

$$\vec{p} + \vec{p} \rightarrow n\, jets + X \qquad (7)$$

with longitudinally polarized protons having helicities h, h'. In Figs.2 and 3 we report the computed asymmetry for 1 jet (n=1) and 2 jet (n=2) production at same possible RHIC energies ($E = \sqrt{s}$); we have two set of curves, corresponding to $\sqrt{Q^2} = 2p_T$ (full) and $\sqrt{Q^2} = p_T$ (dashed).

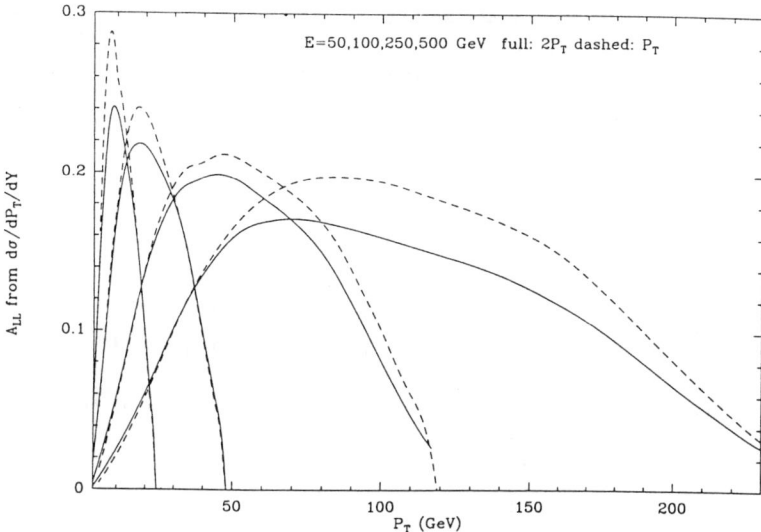

Fig.2 Double helicity asymmetry A_{LL} as a function of p_T at y=0 for 1 jet inclusive production at four different energies: from left to right $\sqrt{s} = 50$, 100, 250 and 500 GeV. Full curves $\sqrt{Q^2} = 2\ p_T$; dashed curves: $\sqrt{Q^2} = p_T$.

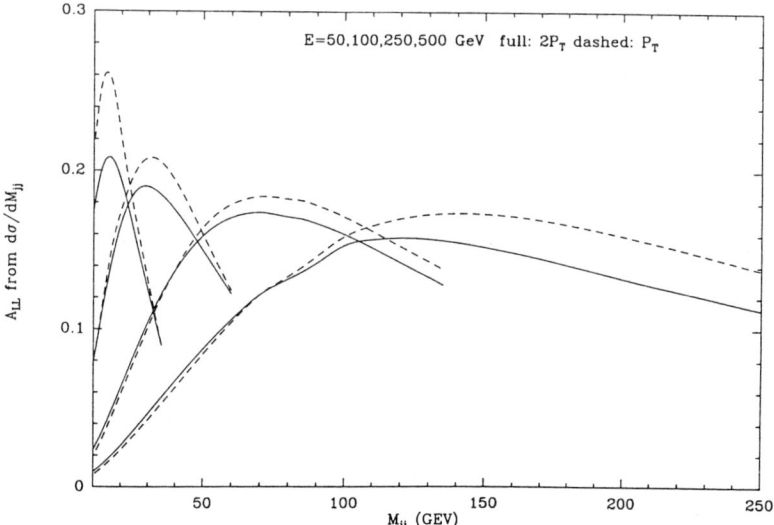

Fig.3 Double helicity asymmetry A_{LL} as a function of the jet pair mass M_{jj} for 2 jet inclusive production at four different energies: from left to right $\sqrt{s} = 50$, 100, 250 and 500 GeV. Full curves: $\sqrt{Q^2} = 2\ p_T$; dashed curves: $\sqrt{Q^2} = p_T$.

In Fig.2 A_{LL} is computed at fixed rapidity y=0 as a function of the p_T jet; in Fig.3 the asymmetry is given as a function of the 2 jet invariant mass M_{jj}. In both cases we obtain sizeable asymmetries, of the order of 10-20%, which would render the RHIC polarized proton beams option a very good stage to observe significant spin effects in jet physics.

REFERENCES

1. J. Ashman et al., Phys. Lett. B206, 364 (1988) and Nucl. Phys. B328, 1 (1989).
2. M. J. Alguard et al., Phys. Rev. Lett. 37, 1261 (1976) and 41,70 (1978); G. Baum et al., Phys. Rev. Lett. 51, 1135 (1983).
3. M. Diemoz, F. Ferroni, E. Longo and G. Martinelli, Z.Phys. C39,21 (1988).
4. P. Chiappetta and G. Nardulli, preprint BARI TH/90-71 (1990).
5. G. Veneziano, Mod. Phys. Lett. A4, 1605 (1989); A. V. Efremov, J. Soffer and N. Tornqvist, Marseille preprint CPT-90/ P.2303.
6. G. Altarelli and G. C. Ross, Phys. Lett. B212, 391 (1988); P. Ratcliffe, Phys. Lett. B192, 180 (1987).

POLARIZED PROTONS AT RHIC

Michael J. Tannenbaum

Brookhaven National Laboratory, Upton, NY 11973*

ABSTRACT

The Physics case is presented for the use of polarized protons at RHIC for one or two months each year. This would provide a facility with polarizations of $\gtrsim 50\%$, high luminosity $\sim 2.0 \times 10^{32}$ cm^{-2} s^{-1}, the possibility of both longitudinal and transverse polarization at the interaction regions, and frequent polarization reversal for control of systematic errors. The annual integrated luminosity for such running ($\sim 10^6$ sec per year) would be $\int \mathcal{L} dt = 2 \times 10^{38}$ cm^{-2}—roughly 20 times the total luminosity integrated in ~ 10 years of operation of the CERN Collider (~ 10 inverse picobarns, 10^{37} cm^{-2}). This facility would be unique in the ability to perform parity-violating measurements and polarization tests of QCD. Also, the existence of $p-p$ collisions in a new energy range would permit the study of "classical" reactions like the total cross section and elastic scattering, etc., and serve as a complement to measurements from $p-\bar{p}$ colliders.

1. INTRODUCTION

The use of RHIC to study the interactions of longitudinally and transversely polarized protons, with a luminosity in excess of 10^{32} cm^{-2} s^{-1}, and c.m. energy in excess of 300 GeV, would open up a totally new field in elementary particle physics and fill a vital gap in the world's accelerators. This facility would be unique in the ability to perform parity-violating measurements and polarization tests of QCD. For many experiments, it would be preferable to run the machine at c.m. energy 300 GeV, rather than the nominal 600 GeV, to obtain the large values of Bjorken x, ($x > 0.4$), required to effectively transmit the polarization of the protons to the constituent quarks and gluons.

---

* This research has been supported in part by the U.S. Department of Energy under Contract DE-AC02-76CH00016

I divide the study of spin effects into 3 classes:

- HIGHBROW—Parity Violation—both the weak interaction effects, which are predicted to be large in this c.m. energy range; and possible new effects in this unexplored realm;
- MIDDLEBROW—-Parity Conserving longitudinal polarization effects, which are fundamental tests of the gauge structure of QCD;
- LOWBROW—Transverse Polarization effects, which are large experimentally, but are not able to be explained theoretically; Polarization effects which QCD predicts to be zero, but which may not be; and polarization of final state particles with unpolarized initial states.

2. PARITY VIOLATING EFFECTS

Two parity violating asymmetries (PVA's) can be measured with longitudinally polarized beams. In the first case, only one beam is polarized, and the cross section difference is measured for the two helicity states of the polarized beam:

$$A_L = (\sigma^+ - \sigma^-)/(\sigma^+ + \sigma^-) \qquad (1)$$

This single spin asymmetry would be suitable for PVA measurements in the CERN collider, if a polarized proton source were available.

The second case involves flipping the helicities of both beams so that they are either left handed ($-$) or right handed ($+$). The two-spin parity-violating asymmetry (A_{PV}) can be defined as

$$A_{PV} = (\sigma^{--} - \sigma^{++})/(\sigma^{--} + \sigma^{++}) \qquad (2)$$

and is about twice as big as A_L, usually, and of opposite sign.

Parity violating effects due to interference between the strong and weak interactions are predicted to be large in the energy range accessible at RHIC. Two examples are: 1) the Total Cross Section for proton-proton interactions, which has a measured PVA (A_L) of $\sim 3 \times 10^{-7}$ at 1.5 GeV/c, 2.6×10^{-6} at 6 GeV/c laboratory momenta, and predictions of $> 10^{-4}$ at RHIC energies[1]; and 2) PVA in inclusive jet production, the leading strong interaction process at large transverse momentum, due to the interference of gluon and W exchange at the constituent level. At $\sqrt{s}$=300 GeV, A_{PV} was estimated to be $\sim 0.8\%$, at jet $p_T = m_W/2$; $\sim 0.5\%$, at $p_T = 50$ GeV/c; 1%, at p_T=70 GeV/c; and 2%, at p_T=95 GeV/c[2,3] (see Fig. 4.1).

In addition to these relatively large interference effects in jet production, the energy and luminosity range of RHIC with polarized protons will open up a totally new regime of hadron physics, a situation in which parity violating effects are dominant. This concerns the direct production of the W and Z bosons of the weak interactions. Although W bosons are predicted to be produced by a parity-violating mechanism, this fact has not yet been demonstrated. The asymmetry A_{PV} is predicted[4,5] to be nearly 70% . RHIC could be the premier and unique laboratory for the study of the PVA in hadroproduction of $W \to e+\nu, \mu+\nu$. In addition to verifying the expected weak interaction PVA effect, this channel will provide the possibility of at least two other important and unique series of measurements:

1) By measuring the PVA for the reaction $W \to e + \nu$ as a function of $\sqrt{s}$, the spin dependent structure functions of the proton can be measured at values of $x \sim m_W/\sqrt{s}$. This brings to mind Val Telegdi's remark: "Yesterday's sensation is today's calibration (and tomorrow's background)."

2) By isolating parity violating production mechanisms, using the PVA subtraction, the rare decays of the W can be studied. The Z will be well studied at LEP and SLC; but there is no comparable facility for the study of W decays until LEPII runs. Thousands of $W \to e + \nu$ events will be produced in each month of running with polarized protons at RHIC. It will be possible to directly measure the leptonic branching ratio of the W, since the dominant hadronic decay, $W \to$jets, will be measurable[5]. Also, the background of electrons from light and heavy quark decays, which prevents electrons from W decay to be observed at transverse momenta below 20 GeV/c, can be substantially reduced (1/100) by the PVA subtraction (see Fig. 4.2), allowing the possibility of the study of rare decays, such as $W \to t + \bar{b}$, in this channel.

The most exciting feature of the study of parity violation in hadron interactions is the possibility of surprises. There are essentially no measurements of, or searches for, parity violation in hard-hadron reactions. THIS FIELD IS TOTALLY UNEXPLORED. In the standard model, no parity violation is expected in hadron reactions. Of course, this is probably a consequence of the fact that nobody ever looked. Recently, some extensions of the standard model have included parity violation. For instance, one possible explanation of the several generations of quarks and leptons is that they are composites

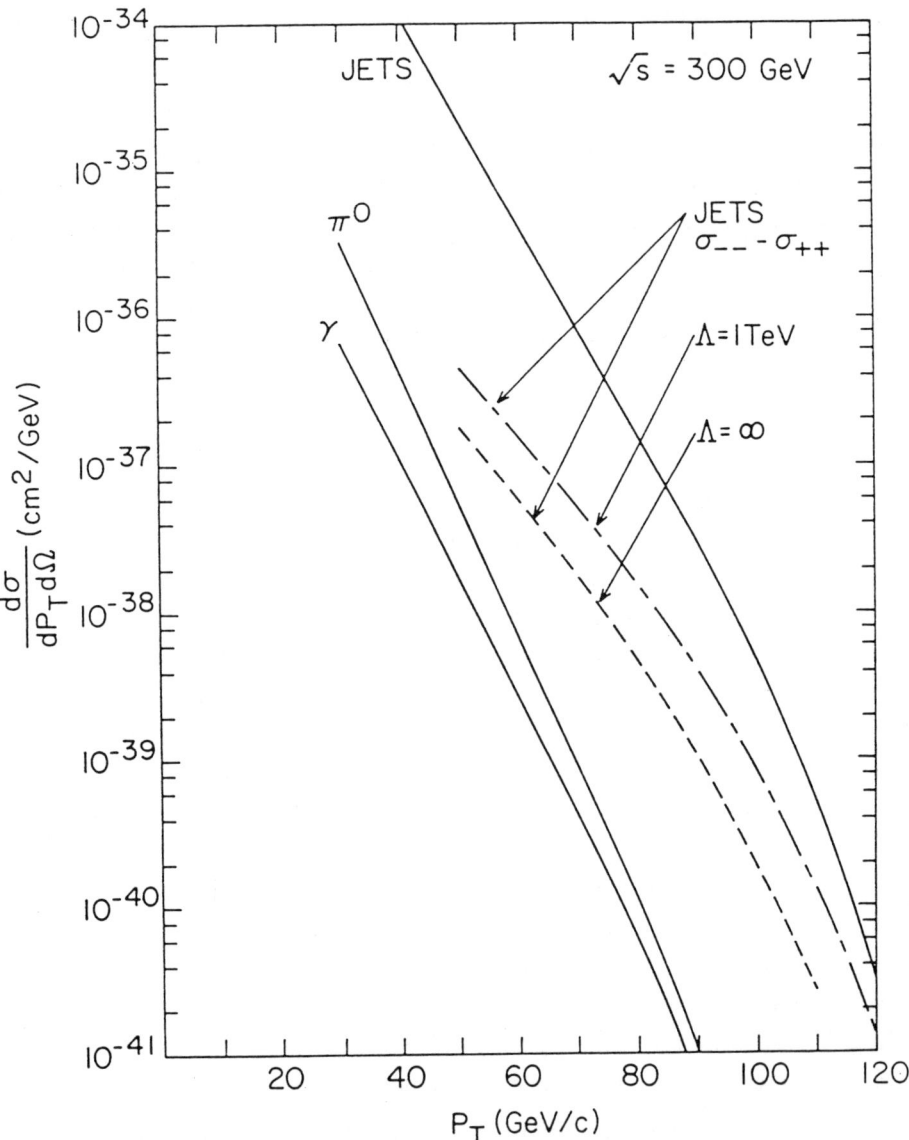

Figure 4.1: Predicted cross sections for representative hadronic inclusive processes at 90° c.m. for $p-p$ collisions at $\sqrt{s}$=300 GeV. The jet cross sections have been computed with ISAJET[3,7]. The π^0 and γ cross sections were estimated using a fit to ISR data[7]. Note that the parity-violating part of the Jet cross section (dashed curve with $\Lambda = \infty$) is much larger than the cross section for inclusive π^0 production. The Λ parameter in the figure refers to a model of quark substructure[6], taken with parameter $A = -1$.

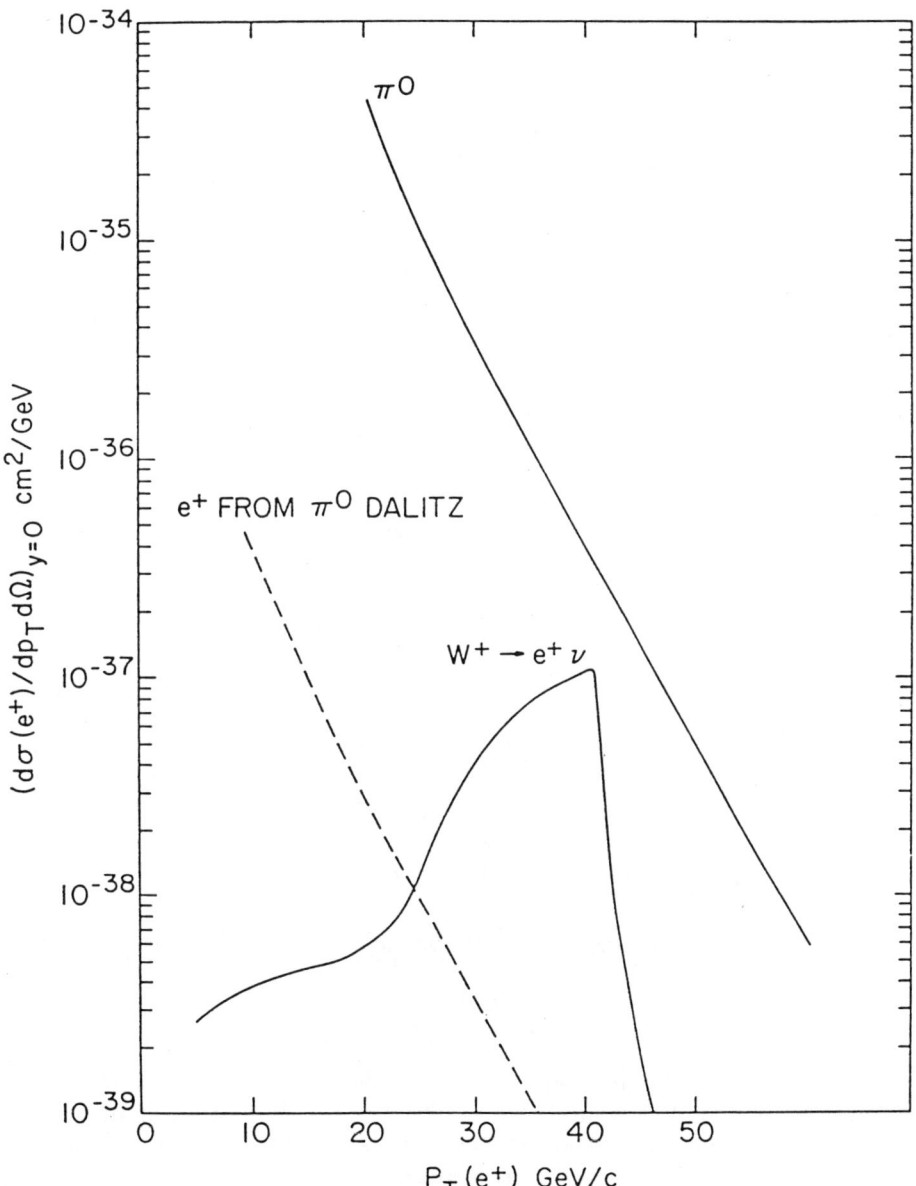

Figure 4.2: Predicted p_T spectrum at $\sqrt{s}=300$ GeV from inclusive π^0, background e^+ from Dalitz decay of π^0, and e^+ from W^+ decay.

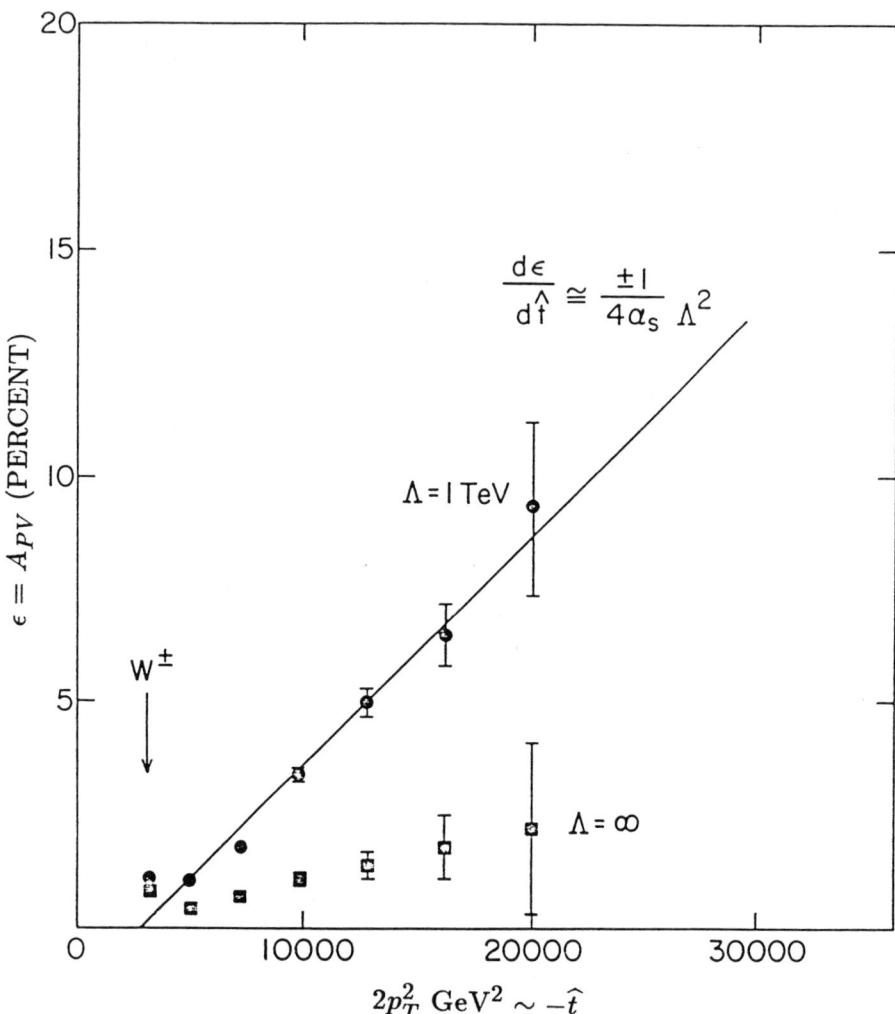

Figure 4.3: Predicted single jet PVA (A_{PV}) for quark substructure model[6] with parameters $\Lambda = 1$ TeV and $A = -1$. The A_{PV} is plotted in percent as a function of the quantity $2p_T^2$ which is a good estimator of $-\hat{t}$, the invariant-four-momentum-transfer-squared of the constituent scattering. The value of Λ can be derived from the observed linear slope, as illustrated. The error bars are calculated assuming an integrated luminosity of 2×10^{39} cm^{-2}, corresponding to 10 months running, and a "nominal" collider detector covering $\Delta\phi = 2\pi$, $\Delta y = \pm 2.5$. This figure indicates that the sensitivity of PVA measurements at RHIC can be much larger than the c.m. energy of the $p-p$ collisions, with sufficient integrated luminosity.

of more fundamental constituents, with a scale of compositenesss $\Lambda \gg 100$ GeV. The intriguing feature of composite models of quarks and leptons is that the interactions generally violate parity, since $\Lambda \gg m_W$. The parity-violating asymmetry then provides direct and much more quantative tests for substructure than other methods. The sensitivity to quark substructure is, of course, model dependent. One model of quark substructure[6] contains an explicitly parity-violating left-left contact interaction between quarks, which results in a PVA in jet production[3,7] proportional to $(p_T/\Lambda)^2$. Without the PVA handle, the CDF detector at the Tevatron is limited to searching for substructure by deviations of jet production from QCD predictions at large values of p_T. The limit is presently[8] at $\Lambda \gtrsim 700$ GeV. Depending on the luminosity, RHIC can produce comparable or better limits. However, the advantage of the PVA signature is that it is a clear indication of new physics and, in addition, Λ can be determined by the dependence of the PVA on p_T—thus, the handedness and other details of any new coupling can be measured (see Fig. 4.3).

There are limitless possibilities beyond the standard model for parity violating effects in hadronic interactions since the subject has hardly been studied. Perhaps the B quark production mechanism is 30% parity violating.....

3. POLARIZATION TESTS OF QCD

QCD is a gauge theory in which helicity plays as fundamental role as "charge". The predicted polarization asymmetry,

$$A_{LL} = \left(\sigma^{++} - \sigma^{+-}\right) / \left(\sigma^{++} + \sigma^{+-}\right) \tag{3}$$

has been given in detail[9]. The $(+)$ and $(-)$ superscripts refer to the helicities of the colliding quarks and gluons. The effect is enormous at the constituent level. However at the particle level, the effect is greatly diluted because the proton polarization is not appreciably transmitted to the constituents, unless $x \gtrsim 0.5$. There is a recent detailed article on this subject[10]. Suffice it to say that the only existing measurement of a polarization effect expected to obey the predictions of QCD, involves the angular distribution of muon pairs produced at large mass and transverse momentum by a π^- beam, in experiment NA10, at CERN[10]. The plane of the lepton pair shows a large azimuthal asymmetry with respect to the production plane—*which is not in accord with QCD predictions.*

A school of thought, led by Jacques Soffer, has claimed for some time that QCD perturbation theory leads to strong polarization of gluons, at large Q^2, independently of any constraint that deep-inelastic lepton scattering data may provide for the distribution of the spin of the nucleon among its constituents. It is therefore important to measure the polarized structure function asymmetry, as directly as possible, in hard processes involving gluons, as well as quarks.

For the case of gluons, the structure function spin asymmetry can be measured directly using single photon production. This reaction is dominated by the subprocess:

$$g + q \rightarrow \gamma + q .$$

The asymmetry A_{LL} in this reaction is proportional to the product of the quark and gluon spin asymmetries, and is predicted to be surprisingly large, in the range ± 10 to 20 %. The charm quark structure function, and its polarization asymmetry, can also be measured in this reaction by tagging events with a lepton in the outgoing jet. Measurements of the spin dependent structure functions of the neutron can be made by colliding polarized protons, or polarized deuterons, against polarized deuterons. RHIC, of course, accomodates beams of different ion species.

To summarize, here is a subject with precise theoretical predictions and no experimental tests. It cries out for measurements—which can only be done using longitudinally polarized proton beams. RHIC is the place.

4. TRANSVERSE POLARIZATION EFFECTS

This subject is the opposite of the preceeding. Large effects have been observed—but there are no theoretical predictions. Examples include elastic scattering at the AGS[11] and a large single spin transverse asymmetry in pion production at large p_T [10]. This is another subject that cries out for a systematic experimental program—to give the theorists some empirical insights into these large polarization effects, which QCD predicts to be small. It is encouraging to note the renewed interest in these effects shown by the participants at this workshop.

5. CONCLUSION

The use of polarized protons at RHIC, with polarizations of $\gtrsim 50\%$, high luminosity $\sim 2 \times 10^{32}$ cm^{-2} s^{-1}, the possibility of both longitudinal and transverse polarization at the interaction regions, and frequent polarization reversal for control of systematic errors, is both possible and desirable. It is time to prepare a detailed case for the implementation of a Polarized Proton Physics program at RHIC.

6. References

1. T. Goldman and D. Preston, Phys. Lett. **168B**, 415 (1986).
2. G. Ranft and J. Ranft, Phys. Lett. **87B**, 122 (1979); F. E. Paige, in Workshop on the Production of New Particles, Madison WI, 1979, (BNL-27066).
3. F. E. Paige and M. J. Tannenbaum , BNL-33119, informal report OG 732 (1983).
4. F. E. Paige, T. L. Trueman and T. N. Tudron, Phys. Rev. **D19**, 935 (1979).
5. G. Bunce, et al., DPF, Snowmass 1982, p 489 (BNL-32150).
6. E. J. Eichten, K. D. Lane and M. E. Peskin, Phys. Rev. Lett. **50**, 811 (1983).
7. M. J. Tannenbaum, internal report OG769, Nov 15, 1983; see also, R. Longacre and M. J. Tannenbaum, BNL-32888, informal report OG 694 (1983).
8. CDF Collaboration, F. Abe, et al., Phys. Rev. Lett. **26**, 3020 (1989).
9. J. Babcock, E. Monsay and D. Sivers, Phys. Rev. **D19**, 1483 (1979).
10. J. Soffer, BNL-41606 (1988), presented at the Symposium on Future Polarization Physics, Fermilab, and to appear in the Proceedings.
11. D. G. Crabb, et al., Phys. Rev. Lett. **60**, 235 (1988).

POLARIZED pp PRODUCTION OF ψ AT LOW p_T AND THE POLARIZED GLUON DISTRIBUTION FUNCTION

M. A. Doncheski
Department of Physics, University of Wisconsin at Madison, Madison, WI 53706

ABSTRACT

It is known that the hadroproduction of ψ is dominated by $gg \to \chi_{0,2} \to \gamma\psi$ and $gg \to g\psi$. The helicity amplitude method allows for the calculation of the necessary polarized parton level processes $gg \to q\bar{q}$ and $gg \to gq\bar{q}$. Techniques exist which can be used to relate the relatively well understood non-relativistic potential model quarkonium wavefunctions to the $gg\chi$ and $ggg\psi$ couplings. Thus it is possible to isolate the polarized gluon contribution to the spin-spin asymmetry from other sources of asymmetry, and study the polarized gluon distribution function directly.

A straightforward analysis of the EMC[1] data indicates that the quarks carry only about 1/8 of the total spin of the proton. The simplest solution is to assume that the gluon carries most of the spin. On the other hand, the parton model predicts that many of the overall properties of the proton (charge, magnetic moment and spin) are carried by the three valence quarks, and that the sea quarks and the gluons have only small effects. A direct measurement of the polarized gluon distribution in the proton would go a long way towards resolving the current difficulties.

Cortes and Pire[2], in 1988, suggested a method to experimentally measure the polarized gluon distribution directly, through the hadroproduction of $\chi_2(c\bar{c})$ at low p_T. A measurement of the photon from the radiative decay $\chi_2 \to \gamma\psi$ is made, which determines the helicity state of the χ_2. They propose a fixed target experiment in which the differential (in x_F and $\cos\theta$) spin-spin asymmetry, in either polarized beam and target or polarized target only, can be used to measure the polarized gluon distribution. The effective $gg\chi$ coupling is to be measured.

Hidaka[3] and more recently Contogouris, Papadopoulos and Kamal[4] perform similar calculations, using partonic semi-local duality ideas to relate the parton level $c\bar{c}$ (or $b\bar{b}$) production to bound state production, below open charm (or bottom) threshold.

Doncheski and Robinett[5] have also recently expanded on the idea of Cortes and Pire. Using a combination of helicity amplitudes[6] and a non-relativistic bound state formalism[7], they can simply relate the polarized $gg\chi_{2,0}$ and $ggg\psi$ couplings to the non-relativistic wavefunction. They also extend the calculation to collider energies, anticipating the polarized hadron collision potential at RHIC and the SSC.

This presentation will review the basic formalism needed, as discussed by Cortes and Pire, followed by an explanation of the improvements suggested by Doncheski and Robinett, applicable to both the fixed target and the collider cases. Finally, results of the polarized collider calculation will be presented.

According to Cortes and Pire[2], the zero p_T production cross section for $pp, p\bar{p} \to \chi_2 X$, to lowest order in α_s, is given by

$$\frac{d\sigma}{dx_F} = \int dx_1 dx_2 f G(x_1, M^2) G(x_2, M^2) \delta(x_1 - x_2 - x_F) \delta\left(x_1 x_2 - \frac{M^2}{s}\right) \quad (1)$$

where f is the effective $gg\chi$ coupling squared, x_F is the longitudinal momentum fraction carried by χ and $G(x, Q^2)$ is the gluon structure function of the proton. The x_1 and x_2 integrals can be carried out, giving

$$\frac{d\sigma}{dx_F} = fG(x_1, M^2)G(x_2, M^2) \quad (2)$$

with

$$x_1 = \frac{x_F + [x_F^2 + 4M^2/s]^{\frac{1}{2}}}{2} \quad (3)$$

$$x_2 = \frac{-x_F + [x_F^2 + 4M^2/s]^{\frac{1}{2}}}{2}. \quad (4)$$

In the electromagnetic decay $\chi_2 \to \gamma\psi$, the angular distribution of the photon depends on the initial helicity state of the χ_2

$$W_\gamma^{(\pm 2)} = \frac{3}{16\pi}(1 + \cos^2\theta) \quad (5)$$

$$W_\gamma^{(0)} = \frac{1}{2\pi} - \frac{3}{16\pi}(1 + \cos^2\theta) \quad (6)$$

where θ is the angle between the z-axis and the photon direction (in the χ_2 rest frame).

If the target and/or beam is polarized, it is advantageous to do the calculation in terms of polarized parton distributions, in this case defining $G_+(x, Q^2)$ ($G_-(x, Q^2)$) as the gluon distribution in the proton, for gluons with same- (opposite-) signed helicity, compared to the helicity of the proton. Also, the effective coupling of $gg\chi$ must be replaced with 'polarized' couplings f_+ (f_-) for production of χ_2 from two gluons with the same- (opposite-) sign helicities. It is possible to calculate cross sections for $p_\lambda p_{\lambda'} \to \chi_2^{(M_J)}$

$$\frac{d\sigma_{+,\pm}^{\pm 2}}{dx_F} = f_- G_+(x_1, M^2) G_\mp(x_2, M^2), \quad (7)$$

and

$$\frac{d\sigma_{+,\pm}^{0}}{dx_F} = f_+[G_+(x_1, M^2)G_\pm(x_2, M^2) + G_-(x_1, M^2)G_\mp(x_2, M^2)]. \quad (8)$$

A spin-spin asymmetry can be defined

$$A_{LL}(x_F, \theta^*) = \frac{d\sigma_{++}/dx_F d\cos\theta^* - d\sigma_{+-}/dx_F d\cos\theta^*}{d\sigma_{++}/dx_F d\cos\theta^* + d\sigma_{+-}/dx_F d\cos\theta^*} \quad (9)$$

$$= \frac{1 - (1 + f_-/f_+)\frac{3}{8}(1 + \cos^2\theta^*)}{1 - (1 - f_-/f_+)\frac{3}{8}(1 + \cos^2\theta^*)} \frac{\Delta G}{G}(x_1, Q^2) \frac{\Delta G}{G}(x_2, Q^2)$$

where θ^* is the angle between the photon and the beam, measured in the χ_2 rest frame and

$$\frac{\Delta G}{G}(x,Q^2) = \frac{G_+(x,Q^2) - G_-(x,Q^2)}{G_+(x,Q^2) + G_-(x,Q^2)}. \tag{10}$$

The ratio f_-/f_+ can be measured in the unpolarized case

$$\frac{d\sigma}{dx_F d\cos\theta^*} = \frac{1 - (1 - f_-/f_+)\frac{3}{8}(1 + \cos^2\theta^*)}{(1 + f_-/f_+)} \frac{d\sigma}{dx_F}, \tag{11}$$

and so $A(x_F, \theta^*)$, at $x_F = 0$, can be used to measure $(\Delta G/G(M/\sqrt{s}, M^2))^2$ directly.

The parton level processes for the production of quarkonium states are known, in terms of non-relativistic potential model wavefunctions:

$$\hat{\sigma}(\hat{s}) = \sigma_0 \delta(\hat{s}/M^2 - 1) \tag{12}$$

where

$$\sigma_0 = \frac{16\pi^2 \alpha_s^2}{M^7} \mid R'_P(0) \mid^2 \tag{13}$$

for $gg \to \chi_2$, and $gg \to \chi_0$ is smaller by a factor of $3/4^8$, and

$$\hat{\sigma}(\hat{s}) = \frac{5\pi \alpha_s^3}{9M^5} \mid R_S(0) \mid^2 I(\frac{\hat{s}}{M^2}) \tag{14}$$

where

$$I(z) = \frac{2}{z^2}\left(\frac{z+1}{z-1} - \frac{2z\ln z}{(z-1)^2}\right) + \frac{2(z-1)}{z(z+1)^2} + \frac{4\ln z}{(z+1)^3} \tag{15}$$

for $gg \to g\psi^9$. The parameter f has been expressed in terms of appropriate factors of wavefunctions and masses. Doncheski and Robinett[5] note that the parameters f_+ and f_- can be similarly calculated. Helicity amplitude calculations[6] give directly the polarized $gg \to c\bar{c}$ and $gg \to gc\bar{c}$ amplitudes, and similar factors of wavefunctions and masses, from the bound state formalism[7], can be used for the parton level cross sections.

For polarized production, following the standard formalism[10], the spin-spin asymmetry can be expressed

$$A_{LL}d\sigma = \int dx_1 dx_2 \Delta G(x_1, Q^2) \Delta G(x_2, Q^2) \hat{a}_{LL} d\hat{\sigma} \tag{16}$$

where

$$\hat{a}_{LL} = \frac{d\hat{\sigma}_{++} - d\hat{\sigma}_{+-}}{d\hat{\sigma}_{++} + d\hat{\sigma}_{+-}}. \tag{17}$$

A measurement of $d\sigma/dy\mid_{y=0}$ or $d\sigma/dx_F\mid_{x_F=0}$ gives (for χ production)

$$A_{LL} = \hat{a}_{LL} \left(\frac{\Delta G(M/\sqrt{s}, M^2)}{G(M/\sqrt{s}, M^2)}\right)^2. \tag{18}$$

$\hat{a}_{LL}$ is found to be -1 $(+1)$ for $gg \to \chi_2$ (χ_0), using the results of Ref. [7].

For $gg \to g\psi$ (with the final gluon integrated out), the parton level asymmetry is found to be

$$\hat{a}_{LL} = \frac{-(z^2-1)F(z) + 2zG(z)\ln z}{(z^2-1)G(z) - 2zF(z)\ln z} \quad (19)$$

where $F(z) = (5z^2 + 2z + 1)$, $G(z) = (2z^3 + z^2 + 4z + 1)$ and $z = \hat{s}/M^2 = x_1 x_2 s/M^2$. Equivalently, this parton level asymmetry can be related to the function $I(\hat{s}/M^2)$ in Eqn. (15)

$$\hat{a}_{LL} = \frac{\tilde{I}(z)}{I(z)} \quad (20)$$

where

$$\tilde{I}(z) = -\frac{2}{z^2}\left(\frac{z+1}{z-1} - \frac{2z\ln z}{(z-1)^2}\right) + \frac{2(z-1)}{z(z+1)^2} + \frac{4\ln z}{(z+1)^3} \quad (21)$$

with $z = \hat{s}/M^2$. The χ_0 and direct ψ production must be included if the three production mechanisms cannot be resolved by experiment.

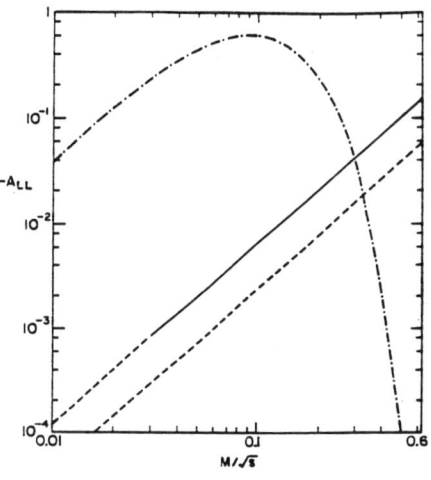

Figure 1 - The negative of the spin-spin asymmetry (in the quantity $d\sigma/dy$ $(y=0)$), $-A_{LL}$, for inclusive χ_2 production via gluon fusion in polarized proton-proton collisions. The solid (dashed, dot-dashed) curve corresponds to the polarized gluon distribution ($\Delta G/G$) of Ref. [11] ([10], [12]).

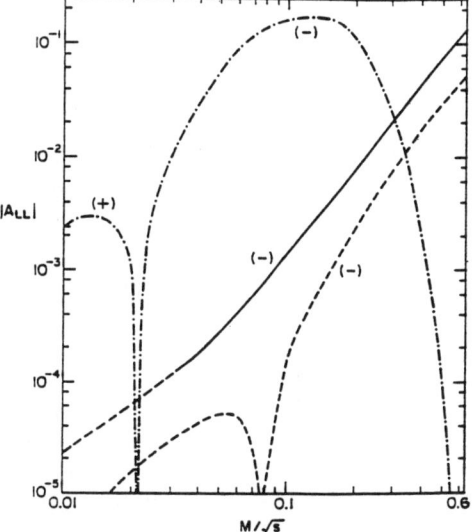

Figure 2 - The spin-spin asymmetry (in the quantity $d\sigma/dy$ $(y=0)$), $|A_{LL}|$, for inclusive ψ production. The solid (dashed, dot-dashed) curve corresponds to the polarized gluon distribution ($\Delta G/G$) of Ref. [11] ([10], [12]).

Some collider results will be shown, using three different parameterizations of $\Delta G(x, Q^2)$. First, is an old (pre-EMC) parameterization by Chiapetta, Guillet and Soffer[11], which is valid for $x \geq 0.03$. The second parameterization is more recent (and EMC inspired) by Bourrely, Soffer, Renard and Taxil[10]. The final parameterization is discussed by Kunszt[12], chosen to satisfy the usual constraints (e.g., $|\Delta G|/G \leq$

1), to be consistent with the interpretation of EMC data suggesting a large gluonic component of proton spin and to exhibit large gluon polarization in the region $x \sim 0.01 - 0.1$.

Figure 1 shows $-A_{LL}$ vs. $M/\sqrt{s}$ for each of these parameterizations, for the case of exclusive χ_2 production. The result for inclusive χ_0 production is identical, but opposite in sign, to the χ_2 result; however, the cross section for ψ production through this channel is small because of the small branching fraction $\chi_0 \to \gamma\psi$. $|A_{LL}|$ vs. $M/\sqrt{s}$ for inclusive ψ production is shown in Figure 2. All figures are for $d\sigma/dy|_{y=0}$.

In conclusion, as suggested by Cortes and Pire, the polarization asymmetry of inclusive χ_2 production is, in principle, an excellent way to probe the polarized gluon distribution of the proton. If the different production mechanisms cannot be resolved experimentally, the asymmetry is reduced somewhat by cancellations arising from the inclusion of direct ψ production. The rapid variation of A_{LL} with $M/\sqrt{s}$ points to the need for a collider with variable center of mass energy, in addition to polarized beams. Finally, the asymmetry is large at relatively large $M/\sqrt{s}$, where the cross sections are small; this points out the need for a high luminosity collider, which may be possible at RHIC[13].

ACKNOWLEDGEMENTS

This research was supported in part by the University of Wisconsin Alumni Research Foundation, and in part by the U. S. Department of Energy under contract DE-AC02-76ER00991.

REFERENCES

[1] EMC Collaboration, J. Ashman *et al.*, Phys. Lett. **B206**, 364(1988).
[2] J. L. Cortes and B. Pire, Phys. Rev. **D38**, 3586(1988).
[3] K. Hidaka, *High-Energy Physics with Polarized Beams and Polarized Targets*, eds. C. Joseph and J. Soffer (Birkhauser Verlag, 1981), p. 77.
[4] A. P. Contogouris, S. Papadopoulos and B. Kamal, Phys. Lett. **B246**, 523(1990).
[5] M. A. Doncheski and R. W. Robinett, Phys. Lett. **B248**, 188(1990).
[6] R. Gastmans, W. Troost and T. T. Wu, Nucl. Phys. **B291**, 731(1987).
[7] J. H. Kühn, J. Kaplan and E. G. O. Safiani, Nucl. Phys. **B157**, 125(1979).
[8] R. Baier and R. Rückl, Z. Phys. **C19**, 251(1983).
[9] V. Barger and A. D. Martin, Phys. Rev. **D31**, 1051(1985).
[10] C. Bourrely, J. Soffer, R. M. Renard and P. Taxil, Phys. Rep. **177**, 319(1989); see also C. Bourrely, J. Soffer and P. Taxil, Phys. Rev. **D36**, 3373(1987).
[11] P. Chiapetta, J. Ph. Guillet and J. Soffer, Nucl. Phys. **B262**, 187(1985).
[12] Z. Kunszt, Phys. Lett. **B218**, 243(1989).
[13] S. Y. Lee and E. D. Courant, Phys. Rev. **D41**, 292(1990), and references therein.

New Bosons in Polarized Hadronic Collisions

John C. Collins And Glenn A. Ladinsky

Department of Physics
104 Davey Laboratory
The Pennsylvania State University
University Park, PA 16802

ABSTRACT

The need for polarized beams at hadron-hadron colliders is reviewed. As an example, lowest order Drell-Yan is examined at the resonance for a new scalar or vector boson. Assuming only scalar and pseudoscalar couplings in the case of a spin zero boson or vector and pseudovector couplings for a new vector boson, it is clear that polarization is required if complete information on the interactions is desired from hadron-hadron collisions. For this case, a study with one beam carrying longitudinal polarization combined with a study where *both* beams have a transverse polarization is required.

1. Introduction

The issues raised in asking how much one gains by polarizing a hadron-hadron collider can be exemplified by the case of the production of a new boson in a Drell-Yan type process. There are plenty of specific calculations that have been done in the context of particular models, however, the justification for a machine should transcend such specifics.

Polarization may or may not provide a greatly increased scope for the *discovery* of new physics, but it surely will be useful in the *measurement* of new interactions after their discovery. We will give an elementary discussion, whose results will not be news to the experts, of what is to be learned by the use of polarized beams.

Given a new particle (like a Higgs, W or Z) produced by a Drell-Yan process, we ask how much extra information is obtained compared with the case of unpolarized beams when:

1) Only one beam is polarized longitudinally
2) Only one beam is polarized transversely
3) Both beams are polarized longitudinally
4) Both beams are polarized transversely.

Furthermore, one can also ask what extra information is gained if one measures the angular distribution of the decay of the boson in addition to the dependence on the polarization of the initial state.

2. The Spin-Dependent Amplitudes for Boson Production

We investigate what can be learned about the couplings for a new scalar boson (X_S) or a new vector boson (X_V) produced in hadron-hadron colliders with or without polarizing the beams. We study this from the perspective of a lowest order Drell-Yan calculation within the resonance of the new particle, where interference effects are minimal. In performing these calculations, we select specific polarizations for the initial quarks by inserting the spin projection operator in the traces which result from squaring the amplitudes.[1] For massless quarks, this operator can be written as

$$\Sigma(s,h) = \frac{1}{2}(1 + \gamma_5 \slashed{s} + h\gamma_5)$$

where the space-like vector $s = (0, \vec{s})$ describes only those polarization directions which are transverse to the fermion's motion and where h is the chirality of the polarization state of interest. Specifically, operating on a bispinor from the left with a positive chirality projection, $\frac{1}{2}(1 + \gamma_5)$, will select a pure state of positive (negative) helicity for the fermion (antifermion) while a negative chirality projection, $\frac{1}{2}(1 - \gamma_5)$, will select a pure state of negative (positive) helicity for the fermion (antifermion). If one wishes to include varying degrees of polarization in these expressions, they may include the appropriate weights in the values for s and h.

We start with two incoming hadrons which are moving along the z-axis. A quark and an antiquark from these hadrons will annihilate, producing the new boson of mass m_X and width Γ_X, which will then decay into two leptons. For a new boson of spin zero, we study the scalar and pseudoscalar couplings, allowing these couplings to be different for each quark and lepton. The two outgoing leptons are taken as massive with momenta q_3 and q_4 while the initial state particles are taken as massless with momenta p_1 and p_2. If the new boson has a narrow width, then at the resonance the invariant mass for the subprocess will be near the mass of the new boson, $\hat{s} = (p_1 + p_2)^2 \approx m_X^2$. An average is taken over quark colors and a sum is performed over all final state spins.

2.1 A new scalar boson

The vertex of the scalar boson with the quarks, $X_S q_1 \bar{q}_2$, is taken as $(\kappa + \kappa' \gamma_5)$ while the vertex with the leptons, $X_S l_1 \bar{l}_2$, is taken as $(\rho + \rho' \gamma_5)$. Here we will assume that κ, κ', ρ and ρ' may be complex. With the incoming quarks moving along the z-axis, we shall define the quark's transverse polarization vectors to have the form[1]

$$s_1^\mu = |\vec{s}_1| (0, \cos \psi_1, -\sin \psi_1, 0) \qquad s_2^\mu = |\vec{s}_2| (0, -\cos \psi_2, -\sin \psi_2, 0).$$

Notice that these vectors are defined to execute a right-handed rotation about the direction of the (anti)fermion's motion as (ψ_2) ψ_1 increases. Using the spin projection operator discussed above, the amplitude squared averaged over colors for the reaction in lowest order reads,

$$|M_S|^2 = \frac{|M_\kappa|^2}{N_c \left[(\hat{s} - m_X^2)^2 + m_X^2 \Gamma_X^2 \right]} \left(\sum_{l_1, \bar{l}_2} |M_\rho|^2 \right)$$

where

$$\sum |M_\rho|^2 = Tr \left[(\slashed{q}_3 + m_3) (\rho + \rho' \gamma_5) (\slashed{q}_4 - m_4) (\rho - \rho' \gamma_5) \right],$$
$$|M_\kappa|^2 = Tr \left[\Sigma (s_2, -h_2) \slashed{p}_2 (\kappa + \kappa' \gamma_5) \Sigma (s_1, h_1) \slashed{p}_1 (\kappa^* - \kappa'^* \gamma_5) \right],$$

and the color factor is $\frac{1}{N_c} = \frac{1}{3}$. Evaluating these traces with the assumption that the incoming particles are moving along the z-axis, we note that the $\vec{s}_i$ ($i = 1, 2$) are transverse to both $\vec{p}_1$ and $\vec{p}_2$. This yields

$$\sum |M_\rho|^2 = 4 \left[(|\rho|^2 + |\rho'|^2) (q_3 \cdot q_4) - m_3 m_4 (|\rho|^2 - |\rho'|^2) \right]$$

and

$$|M_\kappa|^2 = (p_1 \cdot p_2)\left[\left(|\kappa|^2 + |\kappa'|^2\right)(1 + h_1 h_2) + 2\text{Re}\left(\kappa^* \kappa'\right)(h_1 + h_2)\right.$$
$$\left. - \left(|\kappa|^2 - |\kappa'|^2\right)(s_1 \cdot s_2)\right] + 2\text{Im}\left(\kappa^* \kappa'\right)\varepsilon_{\alpha\beta\sigma\tau}p_1^\alpha p_2^\beta s_1^\sigma s_2^\tau.$$

Given that our massless quarks are moving towards each other along the z-axis, the angular dependence on the transverse polarizations may be made explicit with the substitutions $s_1 \cdot s_2 = |\vec{s}_1||\vec{s}_2|\cos(\psi_1 + \psi_2)$ and $\varepsilon_{\alpha\beta\sigma\tau}p_1^\alpha p_2^\beta s_1^\sigma s_2^\tau = |\vec{s}_1||\vec{s}_2|(p_1 \cdot p_2)\sin(\psi_1 + \psi_2)$.

A particular transverse polarization of the initial state is selected in the above amplitudes by taking the four-vectors which describe the direction of transverse spin polarization, s_1 and s_2, as nonzero with $h_1, h_2 = 0$. Purely longitudinal polarizations are obtained when $h_1, h_2 = \pm 1$ while all components of s_1, s_2 are zero. By setting all components of s_1, s_2 to be zero and using $h_1, h_2 = 0$ recovers the unpolarized cross section averaged over initial spin states.

Because we are looking at the decay of a scalar particle, the angular distribution of the final state leptons is isotropic. There is no information to be gained at lowest order by looking at physical quantities like these angular distributions. Integrating over the phase space gives the parton cross section, $\hat{\sigma}_S = |M_S|^2/(16\pi\hat{s})$. It is important to note, however, that varying the angle of the transverse polarizations of the initial state quarks does provide a probe of CP violation as can be seen from the contribution which results from having $\text{Im}(\kappa^*\kappa') \neq 0$.

2.2 A new vector boson

The production of a new vector boson in polarized hadron collisions has been studied in the Drell-Yan case by Ralston and Soper[2] and for the $X_V + jet$ final state by Chaichian, Hayashi, Yamagishi and Soffer.[3] Much work has been done in the context of specific models [4-5] and mass bounds on a massive right-handed vector boson in general $SU(2)_L \times SU(2)_R \times U(1)$ models have been determined by Langacker and Sanker.[6] A more general overview of polarization effects in various models has been written by Bourrely et al.[7]

The vertex with the quarks, $X_V q_1 \bar{q}_2$, is taken as $i\gamma^\mu(g + g'\gamma_5)$ while the vertex with the leptons, $X_V l_1 \bar{l}_2$, is taken as $i\gamma^\mu(\omega + \omega'\gamma_5)$. Here the couplings g, g', ω and ω' are taken to be real, even for nondiagonal quark couplings; we have assumed CP invariance. As with the scalar particles, we take spin projections for the initial state polarizations using $\Sigma(s,h)$.

The relevant amplitude squared for a given flavor of quark and antiquark is

$$|M_V|^2 = \frac{|M_g|^2}{N_c\left[(\hat{s}-m_X^2)^2 + m_X^2\Gamma_X^2\right]}\left(\sum_{l_1,\bar{l}_2}|M_\omega|^2\right)$$

where $1/N_c = 1/3$ and

$$\sum |M_\omega|^2 = Tr\left[(\slashed{q}_3 + m_3)(\omega - \omega'\gamma_5)\gamma^\mu(\slashed{q}_4 - m_4)(\omega - \omega'\gamma_5)\gamma^\nu\right],$$
$$|M_g|^2 = Tr\left[\Sigma(s_2,-h_2)\slashed{p}_2(g - g'\gamma_5)\gamma_\mu\Sigma(s_1,h_1)\slashed{p}_1(g - g'\gamma_5)\gamma_\nu\right].$$

Choosing $h_i = +1$ selects the positive helicity state of the fermion or antifermion. As in the section on the scalar boson, the traces are computed with the initial particles in the subprocess moving along the z-axis. Defining $G_\pm \equiv g^2 \pm g'^2$ and $\Omega_\pm \equiv \omega^2 \pm \omega'^2$ produces

$$|M_V|^2 = \frac{8}{N_c}\left((\hat{s}-m_X^2)^2 + m_X^2\Gamma_X^2\right)^{-1}$$
$$\{m_3m_4\Omega_-(p_1 \cdot p_2)\left[(1-h_1h_2)G_+ + 2(h_1-h_2)gg'\right]$$
$$+(p_1 \cdot q_4)(p_2 \cdot q_3)\left[(1-h_1h_2)(G_+\Omega_+ + 4gg'\omega\omega')\right.$$
$$\left. +2(h_1-h_2)(gg'\Omega_+ + \omega\omega'G_+)\right]$$
$$+(p_1 \cdot q_3)(p_2 \cdot q_4)\left[(1-h_1h_2)(G_+\Omega_+ - 4gg'\omega\omega')\right.$$
$$\left. +2(h_1-h_2)(gg'\Omega_+ - \omega\omega'G_+)\right]$$
$$+G_-\Omega_+(s_1 \cdot s_2)\left[(p_1 \cdot p_2)(q_3 \cdot q_4) - (p_1 \cdot q_4)(p_2 \cdot q_3) - (p_1 \cdot q_3)(p_2 \cdot q_4)\right]$$
$$+2G_-\Omega_+(p_1 \cdot p_2)(q_3 \cdot s_2)(q_3 \cdot s_1)\}.$$

Note there are essentially two terms for the unpolarized states: one with a factor $(g^2 + g'^2)(\omega^2 + \omega'^2)$ and one with a factor $gg'\omega\omega'$. This tells us immediately that knowledge of the spin dependence is required for a complete determination of the vector and pseudovector couplings.

2.3 Angular distributions

We shall now rewrite the amplitude squared for the vector boson production and decay to make the angular distribution of the final state explicit. This has been done in the center of mass of the subprocess such that $\vec{p}_1$ points along the positive z-axis and now all external fermions have been taken to be massless. As before, we define the transverse polarization vectors to have the form

$$s_1^\mu = |\vec{s}_1|\,(0, \cos\psi_1, -\sin\psi_1, 0) \qquad s_2^\mu = |\vec{s}_2|\,(0, -\cos\psi_2, -\sin\psi_2, 0).$$

If we define θ to be the angle between $\vec{p}_1$ and $\vec{q}_3$ in the center of mass for the subprocess then the expression becomes

$$|M_V|^2 = A_1\left(1 + \cos^2\theta\right) + A_2 \cos\theta + A_3 \sin^2\theta \cos\left(\psi_1 - \psi_2 + 2\phi\right)$$

where the A_i are expressed as

$$A_1 = 16 E^4 \Omega_+ \left[(1 - h_1 h_2) G_+ + 2 (h_1 - h_2) gg'\right]/\Delta$$
$$A_2 = 32 E^4 \omega\omega' \left[4 (1 - h_1 h_2) gg' + 2 (h_1 - h_2) G_+\right]/\Delta$$
$$A_3 = -16 E^4 G_- \Omega_+ |\vec{s}_1||\vec{s}_2|/\Delta$$

and $\Delta = N_c((\hat{s} - m_X^2)^2 + m_X^2 \Gamma_X^2)$. In the above formula, E is the energy of the external fermions and the angle ϕ is the azimuthal angle of the momentum vector $\vec{q}_3$.

The information on the scattering of transversely polarized states resides in the term with the coefficient A_3. It is there where we will get deviations in the distribution as compared to the unpolarized differential cross section. The variations from longitudinal polarizations can be found in the terms with the coefficients A_1 and A_2. While the longitudinal polarization affects the angular distribution of the final state at both the order of $\cos\theta$ and $(1 + \cos^2\theta)$, the transverse polarization will affect the distributions only at the level of $(1 - \cos^2\theta)$. The information gained by studying the angular distribution of the leptons would be useful in determining the interaction of the new vector boson. In particular, it may be important to have azimuthal symmetry in the experimental apparatus for studying the asymmetries produced when the initial hadrons are transversely polarized. Integrating over the lepton's phase space yields the parton cross section $\hat{\sigma}_V = A_1/(12\pi\hat{s})$. It is apparent that if any transverse asymmetries are to be observed, one must consider experiments in which the data has not been integrated over the azimuthal angle. This vanishing of the azimuthal dependence in the total cross section is a result of chirality invariance.[8]

3. Discussion

The beams in a hadron-hadron collider may be described as polarized in directions either longitudinal or transverse to their momenta. By using different beam polarizations, it will be possible to extract different information from the interactions of the new boson with the quarks and leptons. As we focus on the problem of completely determining the coupling strengths for the interactions, we will find that a simple examination of the parton level amplitudes will indicate the type of physics we can expect to observe as the beam polarization is varied.

3.1 Unpolarized beams

By setting all the components of the spin polarization vectors, s_1 and s_2, to zero along with the values of h_1 and h_2, we obtain the spin averaged amplitudes for unpolarized initial quarks. Without polarization we can only determine $(|\kappa|^2 + |\kappa'|^2)(|\rho|^2 + |\rho'|^2)$ for the case of the spin zero boson. We can immediately see that it is impossible under these conditions to determine the couplings which describe the interactions. The scalar case looks particularly bad, because there is nothing to vary at the resonance like a final state angle which permits a fit to determine even the *magnitudes* of any of the couplings without losing the generality that the couplings may be different for various initial state quarks and final state leptons. For all practical purposes, that portion of the amplitude which carries the dependence on the couplings is a constant.

The amplitude for the production of a vector boson depends on the couplings in three different ways when the quarks are unpolarized. Unlike the scalar case, we gain crucial information from the angular distribution of the decay products, permiting us to determine, e.g., the value of $|g'/g|$ provided $\omega, \omega' \neq 0$. If the lepton masses are heavy enough, some further information on the couplings may be extracted from the amplitude.

3.2 Only one beam is polarized longitudinally

One longitudinally polarized beam permits a study of parity violation in the lowest order Drell-Yan. When considering the decay of the scalar boson, one longitudinally polarized beam accesses the contribution from $\text{Re}(\kappa^*\kappa')$. If there were no parity violation, then $\kappa' = 0$ or $\kappa = 0$ and this dependence would vanish. Otherwise, by varying the longitudinal polarization, it should be possible to determine (e.g.) the ratio $|\kappa - \kappa'|/|\kappa + \kappa'|$. The symmetry in the amplitude under $\kappa \leftrightarrow \kappa'$ clearly indicates that we will be unable to distinguish between the scalar and pseudoscalar interactions, yet by seeing which helicity polarization gives the maximum cross section, one may determine the sign of $\text{Re}(\kappa^*\kappa')$ which will be a crucial piece of information for getting the relative phase between κ and κ'. To determine the lepton couplings at the same level, however, would require a polarization study on the final state particles as well, but this is unrelated to the issue of polarizing the collider beams. Because the general form for the traces is the same for the $X_S q_1 \bar{q}_2$ and the $X_S l_1 \bar{l}_2$ vertices, the above arguments may be discussed in a similar fashion for the final state.

For vector boson production we find one longitudinally polarized beam will gain access to terms in the amplitude which depend separately on gg' and $\omega\omega'$. Still requiring a fit to the angular distribution of the leptons to extract the values of these terms, it becomes possible to determine the relative signs between the couplings as well as the difference in their magnitudes. The problem that remains is to distinguish between the vector and pseudovector couplings.

3.3 Only one beam is polarized transversely

It is interesting to note that in the lowest order amplitudes presented here, there is no knowledge to be gained by studying the effects of having only one beam with transverse polarization. All contributions due to transverse spin dependences are controlled multiplicatively by both s_1 and s_2. Therefore, if only one beam has a nonzero transverse polarization, these effects vanish. This result is well known.[9]

3.4 Both beams are polarized longitudinally

With two longitudinally polarized beams spin-one bosons may be distinguished from spin-zero bosons. Unfortunately, we do not gain access to any new information on the couplings of the scalar or vector boson when both initial beams are polarized longitudinally as compared to the case when only one beam was longitudinally polarized. For both the scalar and vector bosons the most we can do is to increase the relative effect of different parts of the amplitude by using various degrees and directions of the longitudinal polarization. In the scalar case we may take both helicities to be the same, and by comparing the positive helicities to the negative helicities we will get the maximum variation in the cross section due to the term proportional to $\kappa\kappa'$. Assuming that polarizing a hadron beam will not drastically affect other beam features, a maximum value in the production rate will be reached which was not obtainable by unpolarized scattering. The production cross section for the scalar boson will vanish by helicity conservation when the initial quark and antiquark have pure longitudinal polarizations with opposite helicities and analogously, with quark and antiquark of the same helicity the cross section with the vector boson will vanish.

3.5 Both beams are polarized transversely

It is only with *two* transversely polarized beams that we are able to distinguish between scalar and pseudoscalar couplings or between vector and pseudovector couplings. Furthermore, we find that comparisons where the transverse polarizations of the two colliding hadrons are perpendicular to each other versus parallel or antiparallel provides a probe of the CP invariance of the couplings provided the effects are large enough to be measured. In the scalar amplitude the transverse polarization will extract the contribution due to $(|\kappa|^2 - |\kappa'|^2)$ when the two quark polarizations are parallel or antiparallel. The difference between the squares of the couplings is a necessary piece of information for completely determining the quark couplings. Without it, we would be unable to distinguish between the scalar and pseudoscalar interactions in lowest order Drell-Yan. When the transverse polarizations of the quarks are perpendicular, we may extract the $\text{Im}(\kappa^*\kappa')$ term, which, when combined with the result for $\text{Re}(\kappa^*\kappa')$ from the longitudinal polarization study, will yield the phase difference between κ and κ'. With transversely polarized beams alone, however, the relative phase between κ and κ' is unobtainable since the contribution from the $\text{Re}(\kappa^*\kappa')$ portion of the amplitude is lost, yet, as for the longitudinally polarized quarks, we would be able to determine $|\kappa - \kappa'|/|\kappa + \kappa'|$. The difficulty in going any further is due to our inability to find the lepton couplings. By analogy, this would require polarization studies of the final state. Aligning the two transverse polarizations along the same direction will produce the maximum contribution from the part of the amplitude depending on $(|\kappa|^2 - |\kappa'|^2)$, and by aligning the initial polarizations along perpendicular directions will completely nullify this contribution and instead gain a maximum in the CP violating piece.

We gain a new contribution to the vector boson production with two initial partons carrying transverse polarizations. Namely, a contribution due to $g^2 - g'^2$ appears making it possible to differentiate between vector and pseudovector behaviors. Given that we have the values for $g^2 + g'^2$ and gg' from the previously mentioned polarized scattering, the quark couplings would be completely determined. As in the scalar case a difficulty arises from the factors which depend on the unknown lepton couplings. For heavy leptons we may be able to use the contribution from the dependence on $(\omega^2 - \omega'^2)$, but for the light leptons we are forced to resort to a study of the polarization effects in the final state.

3.6 Further Discussion

There are other issues worth mentioning. As discussed by Bourrely et al.,[7] when one considers dilepton production in hadron-hadron collisions, one must consider the effects of the distribution functions for polarized partons in a hadron. The new boson production studied here in lowest order, depends only on the quark distributions. The quark-antiquark luminosity is generally smaller than the gluon-gluon or gluon-quark luminosities, but here the resonance will yield an increase in the production rate. At a proton-antiproton collider the Drell-Yan could occur through valence quark interactions, however, at a pp collider the lowest order quark-antiquark annihilation will require a sea quark which will not only decrease the luminosity but also dilute any polarization effects when both beams are polarized as compared to the $p\bar{p}$ collisions.

To get an idea of what kind of numbers we are facing, let us do a simple numerical calculation to determine the magnitude of the asymmetries we can observe. Since we have not seen these new bosons, we will have to assume some values to proceed with the estimates. Using the narrow width approximation, the differential cross section with respect to rapidity ($y = \frac{1}{2}\ln(x_1/x_2)$) can be written as

$$\frac{d\sigma}{dy} = f_1\left(\sqrt{\tau}e^y\right) f_2\left(\sqrt{\tau}e^{-y}\right) \hat{\sigma}_{12}(\tau,s) + (1 \leftrightarrow 2)$$

where $\tau = x_1 x_2$, $s = \hat{s}/\tau$ and the x_i are the momentum fractions of the quark and antiquark from the initial hadrons. If the number of Z's expected in one SSC year within $|y| \leq 2.5$ is about 5×10^8, we would expect about $N = 3 \times 10^5$ X's in one SSC year if $m_X = 500\,\text{GeV}$. This includes a reduction factor for the subprocess cross section of $(m_Z^2/m_X^2) = 3 \times 10^{-2}$ and a decrease in the quark-antiquark luminosity of about 2×10^{-2}. Assuming a branching ratio of a few percent gives $\sqrt{N}/N \sim .01$, indicating that we should be able to observe asymmetries down to about a percent given these numbers.

A more general examination of single spin asymmetries when considering longitudinal hadron polarizations indicates that if either the scalar or pseudoscalar couplings vanish, so does the scalar asymmetry. If the two scalar couplings are equal, this asymmetry is 100%. In contrast to the scalar case, we find that the same quantity for the vector boson production has a dependence on kinematics. However, if $|g| = |g'|$ and $|\omega| = |\omega'|$ or if any single coupling is zero, the kinematic dependence disappears.

Finally, it should be mentioned that there is some work which was done in Snowmass 1984 which showed that there is some limited information on the couplings which can be gained using unpolarized beams and examining the decay of the final state leptons.[10] We may also gain similar information at hadron colliders by looking at parity violating asymmetries in inclusive X_V production and using the methods of Hagiwara, Hikasa and Kai.[11] They demonstrate how it is possible to distinguish between $V - A$ and $V + A$ couplings by studying parity violating asymmetries in the production of W + jets using unpolarized hadron beams. Despite this, a complete understanding of the couplings still requires studies with polarized beams.

4. Summary

We have shown, using Drell-Yan production of new bosons as an example, that to make a complete measurement of the couplings for a new physics process requires the use of both transverse and longitudinal polarizations for the hadron beams. If we were to consider a more general process than Drell-Yan, our conclusion would still hold—full polarization information is needed to measure all couplings. We have seen that one longitudinally polarized beam permits a study of parity violation in the lowest order Drell-Yan while with two longitudinally polarized beams spin-one bosons may be distinguished from spin-zero bosons through helicity conservation. Scattering with longitudinal beams alone, however, is insufficient for distinguishing between the scalar coupling and the psuedoscalar couplings or between the vector and the axial vector couplings. Given that we know the boson spin, we also do not gain access to any new information on the couplings of the scalar or vector boson when both initial beams are polarized longitudinally as compared to the case when only one beam was longitudinally polarized.

It is only with *two* transversely polarized beams that were able to distinguish between scalar and pseudoscalar couplings or between vector and axial vector couplings. Furthermore, we found that comparisons where the transverse polarizations of the two colliding hadrons are perpendicular to each other versus parallel or antiparallel provides a probe of the CP invariance of the couplings provided the effects are large enough to be measured. To have only one beam with some transverse polarization gains nothing.

In general, studying the couplings can get more involved than our example. E.g., one may wish to examine other couplings besides those considered here which may enter the effective interactions. With these considerations taken in the light of what has been reviewed in this discussion, polarization is an essential tool for studying the interactions of new bosons in hadron-hadron colliders.

Acknowledgements

This work was supported in part by the U.S. Department of Energy under grant DE-FG02-90ER-40577.

References

1. V.B. Berestetskii, E.M. Lifshitz and L.P. Pitaevskii, *Quantum Electrodynamics*, translated from Russian by J.B. Sykes and J.S. Bell, second edition, Oxford, New York, Pergamon Press, 1982;
   H.A. Olsen, P. Osland and I. Overbo, Nucl. Phys. B171 (1980) 209.
2. J.P. Ralston and D.E. Soper, Nucl. Phys. B152 (1979) 109.
3. M. Chaichian, M. Hayashi, K. Yamagishi and J. Soffer, Il Nuovo Cimento 90A (1985) 327.
4. D. London and J. Rosner, Phys. Rev. D34 (1986) 1530.
5. S. Godfrey, J.L. Hewett and T.G. Rizzo, Phys. Rev. D37 (1988) 643;
   S. Capstick and S. Godfrey, Phys. Rev. D37 (1988) 2466.
6. P. Langacker and S. Uma Sanker, Phys. Rev. D40 (1989) 1569 and references therein.
7. C. Bourrely, J. Soffer, F.M. Renard and P. Taxil, Phys. Rep. 177 (1989) 319.
8. X. Artru and M. Mekhfi Z. Phys. C45 (1990) 669.
9. G.L. Kane, J. Pumplin and W. Repko Phys. Rev. Let. 41 (1978) 1689.
10. H.E. Haber, "Taus-a Probe of New W and Z Couplings," in *Proceedings of the 1984 Summer Study on the Design and Utilization of the Superconducting Supercollider*, 23 June-13 July 1984, Snowmass, Colorado..
11. K. Hagiwara, K. Hikasa and N. Kai, Phys. Rev. Lett. 52 (1984) 1076.

SPIN AND THE INDEPENDENT SCATTERING MECHANISM

John P. Ralston and Bernard Pire
Centre de Physique Théorique - Ecole Polytechnique - 91128 Palaiseau - Cedex.

ABSTRACT

Spin effects observed in proton-proton and meson-proton elastic scattering at large angle show that hadron helicity flip survives at large energies. This may be traced back to the lack of rotationnal symmetry of the independent scattering processes known to dominate at asymptotic energies. We suggest experiments with nuclear targets to test this and the independent scattering contribution to the pp angular distribution.

INTRODUCTION

Elsewhere [1] we have provided evidence that the independent scattering (I.S) mechanism introduced by Landshoff [2] makes a sizable contribution to current data for fixed-angle pp → pp elastic scattering. The oscillations of the data with energy about a power law indicate the I.S contribution is not small. The Brookhaven color transparency experiment [3] is consistent with these contributions being filtered away in a nuclear target.

HADRON HELICITY FLIP

An interesting aspect of the I.S. contributions is that they lack rotational symmetry around any particle axis. The "pinch" power counting shows that the Born amplitude has one power of a hadron transverse size for each pair of independent scatterings. This size scale refers to the direction perpendicular to the scattering plane. The other transverse size scale, in the scattering plane, is small, being of order the inverse energy scale. From this lack of symmetry, we conclude that quark orbital angular momentum can be transmitted through the hard scattering.

Assuming quark helicity is conserved, one nevertheless obtains a mechanism by which hadron helicity is not conserved. A wave function carrying a unit of orbital angular momentum, and the right spin to fit into the hard scattering, can time-evolve into a hadron with its helicity flipped. This does not violate perturbative QCD. Unlike the quark counting (QC) mechanism [4] in which symmetries eliminate all but zero angular momentum, the asymmetry of the I.S mechanism violates the hadron helicity conservation rule.

The size of I.S hadron helicity flips is a detailed dynamical issue. A meson wave function carrying one unit of orbital angular momentum can be made with a factor of $|\vec{b}|\, e^{i\phi}$, where $\vec{b}$ is the transverse separation of quarks. For the case of asymptotically large energies, Sudakov suppression leads to an I.S hadron flip amplitude for meson-meson scattering going like $Q^{-2c\,\ln(1+1/c)-3}$. This can be compared to $Q^{-2c\,\ln(1+1/2c)-3}$ and Q^{-4} for the I.S non-flip and Q.C amplitudes, respectively. Here $c = 8C_F/(11 - 2/3\, n_f)$. The proton case is more complicated and the asymptotic calculation is probabily less relevant. We conclude the I.S mechanism is probably responsible for the observed failure of the hadron helicity conservation rule to hold.

NUCLEAR FILTERING

The idea is testable in a nuclear target. Using nuclear filtering, the asymmetric configurations should disappear for $A \gg 1$. Thus $\pi\, p \to \rho \uparrow p$ in a nuclear target should agree with the perturbative calculation. The power supression in A is calculable. A complementary nuclear target experiment is to measure $d\sigma/dt$ for pp $\to$ pp in the "t^{-8}" region, the t-dependence known for $s \gg |t| \geq \text{GeV}^2$. The t^{-8} dependence, associated with I.S contributions, should fall like a calculable power of A.

ACKNOWLEDGEMENTS

This research has been supported in part under CNRS-NSF International Cooperative Programs Grant INT 8914626 and under DOE Grant OE-FG02-85ER40214.A006. Centre de Physique Théorique is Unite Propre APR0014 du CNRS.

REFERENCES

1. B. Pire and J.P. Ralston, Phys. Lett. 117B (1982), 233.
2. P.V. Landshoff, Phys. Rev. D 10 (1974) 1024 ; J. Botts and
   G. Sterman,Nucl.Phys.B 325 (1989) 62; J. Botts, these proceedings.
3. A.S. Caroll et al, Phys. Rev. Lett. 61 (1988) 1698.
4. S.J. Brodsky and G.P. Lepage, Phys. Rev. D 24 (1981) 2848.
5. J.P. Ralston and B.Pire, Phys. Rev. Lett. 61 (1988) 1823.

Asymptotic Perturbative QCD in Elastic Scattering, Color Transparency and A_{NN}

James Botts

Department of Physics, FM-15, University of Washington, Seattle, WA 98195

Abstract

The effective transverse size of hadrons in leading perturbative contributions to elastic wide angle hadron-hadron scattering is discussed. The leading perturbative contribution to A_{NN}, the spin asymmetry, at $\pi/2$ scattering angle is evaluated as a function of the center of mass energy.

Introduction

Sorting out the various perturbative contributions to wide angle elastic hadron-hadron scattering has been the subject of recent enquiry[1]. Distinguishing the various contributions are the transverse size of the external hadrons and the interaction region and restrictions on the internal momenta flows. For wide angle elastic hadron-hadron scattering, the interaction between two types of perturbative processes, multiple and single scattering, can be the source of interference phenomena and interesting physics. In the following, after a brief description of the leading and non-leading processes, we shall give a picture of what perturbative QCD may have to say about elastic scattering, color transparency and the spin asymmetry, A_{NN}.

Single and Multiple Scattering Processes

The two leading perturbative contributions to wide angle elastic hadron-hadron scattering are the single and multiple scattering processes. A single scattering process[2] is a purely hard process with all the momenta transfer occurring within a distance scale of $\mathcal{O}(1/Q)$. The protons are point-like as well, $1/Q$ in extent. This process obeys dimensional counting and predicts that $d\sigma/dt(pp \to pp) \propto s^{-10}$. To lowest order in α_s, Landshoff[3] showed that diagrams which are like matchsticks aligned out of scattering plane and colliding rather than points have a less severe power law dependence, $d\sigma/dt(pp \to pp) \propto s^{-8}$. The component of quark momenta out scattering plane is constrained to be less than $\mathcal{O}(Q)$ and the constituent quarks are on shell to this order. The inclusion of a nontrivial transverse momenta scale in the Landshoff contribution leads to Sudakov, or double, logarithms when radiative corrections are taken into account that would effectively lower the power law coefficient[4]. The analysis of the Sudakov suppression has been completed and it has been shown[1] that Landshoff scattering is the asymptotically leading process, with a $s^{-9.59}$ power law dependence for baryon-baryon scattering.

A schematic of the region of momenta transfer for single and multiple scattering processes is shown in Fig. 1. Each x represents a hard scattering which to lowest order in α_s is a hard gluon exchange between constituent quarks. In the Landshoff case, the momenta flowing across the x's are $x_1 Q$, $x_2 Q$ and $(1 - x_1 - x_2)Q$, $0 < x_i < 1$ where the $\{x_i\}$ are the momenta fraction of the quarks. The $\{x_i\}$ are constrained to be the same for each proton; this is how the quarks can line up to form three distinct scattering centers. As Q grows large, the distance out of the scattering plane between the hard scatterings of a single scattering process decreases as Q^{-1} and the multiple scattering separation is always larger, like $\Lambda^{-1}(Q/\Lambda)^{-.64}$ for baryons.

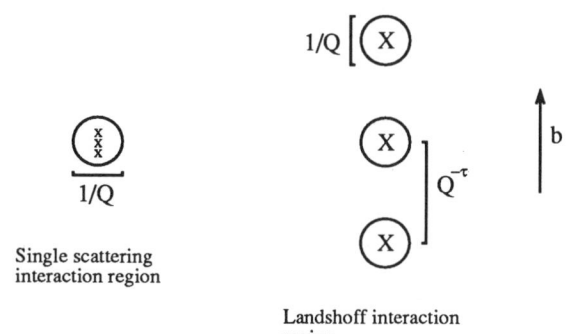

Fig. 1: The interaction regions of single scattering processes and Landshoff processes in baryon- baryon elastic scattering. b is the direction out of the plane of scattering. $\tau_{baryon}=.64$ and $\tau_{meson}=.70$.

Though Landshoff processes result from a relatively small fraction of the internal momenta space necessary for large momenta transfers, its contribution to $d\sigma/dt(pp; \theta = \pi/2)$ for $\ln(s/1GeV^2) \gtrsim 2$, is comparable in size to the data[5]. On the other hand, the dimensional counting contribution has the advantage of a less severe dependence on endpoint, or small x, effects. We define x as small when xQ becomes the order of Λ_{QCD} or a typical hadronic scale of 1 GeV. There is a range in s where the factorization resulting in the purely hard contribution is still trustworthy, its momenta transfers large compared to Λ_{QCD}^2, while the Landshoff is in the endpoint dominated region because it demands three separate scatterings to be hard rather than one. A complete nonasymptotic analysis of wide angle elastic hadron-hadron scattering in the experimentally accessible energy region of $4 \gtrsim \ln(s/1GeV^2)$, would require the inclusion of all the multiple and single scattering contributions, as well as a model for the small x component for each.

Starting from the factorization of Landshoff processes in ref. 1, the leading contribution to $d\sigma/dt(pp; \theta = \pi/2)$ has been numerically evaluated[5] for two choices of nonperturbative hadronic wave function, the King-Sachrajda[6] wave function resulting from QCD sum rules and the asymptotic wave function. In Fig. 2., a typical result is shown for various choices of small x cutoff.

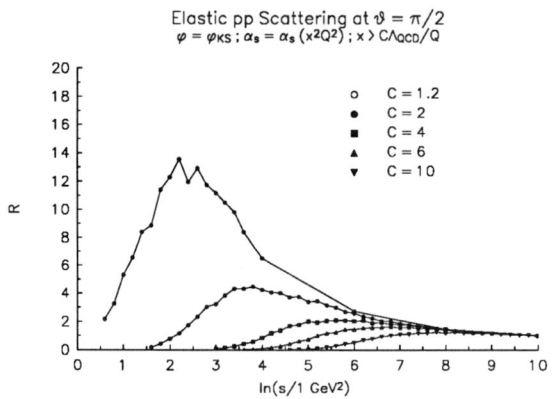

Fig. 2: The cross section for a Landshoff process in elastic pp scattering at center-of mass scattering angle of $\pi/2$ with the dimensional counting power law prediction multiplied out. We have $R \equiv 10^{-8} s^{10} d\sigma/dt(pp \to pp)$.

COLOR TRANSPARENCY

The basic idea of color transparency is that the leading behavior of some elastic hadronic processes comes from contributions where the external hadrons are much smaller in transverse extent than normal[7]. This implies they have much smaller color moments and are much less likely to strongly interact with any background, a nucleus, for example. Sometime later, the quarks and gluons mysteriously hadronize and again behave normally. A useful phenomenology for hadronization within a nuclear background is something lacking in almost every analysis of color transparency, including the treatment herein.

Quasi-elastic pp scattering, $pA \to pp(A-1)$, is a typical example of this paradigm, the transverse separation of the scattered quarks going as Q^{-1} for single scattering or $Q^{-.64}$ for triple. The scattered products of Landshoff processes might be thought to be almost as transparent to the nuclear background as purely hard processes because $Q^{-.64}$ is not so very much different than Q^{-1}. This is in distinction to the original formulation of multiple scattering processes by Landshoff[3] and the corresponding color transparency analysis of Ralston and Pire[8], where the transverse size is considered unconstrained up to the radius of a typical proton.

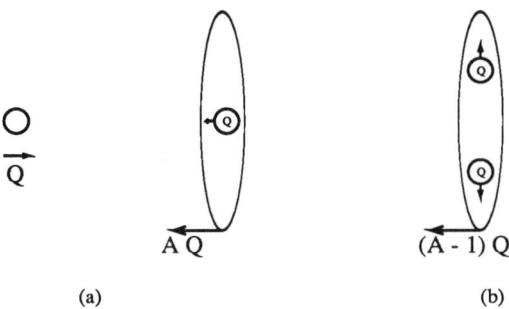

Fig. 3: Quasi-elastic pp scattering at scattering angle $\pi/2$ in the pp center-of-mass frame: a) before scattering, b) after.

The simplest model of color transparency uses a model of hadronization and escape times, $t_{had.}$ and t_{escape}, of the outgoing partons using the uncertainty principle. We have the estimates

$$t_{had.} \sim \frac{r_{transverse}}{<p_{transverse}>} \quad , \quad t_{escape} \sim \frac{r_{longitudinal}}{<p_{longitudinal}>} . \qquad (1)$$

We claim there is transparency for $t_{escape} < t_{had.}$. Consider quasi-elastic proton scattering off a target with A nucleons. We have, using the previous equation and relativistic kinematics,

	multiple scattering	hard scattering
$r_{long.}$	$\sqrt{2}r_h A^{1/3} m_h Q^{-1}$	$\sqrt{2}r_h A^{1/3} m_h Q^{-1}$
r_T	$r_h A^{1/3}$	$r_h A^{1/3}$
$<\vec{p}_T>$	$\Lambda(Q/\Lambda)^{-\tau}$	$\mathcal{O}(\Lambda, m_h)$
$<\vec{p}_{long.}>$	Q	Q
$t_{had.}$	$r_h \Lambda^{-1}(Q/\Lambda)^{-\tau}$	$r_h m_h^{-1}$
t_{escape}	$\sqrt{2}r_h A^{1/3} m_h Q^{-2}$	$\sqrt{2}r_h A^{1/3} m_h Q^{-2}$

(2)

We define the transparency in the usual manner,

$$T(s) = \frac{\frac{d\sigma}{dt}(pA \to p(A-1))}{Z\frac{d\sigma}{dt}(pA \to pp)} \ . \tag{3}$$

Now we shall make an estimate of the transparency of baryon-baryon multiple scattering. The Landshoff amplitude is proportional to an integral over the impact parameter between the hard momenta exchanges, b. We cutoff the b integral in the numerator at $b_{max.}$, where $t_{escape} \approx t_{had.}$. In the leading log approximation and to lowest order in $\alpha_s(Q)$, many terms in the ratio of $T(s)$ drop out and we are left with

$$T(s) \approx \frac{\int_0^{\ln b_{max}\Lambda} d\ln b\Lambda \ e^{-S(b)}}{\int_0^\infty d\ln b\Lambda \ e^{-S(b)}} \ , \tag{4}$$

where e^{-S} is the Sudakov suppression factor. It is roughly a Gaussian as a function of $\ln b\Lambda$ and peaked at $-.64 \ln Q/\Lambda$. The results of this approximation are shown in Fig. 4.

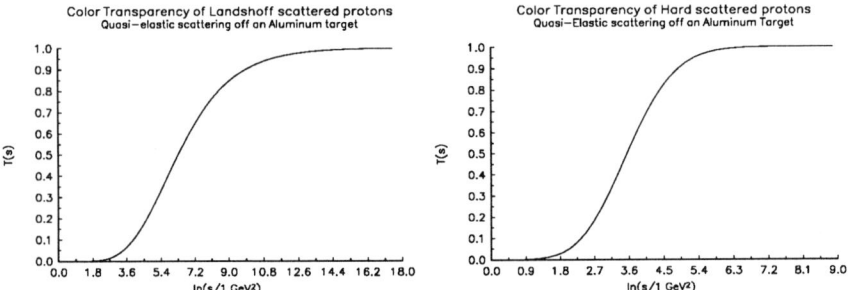

Fig. 4: The color transparency of a Landshoff process and single scattering process in elastic pp scattering at center-of mass scattering angle of $\pi/2$.

In the currently accessible energy region, $\ln(s/s_0) \lesssim 4$, these results would not disagree with Ralston and Pire; the Landshoff contribution is more opaque than the dimensional counting contribution. A typical nuclear target appears to be an efficient filter of Landshoff processes in this energy region. If the oscillations observed in pp scattering is due to interference between these two terms[8], this structure should be mimicked in $T(s)$ because only purely hard scattering would contribute to the numerator. Further details of this work are given in Ref. 9.

A_{NN}

The numerical calculations of the Landshoff contribution to $d\sigma/dt(pp \to pp)$ at center of mass scattering angle of $\pi/2$ can be used to calculate the asymptotic perturbative contribution to the proton spin asymmetry. Except that sums and averages over proton spin are not taken, the numerical calculation is identical.

We define the spin asymmetry, A_{NN}, as a function of the probability for protons to scatter with their incident spins both normal to the scattering plane and parallel and the probability to scatter with spin normal and opposite. We have

$$A_{NN} \equiv \frac{d\sigma(\uparrow\uparrow) - d\sigma(\uparrow\downarrow)}{d\sigma(\uparrow\uparrow) + d\sigma(\uparrow\downarrow)} . \tag{5}$$

Because A_{NN} is a ratio of cross sections, $A_{NN}(Landshoff)$ is independent of overall normalization and the normalization of the hadronic wave functions. A preliminary result is shown in Fig. 5.

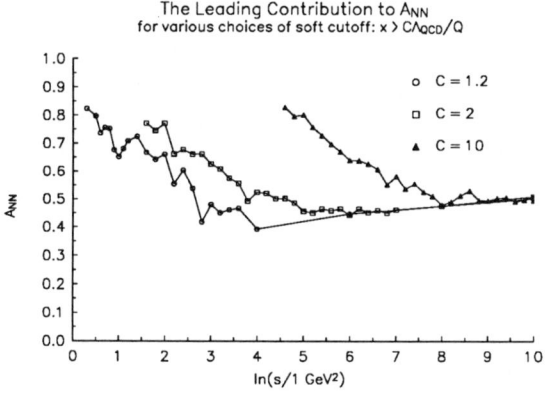

Fig. 5: The spin asymmetry of the pp Landshoff process at $\theta=\pi/2$.

It would appear that at $\ln(s/1\ GeV^2) \gtrsim 5$, $A_{NN} \sim .5$, independent of small x effects. No details of structure can be inferred from the above graph, the errors in the numerical integrations are of the same order as the magnitude of the apparent oscillations at low energy ($\ln(s/1\ GeV^2) \lesssim 3$).

Conclusion

While a full $pQCD$ analysis of wide angle elastic and quasi-elastic hadronic scattering processes at experimentally realizable energies remains for the future, the magnitude of the perturbative contribution has been calculated and estimates of its reliability made. Some of the utility of the asymptotic $pQCD$ results has been shown in regards to the pp elastic cross section, a color transparency analysis and a calculation of A_{NN}. The early results of the perturbative hadronic spin calculation given herein show the potential lies to calculate much more in the perturbative spin realm.

Acknowledgements

The author would very much like to thank George Sterman, John Ralston, Gerry Miller and Leonid Frankfurt for discussions related to the subject to this talk and John Collins, Steve Heppelmann and Richard Robinett for organizing this conference and inviting me. This work was supported in part by the US Department of Energy, Contract No. DE-AS06-88ER40423.

References

1. J. Botts and G. Sterman, Phys. Lett. B224, 201 (1989); Nucl. Phys. B325, 62 (1989); Nucl. Phys. B (Proc. Suppl.) 12, 53 (1989).
2. S. J. Brodsky and G. R. Farrar, Phys. Rev. Lett. 31, 1153 (1973).
3. P. V. Landshoff, Phys. Rev. D10, 1024 (1974); P. V. Landshoff and D. J. Pritchard, Z. Phys. C6, 69 (1980).
4. A. H. Mueller, Phys. Rep. 73, 237 (1981).
5. J. Botts, University of Washington preprint 40423-11 P90, to be published in Nucl. Phys. B.
6. I. D. King and C. T. Sachrajda, Nucl. Phys. B279, 785 (1987).
7. S. J. Brodsky and A. H. Mueller, Santa Barbara preprint NSF-ITP-88-22.
8. J. P. Ralston and B. Pire, Phys. Rev. Lett. 61, 1823 (1988).
9. J. Botts, University of Washington preprint 40423-17 P90.

HADRONIC SINGLE SPIN ASYMMETRIES AT LARGE P_T

P.G. Ratcliffe

INFN, Sezione di Milano, via Celoria 16, 20133 Milano, Italy

ABSTRACT

We present a dynamical explanation of the single transverse-spin asymmetry for the process $pN \to \Lambda^{0\uparrow}X$, with the Λ^0 observed at large angles. The polarisation arises by considering the Λ^0 to be produced either directly or via the virtual dissociation of a parent baryon. Our results reproduce very well both the measured p_T and x_F dependence of the polarisation. We also include a discussion of the left-right asymmetry in pion production off a polarised target: $pN^\uparrow \to \pi X$.

1. INTRODUCTION

Since its discovery in the seventies the large polarisation observed in $pN \to \Lambda^\uparrow X$ at large p_T [1,2] has remained unexplained within the generally accepted approach to high-p_T phenomena, namely perturbative QCD (PQCD). The reason for the failure of PQCD can be traced to two basic requirements for the generation of large polarisations in hadro-production at high p_T:

(i) interference between spin-flip and non-flip amplitudes;

(ii) an imaginary phase between such amplitudes.

In PQCD spin-flip amplitudes are always proportional to *current* quark masses and, since tree diagrams are always purely real, an imaginary phase requires loop contributions. Thus single-spin asymmetries are expected to be doubly suppressed [3]. Although it has been shown that the relevant mass scale may be hadronic [4], the requirement of interference leads to heavy colour and kinematic suppression [4,5]. The general expectation has then been to observe a decrease in polarisation with increasing p_T. Such expectations remain unfulfilled despite attaining high experimental values of $p_T \sim 4\,\text{GeV}/c$ [2], i.e., within the regime of PQCD. Moreover the various phenomenological models adopted to explain this phenomenon [6,7] fail to reproduce quantitatively the x_F dependence and their dynamical basis remains obscure.

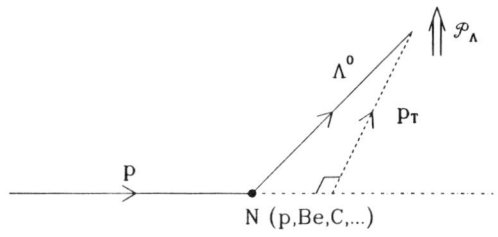

Fig. 1. A schematic representation of the process $pN \to \Lambda^{0\uparrow}X$.

2. HYPERON POLARISATION

In the case of lambda production via proton-hadron scattering the only allowed, parity-conserving, non-zero polarisation of the emerging hyperon is normal to the interaction plane, and thus only for non-zero p_T, fig. 1. As stated above, such polarisations, measured experimentally via the self-analysing decay of the Λ^0, are seen to be large, i.e., of the order of tens of percent, fig. 2.

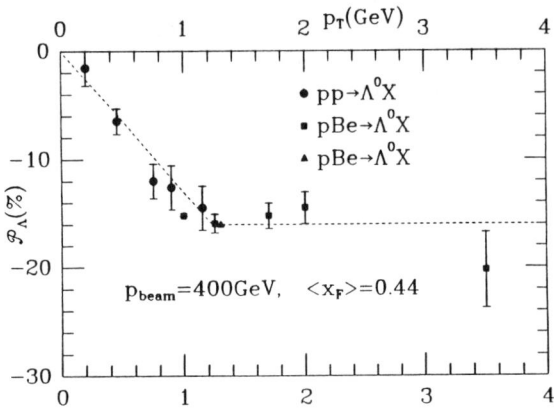

Fig. 2. Data for $pp, pBe \to \Lambda^{0\uparrow} X$ at $p_{lab} = 400$ GeV and $\langle x_F \rangle = 0.44$, the dashed line is merely to guide the eye.

In fig. 3 the observed p_T and x_F dependence of the polarisation is depicted schematically. Note the following features, $\mathcal{P}_{\Lambda^0}$ is:

(i) large and negative (for Λ^0);
(ii) increasing approximately linearly with p_T up to ~ 1 GeV;
(iii) independent of p_T for $p_T > 1$ GeV;
(iv) increasing approximately linearly with x_F;
(v) independent of energy;
(vi) independent of target material.

Moreover $\mathcal{P}_\Sigma$ is opposite in sign and smaller in magnitude, as expected from baryon SU(6) wave-function considerations.

For the inclusive reaction $pN \to B^\uparrow X$, where N is the target nucleon and B the observed final-state hyperon, the polarisation is defined as

$$\mathcal{P}_B = \frac{\sigma^\uparrow - \sigma^\downarrow}{\sigma^\uparrow + \sigma^\downarrow} = \frac{\Delta\sigma}{\sigma}, \qquad (2.1)$$

$\uparrow$ ($\downarrow$) refer to polarisations parallel (anti-parallel) to the scattering-plane normal. Via the optical theorem and transformation to a suitable helicity basis one has

$$\sigma = 2 F^B_{++} \quad \text{and} \quad \Delta\sigma = 2 \operatorname{Im} F^B_{-+}, \qquad (2.2)$$

where
$$F^B_{h'h''} = \sum_{h_p, h_N} \mathrm{disc}_{M^2} \langle p(h_p)\bar{B}(h')N(h_N) | p(h_p)\bar{B}(h'')N(h_N) \rangle.$$
(2.3)

From eq. (2.2) requirements (i) and (ii) above are now immediately obvious.

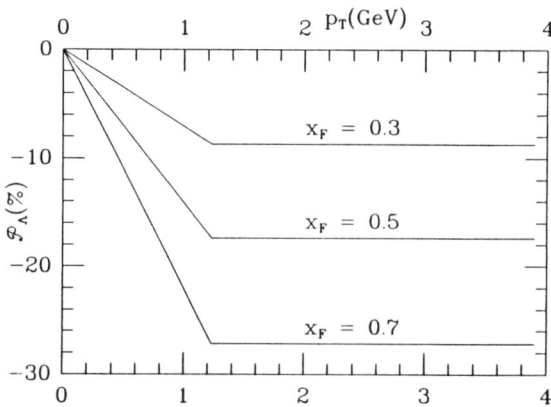

Fig. 3. A schematic representation of the data for $pN \to \Lambda^{0\uparrow}$.

Before proceeding to our calculation let us make a critical examination of a popular approach to explaining this phenomenon: the Lund String Model [6]. In this picture a colour string is stretched between a spin-zero ud pair and the remaining fragments, the string breaks, producing an $s\bar{s}$ pair and the resulting $s(ud)$ system forms the outgoing Λ^0. Since the Λ^0 polarisation is just that of the s-quark the requirement is a polarised $s\bar{s}$ pair. The pair is produced with non-zero p_T and so will have non-zero $L_z = \ell p_T$, where ℓ is the length of string consumed to produce the pair. Angular momentum conservation then demands $s_z = -L_z$. Given $\ell T = E^{\mathrm{pair}} = 2\sqrt{p_T^2 + m_s^2}$, where T is the string tension, one then obtains

$$s_z^s = -p_T \sqrt{p_T^2 + m_s^2}/T. \tag{2.4}$$

At this point one must introduce an *ad hoc* cut-off on the growth of s_z^s, since $|s_z^s| \leq \frac{1}{2}$. Although leading to the correct p_T dependence this contradicts the spirit of the approach, i.e., angular-momentum conservation and no account is given for the x_F dependence of $\mathcal{P}_{\Lambda^0}$.

3. FINAL-STATE INTERFERENCE

It was suggested that the dynamical origin of these large polarisations was to be found in fragmentation effects of the initial nucleon into high-x_F hyperons [8]. Both requirements (i) and (ii) are met by production of the final hyperon via an intermediate step in which the initial nucleon fragments into various members of the baryon octet and decuplet. This approach has been applied in ref. [9] with success to the low-p_T region, although in a somewhat simplified form.

Hadron fragmentation is well described by the triple-Regge model (see fig. 4), which embodies the correct energy and transverse-momentum dependencies [10,11]. In order to describe fragmentation processes into baryon states, which may or may not be the final hyperon, we adopt the following standard triple-Regge parametrisation [5] (h = helicity):

$$\text{disc}_{M^2} \langle p(h_p)\bar{B}'(h')N(h_N) | p(h_p)\bar{B}''(h'')N(h_N) \rangle$$
$$= H(t)\beta_{h_p h'}^{pB'}(t)\beta_{h_p h''}^{pB''}(t)(1-x_F)^{1-2\alpha(t)}, \quad (3.1)$$

where $\alpha(t)$ is the trajectory of the leading Regge-poles, in our case K^* and K^{**}.

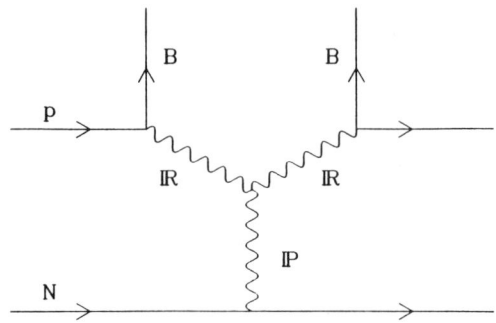

Fig. 4. The triple-Regge exchange mechanism for hyperon production: $B = \Lambda^0$, Σ and Σ^*; $\mathbb{R} = K^*$ and K^{**}; $\mathbb{P}$ = Pomeron.

The Regge residues $\beta_{hh'}^{AB}(t)$ extracted experimentally at small t are in good agreement with theoretical predictions [10-12]. The standard parametrisation:

$$\beta_{hh'}^{AB}(t) = \beta_{hh'}^{AB}(-t'/4m_N^2)^{|h-h'|/2}, \quad (3.2)$$

where $t' = t - t_{min}$ and m_N is the nucleon mass, leads to an unrealistic growth of the spin-flip amplitude for large t so we modify it to

$$\beta_{+-}^{AB}(t) = \beta_{+-}^{AB}\frac{-t'/4m_N^2}{(1-t'/t_0)}, \quad (3.3)$$

with $t_0 \sim 1\,\text{GeV}^2$. In order to reproduce the unpolarised cross-section data we have also found it opportune to modify the Regge trajectories to allow $\alpha(t)$ to tend to a constant for large negative t, as suggested by the quark-parton model. In this way we obtain a satisfactory fit to the Λ^0 cross-section data.

Our model then consists in recognising that the observed hyperon may be produced either directly or via the virtual dissociation of the baryons B' and B'' into $B + \pi$. For Λ^0-production we have the following four processes (see fig. 5):

1. $p + p \to \Lambda^0 + X$ direct production;
2. $p + p \to \Sigma + X \to \Lambda^0 + \pi + X$;
3. $p + p \to \Sigma^* + X \to \Lambda^0 + \pi + X$;
4. $p + p \to \Sigma^0 + X \to \Lambda^0 + \gamma + X$.

The first leads to unpolarised Λ's, while interference between the second and third can give rise to polarisation since the amplitude for the second is imaginary ($m_\Sigma < m_\Lambda + m_\pi$) and that of the third is real ($m_{\Sigma^*} > m_\Lambda + m_\pi$); the phase difference is due to the respective off- and on-shell intermediate propagators. The fourth will also provide polarisation due to the Σ^0 being polarised via the same mechanism.

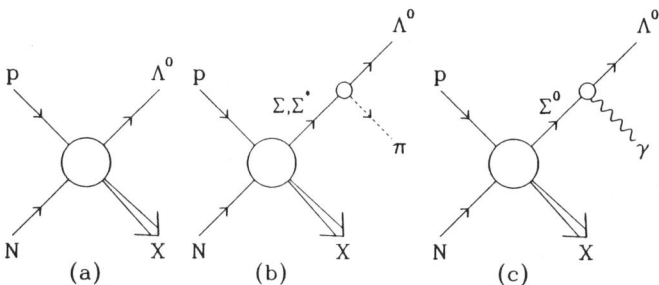

Fig. 5. The production mechanisms for $pN \to \Lambda^0 X$: (a) direct, (b) intermediate baryon dissociation and (c) Σ^0 electromagnetic decay.

4. RESULTS

We should emphasize that the contributions of mechanisms 2. and 3. are comparable in magnitude to that of the direct production throughout the range of p_T; this is a vital precondition for large polarisation. Indeed there is experimental evidence that direct Λ^0 production may be suppressed [135]; this would, of course, slightly enhance the polarization effects we obtain.

It is now straightforward to calculate the desired polarisations by introducing the relevant baryon propagators and couplings as dictated by SU(6) and the known decay widths and branching ratios. In fig. 6 our results [14] for the Λ^0 polarisation are shown versus p_T for various values of x_F. We intend to carry out a calculation for Σ polarisation which, from SU(6) relations, we expect to be consistent with the data. As for $\overline{\Lambda^0}$ polarisation, being based on fragmentation, our model naturally agrees with the negligible value observed [15]. One might ask if this mechanism would give similar results for proton polarisation, which is known to be negligible [16]. In fact this is not the case; the leading Regge exchange for such a process is the Pomeron which leads to domination by the direct channel with consequent zero polarisation. The situation for Ξ production is unclear since it would require an elaborate extension of the Regge model.

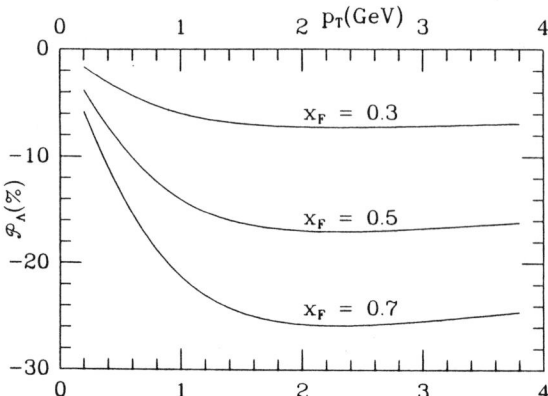

Fig. 6. Our results for the Λ^0 polarisation versus p_T for $x_F = 0.3, 0.5, 0.7$.

5. THE PION ASYMMETRY

A process displaying analogous polarisation effects is pion production in polarised nucleon-nucleon collisions [17]. Here the polarisation (transverse to the beam direction) is in the initial state (either beam or target) and one detects a left-right asymmetry in the π cross-section with respect to the plane defined by the beam and polarisation directions, see fig. 7.

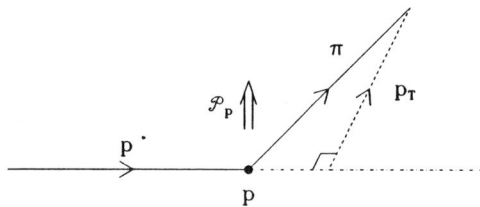

Fig. 7. A schematic representation of the process $pp^\uparrow \to \pi X$.

The diagrams for such a process are analogous to those of fig. 5, where the intermediate resonances are the N and N^* and the trajectories are nonstrange. Although one might imagine *reggeising* the baryons, such contributions are expected to be suppressed with respect to those of the meson trajectories. The computation for this process will be carried out shortly. Prompt-photon production immediately suggests itself as a test of the above mechanism; in this case one would expect zero asymmetry since all the diagrams are relatively real owing to the masslessness of the photon.

6. CONCLUSIONS AND COMMENTS

From the expressions obtained for the cross-sections of the competing processes we find that at small x_F the polarisation is proportional to x_F, while at $x_F \geq 0.8$ the extra factor of $(1 - x_F)$, characteristic of an intermediate decay process, dominates and thus suppresses any polarisation. Thus it is for large x_F (and not p_T) that we expect the polarisation effects to die out. For the pion asymmetry, on the other hand, we expect no such suppression as there is no direct mechanism for π production and thus the effects are undiluted.

In conclusion, as seen from fig. 6, we have produced a full explanation of the hitherto puzzlingly large Λ^0-polarisation observed at high p_T. The description is given entirely in terms of familar aspects of hadron dynamics and without recourse to *ad hoc* assumptions as to magnitudes or kinematical dependence of the dynamics. Moreover we have linked two apparently diverse phenomena: Λ^0 polarisation and the π asymmetry, to one and the same mechanism.

REFERENCES

1. G. Bunce et al., Phys. Rev. Lett. **36** (1976) 1113.
2. K.J Heller, Proc. VIII Int. Symp. on High-Energy Spin Physics, ed. K.J. Heller, AIP conf. proc. 187 (1988).
3. G. Kane, J. Pumplin and W. Repko, Phys. Rev. Lett. **41** (1978) 1689.
4. A.V. Efremov and O.V. Teryaev, Sov. J. Nucl. Phys. **36** (1982) 140; P.G. Ratcliffe, Proc. VI Int. Symp. on High Energy Spin Physics, ed. J. Soffer, J. de Phys. **46** (1985) C2-31.
5. P. Cea, P. Chiappetta and G. Nardulli, Phys. Lett. **209B** (1988) 333.
6. B. Andersson, G. Gustafson and G. Ingelman, Phys. Lett. **85B** (1979) 417.
7. T.A. Degrand and H.I. Miettenen, Phys. Rev. **D24** (1981) 2419.
8. G. Preparata, Proc. High Energy Phys. with Polarised Beams and Polarised Targets, eds. C. Joseph and J. Soffer, Birkhauser, Basel 1981, p. 121.
9. P. Cea, P. Chiappetta, J.-Ph. Guillet and G. Nardulli, Phys. Lett. **193B** (1987) 361.
10. A.C. Irving and R.P. Worden, Phys. Rep. **34** (1977) 117.
11. L.G. Pondrom, Phys. Rep. **122** (1985) 57.
12. S.N. Ganguli and D.P. Roy, Phys. Rep. **67** (1980) 201.
14. R. Barni, G. Preparata and P.G. Ratcliffe, Hyperon Polarization at High p_T Finally Explained (in preparation).
15. K.J. Heller et al., Phys. Rev. Lett. **41** (1978) 607.
16. R.O. Polvado et al., Phys. Rev. Lett. **42** (1979) 1325.
17. J. Antille et al., Phys. Lett. **94B** (1980) 523.

TRANSVERSE SPIN OBSERVABLES IN HADRON-HADRON AND HADRON-NUCLEUS COLLISIONS*

Dennis Sivers[†]

High Energy Physics Division, Argonne National Lab, Argonne, IL 60439

ABSTRACT

Transverse, single-spin asymmetries offer a chance to test QCD at the level of "twist-3" observables. Early suggestions that such asymmetries necessarily vanish as $m_q \to 0$ or involve an extra power of α_s can be refuted with a simple example. Recent experimental results support the interpretation of these data in hard-scattering QCD. The asymmetry in the scattering on nuclear targets can provide new, nontrivial information the space-time structure of the interaction.

TRANSVERSE SPIN

Transverse spin observables can provide considerable insight into the dynamics of composite systems. A simple exercise in spinor algebra can be used to demonstrate the close relationship between spin and mass for spin-1/2 particles. For light like momenta

$$k = E(1, \cos\theta, \sin\theta\cos\phi, \sin\theta\sin\phi) \tag{1}$$

we can define two independent spinors

$$|k, +\rangle = E^{1/2}\begin{pmatrix} \cos\theta/2 \\ \sin\theta/2 e^{i\phi} \end{pmatrix}$$
$$|k, -\rangle = E^{1/2}\begin{pmatrix} \sin\theta/2 e^{-i\phi} \\ \cos\theta/2 \end{pmatrix} \tag{2}$$

For a massive, spin-1/2 particle we can decompose the momentum and "spin" 4-vectors into two light-like vectors, k_1 and k_2.

$$P = 1/2(k_1 + k_2)$$
$$mS = 1/2(k_1 - k_2) \tag{3}$$

with

$$P^2 = m^2 \qquad k_1^2 = k_2^2 = 0$$
$$S^2 = -1 \qquad P \cdot S = 0 \tag{4}$$

In a representation where γ_5 is diagonal, we then combine 2-component spinors defined in (2) to generate the 4-component Dirac spinors

$$u(P, +S) = |k_1, +\rangle + |k_2, -\rangle$$
$$u(P, -S) = |k_1, -\rangle + |k_2, +\rangle \tag{5}$$

[†]Work supported by the U.S. Department of Energy, Division of High Energy Physics, Contract W-31-109-ENG-38.

Let's apply this algebra to the problem of a massive particle (like the proton) made up of massless spin-1/2 constituents neglecting quark masses. These equations can be made Lorentz-covariant, but in order to illustrate the "parton" interpretation it is convenient to use light-cone coordinates and to consider the limit $P^+ \to \infty$.

$$P = (P^+, P^-, \vec{O}) \qquad P^- = m^2/P^+$$
$$k_i = (k_i^+, k_i^-, \vec{k}_i^T) \qquad k_i^2 = O \qquad (6)$$

We now can write,
$$k_i^+ = x_i P^+, \qquad (7)$$

and impose the constraint
$$\sum_i x_i = 1. \qquad (8)$$

We can also impose
$$\sum_1 \vec{k}_i^T = O \qquad (9)$$

but, because of binding, the constituents are off their light-cone "energy" shell and we do not have total conservation of momentum,

$$P^- \neq \sum_i \frac{k_i^{T2}}{x_i P^+}. \qquad (10)$$

By construction, the 4-vector
$$\eta^\mu = P^\mu - \sum_i k_i^\mu \qquad (11)$$

has only a (-) component, and is therefore light-like. Let's now define,

$$D = | \frac{m^2 - \sum_i k_{Ti}^2/x_i}{P^+} | \geq 0, \qquad (12)$$

for definiteness, and write
$$r^\mu = (O, D, \vec{O}) = \pm \eta^\mu. \qquad (13)$$

The 4-vectors
$$p_i^\mu = k_i^\mu + x_i r^\mu$$
$$W_i^\mu = k_i^\mu - x_i r^\mu \qquad (14)$$

are, respectively, timelike and spacelike vectors which differ from k_i^μ only in their (-) component.
$$p_i^\mu p_{i\mu} = x_i^2 P^+ D = -W_i^\mu W_{i\mu}$$
$$W_i^\mu p_{i\mu} = O \qquad (15)$$

The spinor decomposition given in (5) above can therefore be used for the constituents

$$u(p_i, +s) = |k_i, +\rangle + |x_i r, -\rangle$$
$$u(p_i, -s) = |k_i, -\rangle + |x_i r, +\rangle \qquad (16)$$

to define Dirac-spinors which are closely related to the helicity states $|k_i, \pm\rangle$ for the massless quarks. Covariant relationships between these spinors and the proton spinor are well-behaved and we can form "transverse" quark spins in a consistent way, utilizing the fact the P in (6) is timelike and the constituents are bound inside the proton to resolve ambiguities. Note that this construction has built in the connection

$$\vec{W}_i^T = \vec{k}_i^T. \tag{17}$$

which emphasizes that constituent transverse momenta cannot be neglected in a picture of quark transverse-spin observables.

The treatment of transverse spin observables with the neglect of constituent transverse momenta[1] can lead to trouble with single-spin asymmetries in that the approximation artificially enforces a counter-intuitive space-time picture of the production process in which the phase $e^{i k_T \cdot b}$ associated with the finite size of hadrons is ignored compared to phases generated by higher-order terms in the perturbation expansion for hard scattering. There have been many attempts to improve the naive picture.[2,3] In particular, in ref. (4) it was argued to be possible to organize QCD in a way which accounts for transverse momenta effects. In this approach it is assumed that the coherent dynamics can be absorbed into a constituent-level asymmetry in transverse momentum

$$\Delta^N G_{a/p\uparrow}(x, \vec{k}_T; \mu^2) = \sum_h \left[G_{a(h)/p\uparrow}(x, \vec{k}_T; \mu^2) - G_{a(h)/p\downarrow}(x, \vec{k}_T; \mu^2) \right]$$
$$= \sum_h \left[G_{a(h)/p\uparrow}(x, \vec{k}_T; \mu^2) - G_{a(h)/p\uparrow}(x, -\vec{k}_T; \mu^2) \right]. \tag{18}$$

The kinematic dependence of the hard-scattering process on the underlying momenta then gives a production asymmetry

$$A_N \left(E \frac{d^3\sigma}{d^3p}(pp\uparrow \to \pi x) \right) \cong \sum_{ab \to cd} \int d^2\vec{k}_T^1 dx_b \int d^2 k_{Tb} \int \frac{dx_c}{x_c^2}$$
$$\Delta^N G_{a/p\uparrow}(x_a, \vec{k}_{Ta}; \mu^2) G_{b/p}(x_b, k_{Tb}^2; \mu^2) \tag{19}$$
$$D_{\pi/c}(x_c, k_{Tc}^2; \mu^2) \left[\hat{s} \frac{d\sigma}{d\hat{t}}(ab \to cd) \delta(\hat{s}+\hat{t}+\hat{u}) \left(1 + 0(\frac{\alpha_s}{\pi})\right) \right].$$

This new approach leads to an asymmetry which is "higher twist" but does not vanish in the limit that quark masses go to zero and is not of higher-order in the perturbation expansion. The underlying mechanism can be considered a kinematic slingshot effect in that the detected particle or jet is more likely to come from a hard scattering involving a constituent which already had some intrinsic transverse momentum in the right direction. The coherent dynamics associated with the constituent-level asymmetry $\Delta^N G(x, k_T; \mu^2)$ involve the long-range forces which bind quarks and gluons into a proton. The large-p_T limit of (19),

$$A^N \sim \frac{\epsilon \langle k_T \rangle}{P_T}, \tag{20}$$

involves the mean k_T of constituents and a parameter, ϵ, which characterizes the magnitude of $\Delta^N G$.

The identification of a "hard-scattering" mechanism for A^N implies a scaling law of the form

$$\left(\frac{P_T^2 + \mu^2}{\mu P_T} \right) A_N \cong g(x_T) \tag{21}$$

where $x_T = 2P_T/\sqrt{s}$.[5] As shown in Fig. 1, there is support for such scaling behavior from combining experimental data at lower energies[6,7] with recent data[8] from E-704 at Fermilab on $pp\uparrow \to \pi X$.

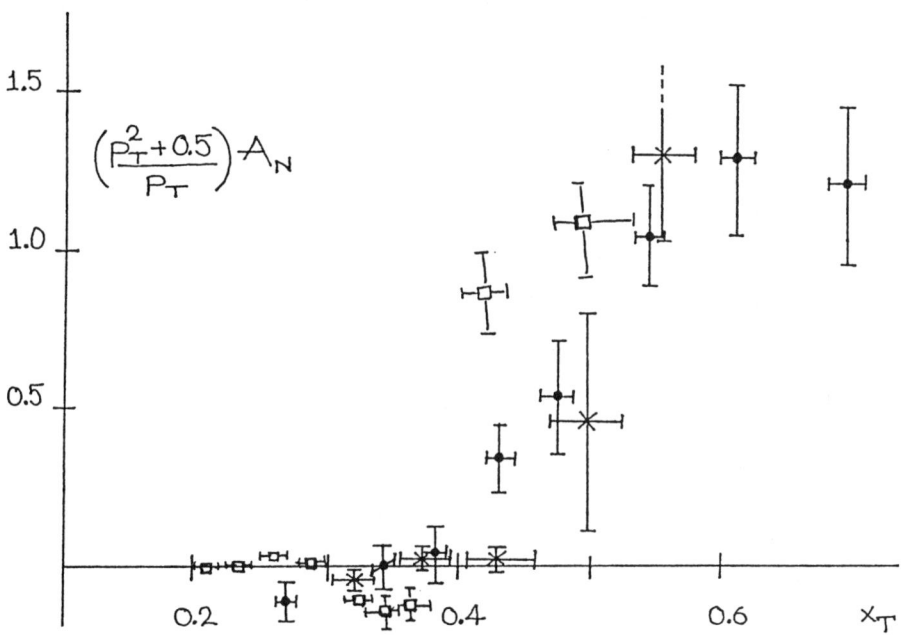

Data from Ref. 6 (●), Ref. 7 (✗), and Ref. 8 (□) on $A_N(p_T^2+0.5)/p_T$ plotted against x_T order to test the scaling law (1.4).

It should be emphasized that establishing the validity of a scaling law of the type (21) provides strong support for the existence of some underlying hard-scattering mechanism. It does not necessarily support the particular version of hard scattering given in eq. (19). The effort to extend factorization in QCD beyond the level of leading twist observables[9] lends credence to the effort to establish the underlying mechanism.

A polarized proton incident on a nuclear target offers a fine opportunity to investigate the nature of the mechanism for the underlying asymmetry. It is not clear that we have an understanding of the dynamics behind the Cronin effect.[10]

$$\frac{d\sigma(pA \to hx)}{d\sigma(pN \to hx)} = A^\alpha \qquad (22)$$

with $\alpha \geq 1$. The most natural explanation of these data involves multiple coherent scatterings. The single-spin asymmetry provides a good probe for multiple scatterings and a program involving polarized protons and nuclear targets would provide groundbreaking information.

REFERENCES

1. G. Kane, J. Pumplin and W. Repko, Phys. Rev. Lett. 41, 1698 (1978).

2. A. V. Efremov and O. V. Teryaev, Sov. Journal of Nucl. Phys. 36, 140 (1982).

3. P. Ratcliffe, *Proceedings of 6th Int. Symposium on High Energy Spin Physics*, 1984 edited by J. Soffer [J. Phys. (Paris) Colloq. 46, C2-31 (1985)].

4. D. Sivers, Phys. Rev. D41, 83 (1990).

5. D. Sivers, Phys. Rev. D43, 261 (1991).

6. J. Antille et al. Phys Lett. 94B, 523 (1980).

7. Data in Proc. of VII Int. Symposium on High Energy Spin Physics.

8. E-704 experiment (A. Yokasawa et al.). See the table of A. Yokasawa, this conference.

9. See, for example J. Qiu, this conference.

10. T. Eichten, et al., Nucl. Phys. B44, 333 (1978); P. Skubic, et al., Phys. Rev. D18, 3115 (1978); L. G. Pondrom, Phys. Reports 122, 457 (1985).

HIGH TWIST EFFECTS IN HADRONIC COLLISIONS

Jianwei Qiu[*] and Geroge Sterman
Institute for Theoretical Physics, State University of New York
Stony Brook, New York 11794-3840

ABSTRACT

We analyze the first nonleading power corrections to inclusive hadron-hadron scattering cross sections. Such corrections decrease as $1/Q$ relative to leading power in the case of polarized hadronic collisions, or as $1/Q^2$ in the case of unpolarized hadronic collisions. We show why such corrections may be treated in perturbative QCD, in terms of generalized factorization theorems.

I. INTRODUCTION

At leading power, perturbative QCD has been very successful in interpreting and predicting high energy scattering processes. The knowledge of the first nonleading power corrections to these processes would make possible new tests of the theory. It would also strengthen our confidence on the predictions given by the leading power formulas.

Such power-suppressed contributions have already received considerable attention. A number of theoretical calculations have been carried out.[1-3] These concentrated on power corrections to structure functions in deeply inelastic scattering (DIS). The results can be summarized in a factorized form, in which perturbatively calculable short-distance coefficient functions are convoluted with nonperturbative matrix elements of parton operators. These operators are different from those in the matrix elements of leading power contributions. Because of these new nonperturbative matrix elements, perturbative QCD alone cannot predict the size of the power-suppressed corrections.

Much of the predictive power of perturbative QCD is contained in factorization theorems.[4] They normally include two assertions. One is that a physically observed quantity can be factorized into some perturbatively calculable short-distance hard parts convoluted with non-perturbative long-distance matrix elements. The other is the universality of the nonperturbative matrix elements. Predictions follow when processes with *different* hard scatterings but the *same* matrix elements are compared. Therefore, we should study power corrections in processes other than deeply inelastic scattering in order to test the universality of the new matrix elements.

Before we start to calculate the power corrections in any process, we should ask whether a factorization theorem applies to the power corrections at all. If not, we will not be able to calculate them consistently. Factorization for the structure functions in deeply inelastic scattering is a natural consequence of operator product expansion (OPE). However, the OPE is not applicable for hard hadronic cross sections, such as the Drell-Yan process. It was originally suggested by Politzer[5] that factorization might be generally true for power corrections to hadronic processes. In Ref. [6], we have given an all-order argument in perturbation theory why the program of factorization can be extended to $1/Q^2$ corrections in a large class of inclusive cross sections of *unpolarized* hadronic scatterings. In this paper, we shall review our arguments and extend them to *polarized* hadronic scattering.

We begin in Section II with a general discussion of how the program of factorization for inclusive hadronic scattering can be carried out. In Section III, we give

---
[*]SSC Fellow

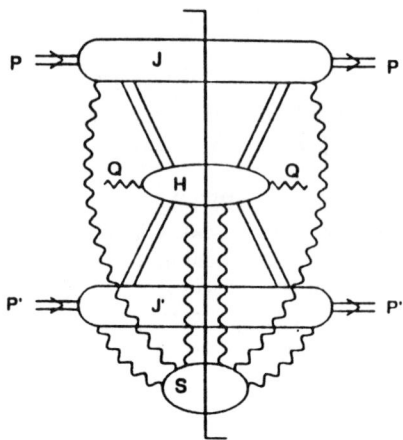

Figure 1: Reduced diagram of a general pinch surface in the Drell-Yan process.

a discussion of how nonfactorizable contributions at the first nonleading order cancel. Such first nonleading corrections decrease as $1/Q$ relative to leading power in the case of polarized hadronic scatterings, or as $1/Q^2$ in the case of unpolarized hadronic scatterings, where Q is a large momentum transfer characterizing the scale of the short-distance hard part. Finally, in Section IV we give our conclusions.

II. FACTORIZATION IN HADRONIC SCATTERING

The aim of factorization is to separate long- from short-distance effects in perturbation theory. The key is to identify the sources of long-distance behavior in Feynman diagrams, and to show that nonfactorizable long-distance contributions are canceled.

In general, a Feynman diagram of a hadronic scattering amplitude can be very complicated. Any number of lines of the diagram can be nearly on-shell and show long-distance behavior. But, in multidimensional complex momentum space, if the loop momenta associated with the on-shell lines are not "trapped" at the on-shell poles, we can deform the contour of momentum integrations far away from the on-shell poles, and make the lines effectively far off-shell. That is, following Refs. [7,8], we need be concerned only with contributions which arise from surfaces in momentum space where momentum integrals are pinched between coalescing singularities associated with Feynman propagators in the zero-mass limit. These surfaces are called the "pinch surfaces" of a Feynman diagram. Different pinch surfaces are associated with different physically realizable processes, so that hadronic scattering processes correspond to different pinch surfaces. To be specific, we will consider the Drell-Yan cross section in the following discussion. A general pinch surface for the Drell-Yan cross section is shown in Fig. 1 where Q is the invariant mass of the lepton pair.[9-11] The arbitrary "jet" subdiagrams J and J' consist of lines which are on-shell and collinear to the incoming hadron momenta p and p', respectively. All internal lines of S are "soft", with momenta which vanish in all components at the pinch surface. Lines in H are "hard", and carry momenta of order Q or larger. The explicit soft gluon lines attaching S to the J's and H represent arbitrary connections, as do the lines connecting the jets

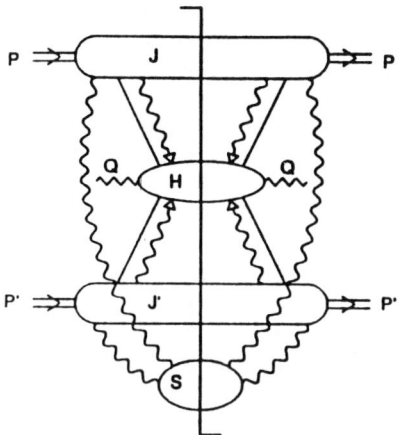

Figure 2: Reduced diagram of a leading pinch surface in the Drell-Yan process. The gluon lines ending in arrows are scalar polarized.

with the hard parts.

Different connections in Fig. 1 represent different pinch surfaces, but not every surface contributes to the cross section at the leading power in Q. In Ref. [8], the "leading" singular surfaces were identified, as shown in Fig. 2. A leading pinch surface has following properties: (i) the soft subdiagram S is connected *only* to the jets, and not directly to the hard subdiagram H. (ii) each jet is connected to the hard subdiagram H by a set of lines of which only *one* carries a physical polarization. This may be a fermion or a physically polarized gluon. In a covariant gauge, there may also be an arbitrary number of unphysical "scalar-polarized" gluons.

Even at the leading power, factorization is not trivial for hadronic processes. In general, we would expect that the leading power factorization formula for hard hadron-hadron scattering is of the form

$$\sigma_{AB}(Q^2) = \sum_{a,b} f_{a/A} \otimes H_{ab} \otimes f_{b/B} + O(\Lambda^2/Q^2), \tag{1}$$

where A and B label two hadrons, and a and b run over all parton flavors, including quarks, antiquarks and gluons. Here $\otimes$ means the convolution of the momentum fraction between hard part H and distribution f. Comparing Eq. (1) with the general *leading* pinch surface shown in Fig. 2, one can see that each of the functions in the factorization form has a direct analog in the *leading* pinch surface. H_{ab} represents the effect of the hard subdiagrams H, while the parton distributions f represent the jet subdiagrams J and J'. The effective statement of factorization is that each parton distribution f depends *only* on the identity of the corresponding *single* incoming hadron, and that the distribution can be defined exactly as in DIS, up to corrections which may be absorbed into the hard part, H_{ab}. The only thing in Fig. 2 which is not in Eq. (1) is the soft gluon subdiagram S. Soft lines which connect the two incoming jets would in general modify the parton distributions before the hard interactions between partons from the two jets can take place. Such soft interactions would destroy the universality of parton distributions,[4] and consequently, factorization would fail.

It has been shown[9-11] that to leading power, the effects of soft gluons cancel in the sum over all final states, and as a result, Eq. (1) holds. To study factorization at the first nonleading power, we have to consider the first nonleading contribution of the leading pinch surfaces, as well as the leading power contributions of the first nonleading pinch surfaces. The first nonleading pinch surfaces can be obtained by relaxing either of the two criterea for identifying the leading pinch surfaces. It is our goal to show that a similar soft gluon cancellation holds in the case of the first nonleading corrections.

III. CANCELLATION OF SOFT GLUONS

Our critereon for *nonfactoring* power-suppressed behavior is that the soft subdiagram S in Fig. 1 should connect lines in *both* incoming jets, and that these lines should be off-shell by no more than some fixed mass squared, denoted as M^2 below, such that at $M^2 = 0$ we encounter a nonleading pinch surface. In pure massless perturbation theory, power corrections are absent due to the lack of fixed masses to set the scales for nonleading terms. In practice, we expect such mass scales to be generated nonperturbatively by the breaking of chiral symmetry.

We must therefore show that after summing over all final states, nonfactoring soft gluon exchanges cancel near *leading* pinch surfaces, not only at leading, but also at first nonleading power. In addition, we must show that the leading nonfactoring behavior from *nonleading* pinch surfaces also cancels.

Two kind momentum regions of soft gluons are of special interest. They are defined by the size of soft gluons' transverse momentum, q_T. The "Central soft" region is defined as $q_T \sim M^2/Q$, while the "Glauber" region is defined as $q_T \sim M$. Both plus and minus momenta of soft gluons are of order M^2/Q. As emphasized in Refs. [12,13], both the Central soft and Glauber regions give leading power contributions. We shall show in the following that the effects of these two kinds of soft gluons cancel at both leading and the first nonleading power for hard hadronic processes.

1) Central Soft Gluons

For leading pinch surfaces, the soft subdiagram S connects *only* to the jets. According to our critereon of nonfactoring behavior above, we need to show that soft gluons can be effectively disconnected between two jets for both leading and first nonleading contributions. For simplicity, we will suppress all unnecessary indices and momentum dependence in all functions in following discussion. Implicitly, all notations are the same as in Ref. [6].

Consider first unpolarized hadronic scattering. For the region of all jet lines off-shell by no more than M^2, we find that

$$\Delta J(q) = J(q) - J^{(+)}(\tilde{q}) = O(M^2/Q^2), \tag{2}$$

where q represents soft gluon momenta. In $\tilde{q}$ only the momentum component opposite to p is kept for each gluon. We have a similar relation for J'. Because soft gluons do not connect to H, we can write the contribution of the leading pinch surface as

$$\begin{aligned} J(q) \otimes S \otimes J'(q') \otimes H &= \big[\Delta J(q) \otimes S \otimes \Delta J'(q') \\ &+ J^{(+)}(\tilde{q}) \otimes S \otimes \Delta J'(q') + \Delta J(q) \otimes S \otimes J'^{(-)}(\tilde{q}') \\ &+ J^{(+)}(\tilde{q}) \otimes S \otimes J'^{(-)}(\tilde{q}')\big] \otimes H \,. \end{aligned} \tag{3}$$

It is clear that the first term on the right-hand side of Eq. (3) is suppressed to $O(M^4/Q^4)$. The second and third terms are suppressed by a factor of M^2/Q^2, while the fourth is the leading term. The final three terms have at least one jet function replaced by its appropriate "soft approximation", $J^{(+)}(\tilde{q})$ or $J'^{(-)}(\tilde{q}')$.[11] Following Ref. [11], each such jet is automatically factorized from the soft lines, and hence from the other jet, once we sum over all connections of the soft lines. As a result, both leading and order of M^2/Q^2 corrections are factorized into three pieces: two jet functions and a hard part.

If both beam and target hadrons are longitudinally polarized, the spin vector s is not an independent vector. The cancellation of central soft gluons in leading pinch surfaces is the same as in the unpolarized case. When the beam or target is transversely polarized, the corresponding ΔJ is of order of M/Q (instead of M^2/Q^2), where the scale M is set by $k_T^{\text{jet}} \cdot s$ with k_T^{jet} the typical transverse momentum of the jet lines. If either beam or target is transversely polarized, the nonfactorizing soft gluons are suppressed by at least the order of M^3/Q^3. If both beam and target are transversely polarized, the nonfactorizing soft gluons are suppressed by at least order M^2/Q^2.

For nonleading pinch surfaces, we have two possible cases, corresponding to relaxing one of the two properties of leading pinch surfaces. Thus we have either (1) a soft gluon attaching to H, or (2) more physically polarized partons attached between the jets and H.

For the first case, the hard part H in Eq. (3) can now connect to S. The first term is again suppressed by at least $O(M^4/Q^4)$. The fourth term can be factorized into three pieces because of $J^{(+)}$ and $J'^{(-)}$. The second and third terms can be thought as the same terms as for a leading pinch surface, after moving a number of soft gluons from ΔJ (or $\Delta J'$) to H. In the case of unpolarized scatterings, moving a soft gluon from a jet function to hard part H produces a factor M^2/Q^2, because of the difference of momentum scales in the jet function and the hard part. Therefore, these two terms are suppressed by at least the order of $(M^2/Q^2)^{N+1}$, relative to the leading power, with N being the number of soft gluons attached to H. It is clear that nonfactorizing soft gluons are suppressed by at least $O(M^4/Q^4)$. Again, the conclusion is the same for the case when both beam and target are longitudinally polarized. But, when either beam or target (or both beam and target) is (are) transversely polarized, the nonfactorizing soft gluons are suppressed by at least order of $(M/Q)^{N+1}$.

For the second case, we limit ourself to the situation in which only one jet has more physically polarized partons attaching to H. The jet function with more physically polarized partons is a higher twist matrix element, and the fourth (or leading) term in Eq. (3) is already suppressed by a factor $(M/Q)^N$ compared with the same term of a leading pinch surface. Here $N = 2$ when the jet is unpolarized or longitudinally polarized, and $N = 1$ when the jet is transversely polarized. Applying the soft approximation to the other jet, one can separate the other jet from S, and hence from the jet with more physically polarized partons. The fourth term is thus factorized.

2) Glauber Exchange

From Ref. [6], we need consider only the spectator-spectator interactions in this region. The arguments for cancellation are based on analyticity and unitarity, and do not depend on symmetry arguments. Therefore, the cancellation of Glauber exchange at the first nonleading order in polarized hadronic processes should work the same way as in unpolarized case.

We conclude that central soft gluons, as well as Glauber soft gluons, will not destroy factorization, at least at the first nonleading contributions to cross sections of hadronic scatterings. Such contribuions are order $(M/Q)^N$ corrections relative to the leading power, where $N = 2$ for unpolarized or longitudinally polarized hadronic scattering, and $N = 1$ for transversely polarized scattering.

IV. CONCLUSIONS

Having shown that both kinds of leading soft gluons do not introduce any non-factorizing contribution at the first nonleading order to cross sections for hard hadronic scattering, we see no major obstacle for the factorization of such first nonleading corrections. Our conclusion is consistent with the known failure of naive factorization at order of $1/Q^4$ by noncancelling infrared divergences at two loops.[14,15] It is also consistent with heuristic arguments based on classical eletrodynamics.[15] Having gained confidence in factorization, reliable and consistent calculations and predictions can be obtained.[3]

In closing, we give the factorization formula for cross sections of hadronic scatterings

$$\sigma(Q) = H^0 \otimes f_2 \otimes f_2 + \left(\frac{1}{Q}\right)^N H^1 \otimes f_2 \otimes f_{2+N} \qquad (4)$$

where $N = 2$ for unpolarized and longitudinally polarized scatterings, while $N = 1$ for transversely polarized scatterings. In Eq. (4), H^0 and H^1 are perturbatively calculable coefficient functions, the f_n's, with $n = 2, 3, 4$, are the twist-n matrix elements measured in DIS.[3]

Acknowledgement

The research was supported in part by the National Science Foundation under grant No. PHY-89-08495.

REFERENCES

1. R.L. Jaffe and M. Soldate, Phys. Lett. 105B, 467 (1981), Phys. Rev. D26, 49 (1982); R.K. Ellis, W. Furmanski and R. Petronzio, Nucl. Phys. B207, 1 (1982), B212, 29 (1983); R.L. Jaffe, Nucl. Phys. B229, 205 (1983); J. Qiu, Phys. Rev. D42, 30 (1990).
2. E.L. Berger and S.J. Brodsky, Phys. Rev. Lett. 42, 940 (1979).
3. J. Qiu and G. Sterman. Stony Brook Preprint, ITP-SB-90-49, Nucl. Phys. (in press).
4. J.C. Collins, D.E. Soper and G. Sterman, in "*Perturbative Quantum Chromodynamics*", A.H. Mueller Editor, World Scientific, Singapore, 1989.
5. H.D. Politzer, Nucl. Phys. B172, 349 (1980).
6. J. Qiu and G. Sterman. Stony Brook Preprint, ITP-SB-90-50, Nucl. Phys. (in press).
7. S. Coleman and R.E. Norton, Nuovo Cim. 38, 438 (1965); G. Sterman, Phys. Rev. D17, 2773 (1978), D17, 2789 (1978).
8. S. Libby and G. Sterman. Phys. Rev. D18, 3252 (1978), D18, 4737 (1978).
9. G. Bodwin, Phys. Rev. D31, 2616 (1985).
10. J.C. Collins, D.E. Soper and G. Sterman, Nucl. Phys. B261, 104 (1985).
11. J.C. Collins, D.E. Soper and G. Sterman, Nucl. Phys. B308, 833 (1988).
12. J.C. Collins and G. Sterman, Nucl. Phys. B185, 172 (1981).
13. G. Bodwin, S.J. Brodsky and G.P. Lepage, Phys. Rev. Lett. 47, 1799 (1981).
14. R. Doria, J. Frenkel, and J.C. Taylor, Nucl. Phys. B168, 93 (1980); C. Di'Lieto, S. Gendron, I.G. Halliday and C.T. Sachrajda, *ibid*. B183, 223 (1981); F.T. Brandt, J. Frenkel, and J.C. Taylor, *ibid*. B312, 589 (1989).
15. R. Basu, A.J. Ramalho and G. Sterman, Nucl. Phys. B244, 221 (1984).

COLOR TRANSPARENCY IN NUCLEUS–NUCLEUS COLLISIONS.

Leonid Frankfurt and Mark Strikman
Department of Physics, University of Illinois at Urbana-Champaign
1110 W. Green Street, Urbana, IL 61801 [†,±]

ABSTRACT

We predict that probability for a nucleon to traverse nucleus in the central collisions should noticeably increase with energy increase between BNL and CERN energies due to color screening phenomenon. At RHIC energies this effect should practically disappear if perturbative QCD scenario is applicable.

It is widely discussed now that QCD exhibits the following fundamental property: the smaller the spatial radius (transverse size) of white configuration, the smaller its interaction with the target. (This can be considered as another form of the renormalization group invariance of the strong interactions.) The recent theoretical papers predominantly centered on the options for studying this property of QCD via color transparency phenomenon[1-2], which predicts that cross section of large angle quasielastic reaction $1(h) + A \to 1(h) + N + (A-1)^*$ should be proportional to A provided elementary hard amplitude is dominated by the contribution of small size (point-like) configurations, PLC, in colliding hadrons. Here $(A-1)^*$ is a system of nucleons and nuclear fragments which contains no pions. However, it was demonstrated in [3] that expansion of PLC to normal size configurations would mask to large extent the color transparency phenomenon up to very large energies (momentum transfer). Basically, this is because the momentum of the ejected nucleon is about $-t/2m_N$ for large momentum transfer to the target nucleon $|t|$, while condition that hadronic configuration with mass M_{PLC} can be considered as frozen during passage of the nucleus is

$$2p_N/(M_{PLC}^2 - m_N^2) > 2 R_A \tag{1}$$

Since PLC is superposition of hadronic states $|n\rangle$ of different masses:

$$\Psi_{PLC} = \sum_n \Psi_n \exp\left(i \frac{M_n^2 - m^2}{2p} t\right) \tag{2}$$

evidently $M_{PLC} > M_{N^*}$, where N^* is the resonance with nucleon quantum numbers. This corresponds to $-t > 12$ GeV^2 A$^{1/3}$.

© 1991 American Institute of Physics

Consequently, when studying the color transparency phenomenon question of presence of PLC in hadrons has been intermixed with two interesting but different questions: 1) whether the elementary reaction is dominated by PLC; ii) whether the hadron energies are high enough to consider PLC as frozen during the interaction process. Therefore, study of these reactions would allow, at best, to reveal existence of the tiny fraction of all small size (weakly interacting) configurations in hadrons. At the same time in ultrarelativistic heavy ion collisions (CERN, RHIC) the collision energy is sufficiently large so PLCs in nucleons are really frozen in the process of collision.

The main aim of this talk is to suggest an experimentally feasible method of measuring the total probability of PLC in nucleons via study of spectator nucleon production in untrarelativistic central collisions of light nuclei with heavy nuclei at the energy range available now in CERN. We will argue also that studies of this reaction at higher energies to be available at RHIC would give unique information on the origin of Pomeron. We also point out that nucleus-nucleus collisions allow to enhance sensitivity to the color transparency effects.

Our discussion stems from the experimental observation[4] that it is possible to device effective triggers for events corresponding to central collisions of light (A) and heavy (B) nuclei. In the traditional approach which ignores PLC in hadrons probability to produce in such a collision, a nucleon spectator in the light nucleus fragmentation region (i.e. nucleon with momentum fraction $\alpha = Ap_N/p_A \geq 1$) is very small. In the Glauber model approach it is as follows:

$$P_{spect}/A \approx \exp(-\sigma_{inel}^{NN} \rho_B 2R_B), \text{ i.e. } P_{spect}/A \sim 6.10^{-4} \text{ for } B = 240 \quad (3)$$

where $\rho_B \approx \rho_O$, is the mean nuclear density and R_B is the radius of the nucleus B. However, if inelastic nucleon-nucleon cross section arises as a superposition of interactions of nucleon in configurations of different transverse size:

$$\sigma_{inel}^{NN} = \int \Psi_N^2(r_t^2) \sigma_{inel}^{NN}(r_t^2) d^2r_t \quad (4)$$

P_{spec}/A may be noticeably larger. Indeed in the eikonal approximation the total probability for a transition of a nucleon to a state X while passing through the nucleus at small impact parameter b without inelastic interaction is given by:

$$W^{N \to X}(b) = \left\{ \int \Psi_N(r_t^2) \Psi_X(r_t^2) \exp\left[-\sigma_{inel}^{hN}(r_t^2) T(b)/2\right] d^2r_t \right\}^2 \quad (5)$$

where $T(b) = \int_{-\infty}^{\infty} \rho_B(z,b)\,dz \approx 2.2\,\rho_0 B^{1/3}$ for $b \approx 0$, and $B \gg 1$.

For the wave function of the nucleus B, the approximation of independent particles has been used. $W^{N \to X}(b)$ is the probability for the propagation of configuration of the transverse size r_t^2 without inelastic interactions and it accounts for the overlapping integral between outgoing system and state X (Evidently X is a state with longitudinal momentum $|Ap/p_A - 1| \leq k_F/m_N$ and with nucleon quantum numbers.) All other nucleons in these events interact inelastically. In particular the probability of transition to a nucleon is:

$$W^{N \to N}(b) = \left[\int \Psi_N^2(r_t^2) \exp\left[-\sigma_{inel}^{NN}(r_t^2)\,T(b)/2\right] d^2 r_t \right]^2 \quad (5')$$

The total probability of the diffraction can be derived from eq. 4 using closure over the states X:

$$W_{tot}^{N \to X}(b) = \int \Psi_N^2(r_t^2) \exp\left[-\sigma_{inel}^{NN}(r_t^2)\,T(b)\right] d^2 r_t \quad (6)$$

Let us first consider several qualitative consequences of eqs. (5), (6).

i) Eq. 5' predicts a noticeably larger number of spectators than eq. 2. (We will give some numerical estimates below.)

ii) In line with eq. 1, one should expect increase of the spectator yield at energies where

$$2p_A/\{A(M_{PLC}^2 - m_N^2)\} > 2R_B, \text{ i.e. } p_A/A \geq 9 \text{ GeV/c } B^{1/3} \quad (7)$$

in the reference frame where nucleus B is at rest. This increase should be saturated at the energies where essential small size (large mass configurations) in nucleon can be considered as frozen. Since the probability of PLC in a nucleon rapidly decreases with decrease of r_t^2 one may expect that configurations with $M_{PLC} > 2\text{-}3$ GeV don't contribute significantly to $W^{N \to N}$. Thus the saturation should be reached at $p_A/A \leq 40$ GeV/c $B^{1/3}$.

Therefore the presence of PLC in nucleons should lead to small modifications of Eq. 2 at the energies currently studied at BNL, though it may be observed in the energy range available now at CERN. Evidently the study of the energy dependence of the number of spectators would provide by product fairly direct information on the structure of the nucleus wave function.

iii) It is evident from the comparison of eqs. 4 and 5 that $W^{N \to X}_{tot} \gg W^{N \to N}$, i.e. cross section of diffraction production of nucleon resonances and continuum with nucleon quantum numbers should be much larger than cross section of leading nucleon production. (This enhancement of diffraction resembles phenomenon of hard pion diffraction off nuclei first discussed in Ref. 5.) Though the measurement of the production of nucleon resonances in the central nucleus-nucleus collisions seems to be a much more difficult task than studying of nucleon production, it is not hopeless because production of leading pions in the central nucleus-nucleus collisions is strongly suppressed. Estimate of this background is out of the scope of this talk.

To be more specific let us make some simple numerical estimates. In any quark model of a hadron h (consisting of n quarks) for small r_t^2

$$\Psi_h^2(r_t^2) \sim (r_t^2)^{n-2},$$

mostly due to the phase volume factor and lack of singularity of the wavefunctions of a hadron at $t \to 0$. To take into account contribution of configurations which size is comparable to the average size we assume that transition from average to PL configurations is being smooth. Since at large r_t^2 $\Psi_N^2(r_t^2)$ should decrease rather fast we choose

$$\Psi_N^2(r_t^2) = \frac{1}{\pi} \exp(-r_t^2/r_c^2) \, r_t^2/r_c^4, \text{ with } r_c \sim 0.5f \text{ (to get } <r_t^2>^{1/2} \sim 0.65f)$$

leading to

$$W^{N \to N}_{tot} = 4/(z+2)^2 \text{ and } W^{N \to N} = 16/(z/2+2)^4,$$

where $z = T(b)\sigma_{inel}\rho_0$. For B = 240 this equation leads to $W^{N \to X} = 0.04$ and $W^{N \to X} = 0.015$. The last number should be compared with the BNL experimental number 4 for $^{28}S + ^{208}Pb$ central collisions of $P_{spect}/A \leq 2 \cdot 10^{-3}$. (Obviously the background from the processes when one of the nucleons of nucleus A traverses through the edge of heavy nucleus B increases with increase of A). Thus, this estimate indicates that quite significant enhancement of the spectator nucleon yield is possible between BNL and CERN energies.

Note that similar effects are present in reaction like

A + B → leading nucleon + (massive 1^+ 1^- pair),

since the Dress-Yan trigger also enhances central collisions. However, in this case peripheral collisions give noticeable contribution and therefore the predicted deviation from the Glauber picture is considerably smaller.[6] Besides, one loses also on the small cross section of the Dress-Yan process.

When one would go to yet another energy scale of $s^{1/2} > 100$ GeV to be available at RHIC a new change of the pattern may be expected. Since the σ_{inel} increases very significantly at these energies (~60 mb) the soft interaction of nucleon would be more filtered away. It may help to observe another property of QCD. Really the effective strong coupling constant decreases with the size of PLC (asymptotic freedom). So the interaction of a nucleon in PLC with nuclear matter becomes more close to the perturbative regime. However, the distinctive property of pQCD as understood now (cf. discussion in [7]) that the cross section rather rapidly increases with initial energy ν: $\sigma \sim \sigma_0 (\nu/\nu_0)^{12\alpha_s/\pi}$.[8] The net result decreases with the increase of ν of $W^{N \to X}_{tot}$ and $W^{N \to N}$ at RHIC energies.

The method discussed in this paper could be used also to reach better sensitivity in the "classical" color transparency experiment $p+A \to p+p+(A-1)^*$ by using as a projectile light nucleus, $A+B \to pp +X$, measuring the pp in the same quasielastic kinematics and looking for events corresponding to the central collision trigger. In this way the peripheral contribution to $pA \to pp(A-1)$ which dominates in a wide range of incident energies[3] will be strongly suppressed and small color transparency effects would be enhanced.

It is worth also noting that the enhancement of the resonances production in the diffraction processes off heavy nuclei due to filtering of PLC is quite general phenomenon. For example, it shouldreveal itself also in diffractive photoproduction as an increase with atomic number of the ratio of the yields of ϕ'/ϕ, Ψ'/Ψ, etc. It follows from Eq. 2 that in any reactions where PLC dominates

$$\frac{\sigma(N^*)}{\sigma(N)} = \left[\frac{\Psi_N^*(0)}{\Psi_N(0)}\right]^2 \qquad (8)$$

We use here decomposition of δ-function over eigen states of the hamiltonian $\delta(r) = \sum_n \Psi_n(0) \Psi_n(r)$.

To summarize, the study of nucleon diffraction and spectator nucleon production in the central light nucleus–heavy nucleus collisions over all energy range to be available in the next decade could give valuable new information on the strong interaction. It appears that these experiments would be feasible both at CERN and at RHIC (though in the later case a better coverage of small angles than currently discussed would be necessary.)

We would like to thank J. J. Stachel for useful questions which stimulated us to return to this problem. We are thankful to A. Mueller and G. Farrar for stimulating discussions. One of us (M. S.) would like to thank the theory division of BNL for hospitality during the period when part of this work has been done.

REFERENCES

† On leave of absence from Leningrad Nuclear Physics Institute.

± The work has been supported by the National Science Foundation, Grant No. PHY 89-21025.

1. A. Mueller, in Proceedings of the Seventeenth Recontre de Moriond, Les Arcs, France (1982), edited by J. Tran Thanh Van (Editions Frontieres, Gif-sur-Yvette, 1982), and Columbia University Report No. CU-TP-232, 1982 (unpublished).
2. S. Brodsky, in Proceedings Thirteenth International Symposium on Multiparticle Dynamics, Volendavm, The Netherlands, 6-11 June 1982, edited by E. W. Kittle, W. Metzger, and A. Stergion (World Scientific, Singapore, 1982).
3. G. R. Farrar, H. Liu, L. L. Frankfurt, M. I. Strikman, Phys. Rev. Lett. 61, 686 (1988).
4. J. Barrette er al., Phys. Rev. Lett., 64, 1219 (1990).
5. J. Bertch, S. J. Brodsky, A. S. Goldhaber and J. G. Gunion, Phys. Rev. Lett. 47, 297 (1981).
6. L. L. Frankfurt and M. I. Strikman, Nucl. Phys. B250, 143 (1985).
7. A. H. Mueller, Columbia University Report CU-TP-475, (1990) (to be published).
8. E. A. Kuraev, L. N. Lipatov and V. S. Fadin, Sov. Phys. JETP 45, 199 (1977).

The Proton Spin and the Gluon Anomaly[1]

Jeffrey E. Mandula
Department of Energy
Division of High Energy Physics
Washington, DC 20545

Everyone at this conference is surely aware of the important experiment on polarized muon-proton scattering carried out by the European Muon Collaboration (EMC).[2] The experiment measured the spin asymmetry in deep inelastic muon scattering from protons. The key result was interpreted as indicating that essentially none of the proton's spin was carried by the spins of its quarks. A reexamination of this interpretation held that the apparent vanishing of the quark spin contribution could be due to a cancellation between an intrinsic quark contribution and a gluonic contribution induced by the triangle diagram that gives rise to the Adler-Bell-Jackiw anomaly.

In this talk we review the experimental and theoretical situation, and then describe a calculation of the induced, anomalous gluonic contribution by means of a lattice QCD simulation. We will discuss some necessary theoretical and lattice preliminaries to the calculation, and the algorithm that was used. We will also mention some important issues having to do with gauge invariance. The main conclusion that we report here is that, based on the lattice calculation, the anomalous gluonic contribution is much too small to change the interpretation of the EMC experiment. Finally we will address some of the limitations of the calculation, and provide an estimate of its accuracy.

The principal result of the EMC experiment is that the integral of the polarized structure function g_1 is

$$\int_0^1 g_1(x,q^2 = -10.7 \ GeV) \, dx = \frac{1}{2}(\frac{4}{9}\Delta u + \frac{1}{9}\Delta d + \frac{1}{9}\Delta s) = .126 \pm .010 \pm .015 \qquad (1)$$

where the statistical and systematic errors are shown separately. The quoted systematic error of $\pm .015$ includes an estimate of the uncertainty due to the need to extrapolate the measured value of the structure functions to the regions above $x = .7$ and below $x = .01$.

The scaling behavior of g_1 and its connection to polarized proton quark spin fractions is known from operator product expansion analysis.[3] In particular, g_1 has at most a logarithmic dependence on q^2, and in the deep inelastic limit its integral is a sum of axial current matrix elements between proton states of polarization s:

$$4 m s_\mu \int_0^1 g_1(x) \, dx = \sum Q_i^2 \langle ps | \bar{q}_i \, i \gamma_\mu \gamma_5 \, q_i | ps \rangle \qquad (2)$$

The axial currents are the canonical spin operators for each quark, so their forward matrix elements are the fractions of the proton's spin carried by each kind of quark,

$$\langle ps | \bar{q}_i \, i \gamma_\mu \gamma_5 \, q_i | ps \rangle = 2 m s_\mu \Delta q_i \qquad (3)$$

The significance of the EMC measurement stands out when combined with information from semi-leptonic axial weak decays of neutrons and hyperons. Specifically, neutron β-decay plus isospin symmetry gives the Bjorken sum rule[4]

$$\Delta u - \Delta d = g_A = 1.254 \pm .006 \qquad (4)$$

while strangeness changing hyperon decay and flavor SU(3) symmetry gives[5]

$$\Delta u + \Delta d - 2 \Delta s = .60 \pm .12 \qquad (5)$$

Putting this all together gives

$$\begin{aligned} \Delta u &= +.74 \pm .05 \\ \Delta d &= -.51 \pm .05 \\ \Delta s &= -.19 \pm .07 \end{aligned} \qquad (6)$$

and

$$\Delta u + \Delta d + \Delta s = +.04 \pm .16 \qquad (7)$$

The statistical and systematic errors from the EMC result have been combined in quadrature. The errors in Eqs. (6) and (7) are all dominated by the error on the EMC result.

The result contained in Eq. (7), which can be phrased as saying that only a small fraction, if any, of the proton's spin is carried by the spins of its constituent quarks was completely unanticipated, as was the conclusion that the strange quark's contribution was comparable to that of the down quark.

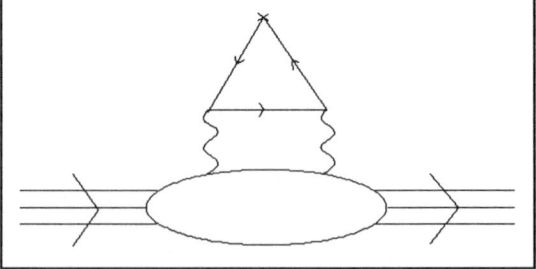

Figure 1 The triangle diagram, which induces an effective gluonic contribution to each quark spin distribution in the proton.

The interesting suggestion for understanding or reinterpreting) the EMC result by separating out an anomalous gluonic contribution was made by Efremov and Teryaev,[6] Altarelli and Ross,[7] and Carlitz, Collins, and Mueller.[8] They observed that there is a contribution to each Δq_i associated with the gluonic spin distribution inside the proton which comes from the triangle diagram (see Figure 1) that is the source of the Adler-Bardeen-Jackiw anomaly.[9] The evaluation of this contribution requires a choice of gauge and a careful treatment of high momentum limits. In axial gauges, at least, it comes entirely from infinite loop momenta around the triangle, so that the gluon vertices are effectively contracted to a point. In the (canonical) $A_0=0$ gauge, this gluonic spin contribution to each quark axial current matrix element is the matrix element of the anomalous current

$$\langle ps|[j^5_\mu]_{Triangle}|ps\rangle = \frac{\alpha}{2\pi}\langle ps|K_\mu|ps\rangle = -2m\, s_\mu\, \frac{\alpha}{2\pi}\Delta g \tag{8}$$

which is the canonical gluon spin operator in that gauge.[5] We identify $-\frac{\alpha}{2\pi}\Delta g$ as the gluonic part of each quark spin fraction.

The fact that neither the isolation of this gluonic contribution nor even the final formula can be expressed in gauge invariant terms has been criticized and its physical interpretation questioned.[5,10,11] An important issue in that connection is the possible presence of gauge-variant singularities in the matrix element $\langle p+q/2\; s|K_\mu|p-q/2\; s'\rangle$.

In fact, such singularities are not likely to be present. The reason is that the conditions that are known to allow singularities in gauge-variant matrix elements, for example in the Landau-gauge Schwinger model, are absent from axial gauges. Specifically, the phenomenon of having singularities that contribute to gauge-variant amplitudes but cancel in gauge-invariant amplitudes, known as "Kogut-Susskind dipoles",[12] requires that there be ghost (negative probability) states. In the Schwinger model, which is the paradigm for this phenomenon, there is a marked contrast between the Landau and axial gauges. In the Landau gauge, which has ghosts, there are Kogut-Susskind dipoles, and they make a singular contribution to the anomalous current at zero momentum transfer. In the axial gauge, which is ghost free, there are no Kogut-Susskind dipoles, and no singularities in the anomalous current. In QCD, the Landau gauge does contain ghosts, but axial gauges, and most obviously the temporal axial gauge, do not have ghost states.

This is an important issue because the forward matrix element $\langle ps|K_\mu|ps'\rangle$ can depend on the gauge parameter η only if there are $q\to 0$ singularities. The reason is that under an infinitesimal gauge transformation ω, the change in the operator K_μ is

$$\delta K_\mu = \frac{2}{g}\epsilon_{\mu\nu\lambda\sigma}\partial_\nu[Tr\,\omega\,\partial_\lambda A_\sigma] \tag{9}$$

Barring $q\to 0$ singularities, forward matrix elements of this operator vanish because of the explicit factor of q_ν coming from the overall derivative. Note that an infinitesimal gauge transformation ω can always be found which effects an infinitesimal change $\delta\eta$ in the gauge direction parameter η.

These arguments can be explicitly illustrated by calculating the lowest order contribution to the forward matrix element of the anomalous current in a general axial gauge. That is the triangle diagram shown in Figure 2. Using the axial gauge gluon propagator,

$$S_{\mu\nu}(k) = \frac{1}{k^2}\left[g_{\mu\nu} - \frac{k_\mu \eta_\nu + \eta_\mu k_\nu}{\eta \cdot k} - \frac{\eta^2 k_\mu k_\nu}{(\eta \cdot k)^2}\right] \quad (10)$$

this contribution to the forward matrix element,

$$\epsilon_{\mu\nu\sigma\tau}\int \frac{d^4k}{(2\pi)^4} S_{\nu\alpha}(k) k_\sigma S_{\tau\beta}(k) \bar{u}_{ps}\gamma_\alpha \frac{\not{p}+\not{k}+m}{(p+k)^2-m^2}\gamma_\beta u_{ps} \quad (11)$$

can be seen by direct calculation to be independent of the gauge parameter η. This is also not a proof, but it reinforces the conclusion of the previous remarks that the forward matrix element $\langle ps|K_\mu|ps'\rangle$ is the same in all axial gauges, whatever the direction of the gauge parameter η.

That the triangle diagram meaningfully isolates a gluonic contribution hidden in a quark operator matrix element has not been rigorously proved in any sense. However, the fact that the sort of singular behavior that would interfere with such an interpretation seems most unlikely to be present, and so we regard the meaningful isolation of a gluonic component of the proton's spin by the triangle diagram as more than plausible. We will accept it and use it as the basis of the following calculation and analysis.

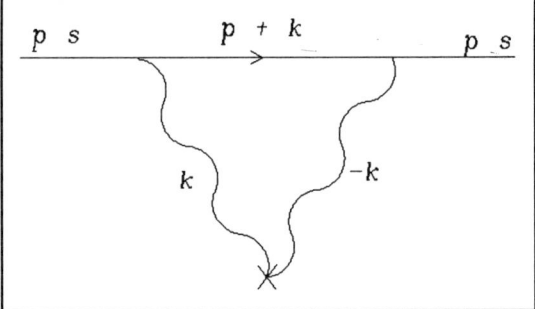

Figure 2 The lowest order contribution to the forward matrix element of the anomalous current between fermion states.

In the remainder of this lecture we describe a lattice gauge simulation of Δg by evaluating the $\langle ps|K_\mu|ps\rangle$ matrix element. We use the standard formulation: Wilson's SU(3) lattice action on a periodic Euclidean lattice to describe the gauge fields and (4-spin × 3-color component) Wilson fermions to describe the quarks. We work in the quenched approximation, which omits closed quark loop effects. Our calculation uses lattices generated by Bernard, Hockney, and Soni for a different purpose — the calculation of hadronic matrix elements that enter into weak interaction rates.[13] They made available for this purpose an ensemble of 204

$6^3 \times 10$ lattices with $\beta = 5.7$ together with quark propagators computed through 8 units of Euclidean time both forward and backwards. The quark propagators satisfy open (Neumann) boundary conditions with hopping constant $\kappa = .162$.

A key issue which arises in this calculation is the formulation of a suitable lattice approximation to the anomalous current. For the reasons discussed earlier, we work in the gauge $A_4=0$, which we implement on each lattice by gauge transforming all timelike link variables (except at a single reference times outside the quark propagator range) to $U_4=1$. In the continuum, the anomalous current simplifies in this gauge to

$$K_i = 2\, \epsilon_{ijk}\, \mathrm{Tr}\, A_j\, F_{k4} \qquad (12)$$

A convenient lattice approximation to F_{k4} is[14])

$$F_{k4}{}^{Latt}(n) = \frac{1}{8i}\,[U_k(n)U_k{}^\dagger(n+\hat{4}) + U_k{}^\dagger(n+\hat{4}-\hat{k})U_k(n-\hat{k})$$
$$+ U_k{}^\dagger(n-\hat{k})U_k(n-\hat{4}-\hat{k}) + U_k(n-\hat{4})U_k{}^\dagger(n)] \qquad (13)$$

$$- \text{Hermitean Conjugate}]_{Traceless}$$

It has the correct continuum limit, transforms simply under the lattice hypercubic symmetry group, and depends under gauge transformation only on the value of the gauge transformation at site n. With this expression for the field strength a convenient lattice approximation to the anomalous current is

$$K_i{}^{Latt}(n) = \mathrm{Im}\, \epsilon_{ijk}\, \mathrm{Tr}\, U_j(n\hat{j})\, F_{k4}{}^{Latt}(n)\, U_j(n) \qquad (14)$$

Like $F_{k4}{}^{Latt}(n)$, it has the correct continuum limit and transforms simply under the hypercubic group. Under infinitesimal gauge transformations its change depends on the difference of the values of the gauge transformation at sites $n + \hat{j}$ and $n - \hat{j}$ ($j \neq i$).

This is not a perfect approximation to the anomalous current, however, and some important properties are only recovered in the continuum. These include its response to x_4 independent gauge transformations ω,

$$\delta K_i \to \partial_4\, \epsilon_{ijk}\, Tr\, [\partial_j \omega\, A_k] \qquad (15)$$

and the implied vanishing of the proton expectation value of the gauge transformation in K_i

$$\langle ps | \delta K_i | ps' \rangle \to 0 \qquad (16)$$

Other important properties of K_μ that are only recovered in the continuum are the gauge invariance of its divergence and its relation to topology changing gauge transformations.

An alternative method for evaluating Δg has the advantage that it is (almost) manifestly gauge invariant on the lattice. This algorithm also needs zero spacing limit to recover all the continuum properties. It has the disadvantage that its statistical noise is larger than that of the axial gauge algorithm. The basic idea of the algorithm is that the space components of the forward matrix elements of K_μ can be inferred from those of $\partial_\mu K_\mu$ by means of the limit

$$\langle \vec{0} s | K_i | \vec{0} s' \rangle = \lim_{\vec{p}=p\hat{i} \to \vec{0}} \frac{i}{p} \langle \vec{0} s | \partial_\mu K_\mu | \vec{p} s' \rangle \tag{17}$$

where

$$\partial_\mu K_\mu = \frac{1}{2} \epsilon_{\mu\nu\lambda\sigma} Tr\, F_{\mu\nu} F_{\lambda\sigma} \tag{18}$$

Only space components are recovered, but K_0 does not enter into the spin fraction, so this is not a problem.

To implement this on the lattice we define

$$\Delta_\mu K_\mu^{Invariant} = \frac{1}{2} \epsilon_{\mu\nu\lambda\sigma} Tr\, F_{\mu\nu}^{Latt} F_{\lambda\sigma}^{Latt} \tag{19}$$

and take the momentum p to be the smallest possible on the finite lattice along each direction

$$\langle \vec{0} s | K_i | \vec{0} s' \rangle = \frac{i}{2p}(\langle \vec{0} s | \Delta_\mu K_\mu | \vec{p} s' \rangle - \langle \vec{0} s | \Delta_\mu K_\mu | -\vec{p} s' \rangle) \tag{20}$$
$$+ O(p^2) \qquad (\vec{p} \parallel \hat{i})$$

The two definitions of K_μ are not equal on the lattice, although they become equal in the continuum limit.

Of course, since K_μ is not perfectly gauge invariant, there must be some gauge dependence remaining in the covariant algorithm. That residual gauge invariance is contained in the tacit assumption that there are no unphysical singularities in matrix elements of K_μ which would interfere with the momentum limit. As we discussed earlier, such singularities are expected to be absent from axial gauges, and so what we have called a gauge invariant algorithm is more properly called a general axial gauge algorithm.

We project out the polarization asymmetry by tracing with $i\gamma_5 \rlap{/}s$, and divide by the trace of the propagator to remove the leading exponential decay, giving the convenient expression

$$\Delta g = \frac{Tr\, P_+ i\gamma_5 \rlap{/}s \langle 0 | \Psi(x_4)\, \vec{s}\cdot\vec{K}(y_4) \bar{\Psi}(z_4) | 0 \rangle}{Tr \langle 0 | \Psi(x_4) \bar{\Psi}(z_4) | 0 \rangle} \quad (21)$$

which is independent of the normalization of the proton field. It is from this formula that we extract the value of Δg in computer simulations. To improve statistics we average over the three principal axis choices for $\vec{s}$.

In a lattice numerical simulation, the numerator of Eq. (21) measures the correlation, lattice by lattice, of two quantities which are fluctuating about 0. One is the polarization asymmetry of the proton propagator, measured by

$$Tr\, P_+ i\gamma_5 \rlap{/}s\, \Psi(x_4) \bar{\Psi}(z_4) \quad (22)$$

which is composed only of fermion operators, while the other is the $\vec{s}$ component of the anomalous current,

$$\vec{s}\cdot\vec{K}(y_4) \quad (23)$$

which is composed only of the bosonic, gauge field variables. The correlation between these fluctuations arises because, lattice by lattice, the polarization asymmetry of the proton propagator is determined by the values of the same gauge variables, through the lattice Dirac equation, that comprise the lattice anomalous current operator.

It should be noted that this aspect of the calculation is the same in both the quenched approximation and the full theory. The effect of including quark loops is only to modify the relative weighting of different gauge configurations. The core of the calculation is still the computation of gluon variable with quark variable fluctuation correlations.

We "measured" the gluon spin fraction Δg of Eq. (21) using the sample of 204 $6^3 \times 10$ lattices provided by the Bernard-Soni collaboration. With the parameters of those lattices, $\beta = 5.7$ and $\kappa = .162$, the lattice spacing is about 1.0 GeV^{-1} and the proton mass about 1.4 inverse lattice spacings.

We constructed a color singlet proton field from colored quark fields by projecting on the appropriate spin and symmetry. To extract Δg using Eq. (21), we should take $x_4 - y_4$ and $y_4 - z_4$ as large as possible. It is also desirable to avoid the largest values of $x_4 - z_4$, since the quark propagator will be most affected by the boundary condition in that case. The best compromises available are $x_4 - z_4 \leq 7$ with $x_4 - y_4 \geq 3$ and $y_4 - z_4 \geq 3$. The values of Δg are extracted from the forward and backward propagation separately, and are given along with their statistical errors in the two Tables.

Table I. Δg Using Axial Gauge Construction

Separations		Δg	
$x_4 - y_4$	$y_4 - z_4$	Forward	Backward
4	3	−.158 ± .671	+.008 ±.191
3	4	−.292 ± .367	+.009 ±.124
3	3	−.097 ± .397	+.001 ±.114

Table II. Δg Using Gauge Invariant Construction But Finite Momentum Approximation

Separations		Δg	
$x_4 - y_4$	$y_4 - z_4$	Forward	Backward
4	3	−0.83 ± 2.58	+0.07 ± .80
3	4	+2.89 ± 3.58	+0.07 ± .72
3	3	−0.15 ± 1.42	−0.08 ± .46

Several comments are in order regarding the quality of these results. One is that they are all mutually consistent, and consistent with 0. That is, we have not arrived at a value for Δg, but rather a bound, $|\Delta g| \leq .5$. Another is that the finite lattice spacing and lattice size are serious but probably not fatal limitations. The proton mass is almost 50% too large on these lattices, for example. In the zero spacing and large time limits, the value of Δg gotten from the simulation would be independent of the exact values of $x_4 - y_4$ and $y_4 - z_4$, so long as both were large enough. Within the statistics, there is no evidence that this is failing, and the much closer than statistics self-agreement of the forward and backward results separately indicates that the finite spacing might not be much too coarse. Similarly, quenched approximation results in other calculations, which like this one would not be structurally different in the full theory, are usually reasonable, and have not yet been seen to be qualitatively wrong.

Finally, at the fairly small values of $x_4 - y_4$ and $y_4 - z_4$ at which we worked (either one or both were $\Delta t = 3$), one should ask how significant is the contribution of high mass states with the same quantum numbers as the proton to the three-point Green's function from which we extracted Δg. We can estimate this by examining an extrapolation of the simulated proton propagator from the largest values of Δt, at which the single proton contribution is purest, back to $\Delta t = 3$ or 4, and seeing by how much of the simulated propagator exceeds the extrapolation. This is shown in Figure 3.

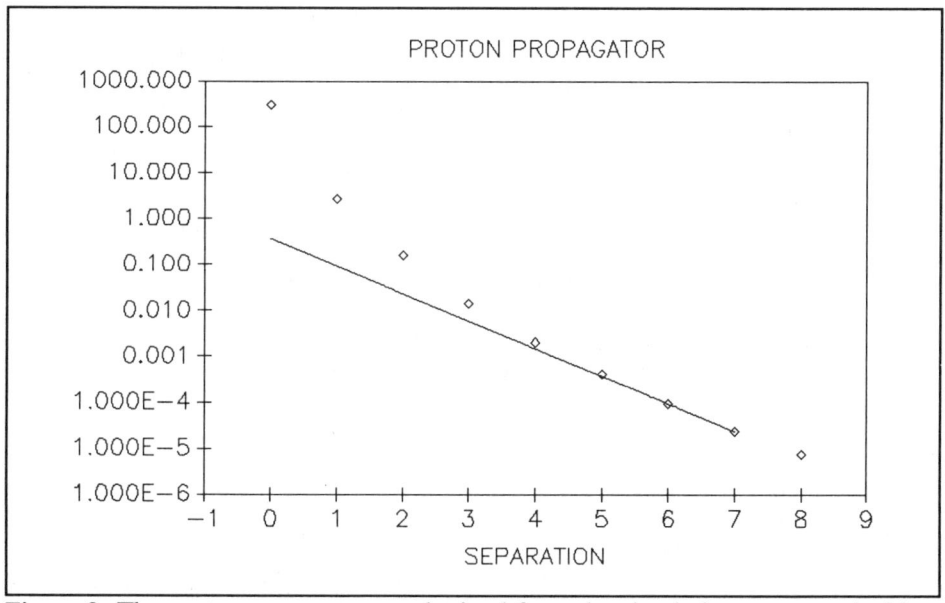

Figure 3 The proton propagator as obtained from the simulation, compared with the extrapolation back from $\Delta t = 6$ and $\Delta t = 7$.

At $\Delta t = 4$ the admixture of higher states is about 14%, and at $\Delta t = 3$ it is about 40%. This is comparable to the other sources of error. The overall effect of all these considerations is to weaken the bound coming from the simulation by about a factor of 2.

The basic conclusion of the foregoing analysis is that the gluon spin fraction Δg is apparently less than 1 in magnitude. Since at $\beta = 6/g^2 = 5.7$ the QCD coupling is $\alpha = g^2/(4\pi) \approx .08$, this gives a bound on the correction to the total quark spin fraction of the proton's spin of $(3\alpha/2\pi)|\Delta g| \leq .05$. If, following Refs. 7-9, we modify Eq. (7) by removing the anomalous gluonic contribution to the total quark spin fraction of the proton, the result would be changed by at most .05, and each component of Eq. (6) by a third of that.

The analysis described here was undertaken to evaluate the extent to which the anomalous contribution of gluons to the spin dependent quark structure functions would modify the conclusion of the EMC experiment that the net quark contribution to the proton's spin is very small. In so far as the results using such small lattices can be trusted, the conclusion of this simulation is that the effect is an order of magnitude too small to change the thrust of the EMC conclusion.

Acknowledgements
The author wishes to express his appreciation to the Bernard-Soni collaboration for generously making available the lattices that form the basis for this analysis. The author has also greatly benefitted from valuable discussions and encouragement to all stages of this work with Carl Bender, Robert Carlitz, Robert Jaffe, Marek Karliner, James Labrenz, Alfred Mueller, Michael Ogilvie, Amarjit Soni, and Vigdor Teplitz.

References
1. Some of the material presented here is contained in J.E. Mandula, Phys. Rev. Letts. **65**, 1403 (1990), and is included here for completeness.
2. The EMC Collaboration, Nucl. Phys **B328**, 1 (1989)
3. J. Kodaira, Nucl. Phys., **B165**, 129 (1979)
4. J.D. Bjorken, Phys. Rev. **148**, 1467 (1966)
5. R.L. Jaffe and A. Manohar, Nucl. Phys. **B337**, 509 (1990)
6. A.V. Efremov and O.V. Teryaev, JINR report E2-88-278 (1988)
7. G. Altarelli and G.G. Ross, Phys. Letts. **B212**, 391 (1988)
8. R.D. Carlitz, J.C. Collins, and A.H. Mueller, Phys. Letts. **214B**, 229 (1988)
9. J.S. Bell and R. Jackiw, Nuovo Cimento **A51**, 47 (1967); S.L Adler, Phys. Rev. **177**, 2426 (1969)
10. G. Bodwin and J. Qiu, Phys. Rev. D **41**, 2755 (1990)
11. A. Efremov, J. Soffer, and N. Törnqvist, to be published; A. Manohar, to be published
12. J. Kogut and L. Susskind, Phys. Rev. D **11**, 3594 (1975)
13. See, for example, C. Bernard, T. Draper, G. Hockney, A. Rushton, and A. Soni, Phys. Rev. Letts. **55**, 2770 (1985)
14. J.E. Mandula, J. Govaerts, and G. Zweig, Nucl. Phys. **B228**, 109 (1983)

DOES PARTON MODEL NEED ANOTHER GLUON OPERATOR ?

Zbigniew Ryzak
Lyman Laboratory of Physics
Harvard University, Cambridge, Mass. 02138

Abstract

I address the issue of the properties of the first moments of the polarized structure functions. I demonstrate that consistency with the Altarelli-Parisi analysis requires an introduction of a new gluon operator which is not present in the OPE expansion. The proton matix element of this operator normalizes the first moment of the polarized gluon density.

Parton Densities in the QCD Parton Model

The analysis of the inclusive deep inelastic scattering requires a calculation of the forward Compton amplitude for the photon (W) scattering off the proton. This is a problem with three energy scales: a) Q^2 which corresponds to the (minus) invariant mass of the probe; b) μ^2 which is the renormalization scale; c) Λ^2_{con} which is the confinement scale associated with the dynamics of the proton's constituents i.e., Λ^2_{con} is the scale associated with a representation of the proton as a bound state. The QCD parton model description of the forward Compton amplitude is based on the factorization property of QCD[1]. Factorization assumes a particularly simple form for the moments of the proton's structure functions:

$$M_n(Q^2) = \int_0^1 dx\, x^{n-1} \mathcal{F}(x) \quad (\mathcal{F} = F_1; \frac{1}{2x} F_2; F_3; g_1; ...), \quad (1)$$

where $\mathcal{F}$ corresponds to any of the measured structure functions. In this talk I am particularly interested in the singlet combinations of the structure functions $\mathcal{F} = \mathcal{F}^S$. The parton model predicts that the fundamental scattering occurs on both quark and gluon constituents of the proton and the formula for the n-th moment of $\mathcal{F}^S$ is:

$$M_n^S(Q^2) = S_n^{(q)}\left(\frac{Q^2}{\mu^2}, \frac{\Lambda^2_{con}}{\mu^2}, \alpha\right) A_n^S + S_n^{(G)}\left(\frac{Q^2}{\mu^2}, \frac{\Lambda^2_{con}}{\mu^2}, \alpha\right) B_n^G. \quad (2)$$

A_n^S is the n-th moment of a singlet combination of the bare quark (and antiquark) densities, and B_n^G is the n-th moment of the bare gluon density. $S_n^{(q)}$ and $S_n^{(G)}$ correspond to the n-th moment of the photon-quark (or W-quark) and photon-gluon (or W-gluon) forward Compton amplitudes respectively.

We can use the OPE analysis to represent the moments $S_n^{(q)}\left(\frac{Q^2}{\mu^2}, \frac{\Lambda^2_{con}}{\mu^2}, \alpha\right)$ together with $S_n^{(G)}\left(\frac{Q^2}{\mu^2}, \frac{\Lambda^2_{con}}{\mu^2}, \alpha\right)$ in the high-Q^2 regime:

$$\left(S_n^{(q)}\left(\frac{Q^2}{\mu^2}, \frac{\Lambda^2_{con}}{\mu^2}, \alpha\right), S_n^{(G)}\left(\frac{Q^2}{\mu^2}, \frac{\Lambda^2_{con}}{\mu^2}, \alpha\right)\right) = \\ \left(C_n^{(q)}\left(\frac{Q^2}{\mu^2}, \alpha\right), C_n^{(G)}\left(\frac{Q^2}{\mu^2}, \alpha\right)\right) \begin{pmatrix} V_n^{(qq)}\left(\frac{\Lambda^2_{con}}{\mu^2}, \alpha\right) & V_n^{(qG)}\left(\frac{\Lambda^2_{con}}{\mu^2}, \alpha\right) \\ V_n^{(Gq)}\left(\frac{\Lambda^2_{con}}{\mu^2}, \alpha\right) & V_n^{(GG)}\left(\frac{\Lambda^2_{con}}{\mu^2}, \alpha\right) \end{pmatrix}. \quad (3)$$

$C_n^{(q)}\left(\frac{Q^2}{\mu^2},\alpha\right)$ and $C_n^{(G)}\left(\frac{Q^2}{\mu^2},\alpha\right)$ are the Wilson coefficients of the singlet spin-n twist-2 quark and gluon operators: $\mathcal{O}_n^{(q)}$ and $\mathcal{O}_n^{(G)}$ respectively. $V_n^{(qq)}$ ($V_n^{(qG)}$) describes the quark (gluon) two-point function of the operator $\mathcal{O}_n^{(q)}$. $V_n^{(Gq)}$ ($V_n^{(GG)}$) describes the quark (gluon) two-point function of the operator $\mathcal{O}_n^{(G)}$. The operators are renormalized at the scale μ^2 ($\mathcal{O}_n^{(q)} \equiv \mathcal{O}_n^{(q)}|_{\mu^2}$, $\mathcal{O}_n^{(G)} \equiv \mathcal{O}_n^{(G)}|_{\mu^2}$) i.e., they are defined at the kinematical point $p^2 = -\mu^2$ by relations:

$$\begin{aligned}
V_n^{(qq)}(1,\alpha) &\equiv v_n^{(qq)}(\alpha) = 1 + \mathcal{O}(\alpha)\\
V_n^{(qG)}(1,\alpha) &\equiv v_n^{(qG)}(\alpha) = \mathcal{O}(\alpha)\\
V_n^{(Gq)}(1,\alpha) &\equiv v_n^{(Gq)}(\alpha) = \mathcal{O}(\alpha)\\
V_n^{(GG)}(1,\alpha) &\equiv v_n^{(GG)}(\alpha) = 1 + \mathcal{O}(\alpha).
\end{aligned} \qquad (4)$$

where we have indicated the required asymptotic behavior of the functions $v_n^{(mn)}(\alpha)$ ($m,n = q,G$) at $\alpha \to 0$. Aside from those asymptotic requirements the functions $v_n^{(mn)}(\alpha)$ are arbitrary in a sense that any choice corresponds to an acceptable renormalization prescription. Now we can use the RG analysis to express the n-th moment as:

$$M_n^S(Q^2) = \left(c_n^{(q)}(\alpha(Q^2)),\ c_n^{(G)}(\alpha(Q^2))\right) \mathcal{V}_n\left(\frac{Q^2}{\mu^2},\frac{\Lambda_{con}^2}{\mu^2},\alpha\right)\begin{pmatrix} A_n^S \\ B_n^G \end{pmatrix}, \qquad (5)$$

where $\mathcal{V}_n$ is a matrix which is a product of two matrices that connect the scales and of a matrix of the quark and gluon forward matrix elements of the twist-2 operators:

$$\mathcal{V}_n = \mathrm{T}e^{-\int_0^{tQ} dz\,\gamma_n^S(\bar{\alpha}(z))}\,\tilde{\mathrm{T}}e^{\int_0^{t\Lambda} dz\,\gamma_n^S(\bar{\alpha}(z))}\begin{pmatrix} v_n^{(qq)}(\alpha(\Lambda_{con}^2)) & v_n^{(qG)}(\alpha(\Lambda_{con}^2)) \\ v_n^{(Gq)}(\alpha(\Lambda_{con}^2)) & v_n^{(GG)}(\alpha(\Lambda_{con}^2)) \end{pmatrix}. \qquad (6)$$

In eqn.(6) T ($\tilde{\mathrm{T}}$) denotes ordering (anti-ordering) and γ_n^S is the anomalous dimension matrix of the pair of operators $\mathcal{O}_n^{(q)}$ and $\mathcal{O}_n^{(G)}$. The coefficients $c_n^{(q)}$ and $c_n^{(G)}$ are defined using the Wilson coefficients: $c_n^{(q)}(\alpha) \equiv C_n^{(q)}(1,\alpha)$ and $c_n^{(G)}(\alpha) \equiv C_n^{(G)}(1,\alpha)$. Eqn.(5) shows that the bare densities A_n^S and B_n^G are in principle measurable, provided perturbative QCD returns full expressions for the functions $\gamma_n^S(\alpha)$, $c_n(\alpha)$ and $v_n^{(mn)}(\alpha)$. Of course, in practice this result is not available. A general procedure of the parton model suggests that one should "hide" all the dependence on the soft physics inside conveniently defined: the singlet quark density $q^S(x,Q^2)$ and the gluon density $G(x,Q^2)$. The moments of those densities we denote as: $q_n^S(Q^2) = \int_0^1 dx\, x^{n-1} q^S(x,Q^2)$, and $G_n(Q^2) = \int_0^1 dx\, x^{n-1} G(x,Q^2)$. An appropriate normalization for the moments corresponds to:

$$\begin{pmatrix} q_n^S(Q^2) \\ G_n(Q^2) \end{pmatrix} = \mathcal{D}_n(\alpha(Q^2))\mathcal{V}_n\begin{pmatrix} A_n^S \\ B_n^G \end{pmatrix}, \qquad (7)$$

where $\mathcal{D}_n$ is a non-singular matrix which goes to identity in the limit $Q^2 \to \infty$ i.e.: $d_n^{11}(\alpha) = 1 + \mathcal{O}(\alpha)$, $d_n^{22}(\alpha) = 1 + \mathcal{O}(\alpha)$ and $d_n^{21}(\alpha) = \mathcal{O}(\alpha)$, $d_n^{12}(\alpha) = \mathcal{O}(\alpha)$. We see that the running moments $q_n^S(Q^2)$ and $G_n(Q^2)$ are linear combinations of the bare moments A_n^S and B_n^G. In this sense the information about the proton contained in the bare and the running moments is equivalent.

Now we can write an expression for $M_n^S(Q^2)$ in terms of the running quark and gluon densities:

$$M_n^S(Q^2) = \left(c_n^{(q)}(\alpha(Q^2)),\ c_n^{(G)}(\alpha(Q^2))\right)\mathcal{D}_n^{-1}(\alpha(Q^2))\begin{pmatrix} q_n^S(Q^2) \\ G_n(Q^2) \end{pmatrix}. \qquad (8)$$

Note that the deep-inelastic scattering can be consistently discussed with any choice of the matrix $\mathcal{D}_n$. Of course some choices may be especially convenient. In particular, one can choose a given singlet structure function $\mathcal{F}_a^S$ and define $\mathcal{D}_n$ with a help of the appropriate Wilson functions: $d_n^{11}(\alpha) = c_n^{(q)}|_{\mathcal{F}_a^S}(\alpha)$ and $c_n^{(G)}|_{\mathcal{F}_a^S}(\alpha)$ which leads to the identification

$$M_n^{\mathcal{F}_a^S}(Q^2) \equiv q_n^S(Q^2). \quad (9)$$

In general, another singlet structure function $\mathcal{F}_b^S \neq \mathcal{F}_a^S$ leads to a different set of Wilson coefficients $c_n^{(q)}|_{\mathcal{F}_a^S}(\alpha) \neq c_n^{(q)}|_{\mathcal{F}_b^S}(\alpha)$ and $c_n^{(G)}|_{\mathcal{F}_a^S}(\alpha) \neq c_n^{(G)}|_{\mathcal{F}_b^S}(\alpha)$ (for example in ref.[2] the case where $\mathcal{F}_b = 2F_1(x, Q^2)$ and $\mathcal{F}_a = \frac{1}{x}F_2(x, Q^2)$ was discussed). If one identifies the running quark density with $\mathcal{F}_a^S$ as in eqn.(9), the other structure functions like $\mathcal{F}_b^S$ will correspond to linear combinations of $q^S(x, Q^2)$ and $G(x, Q^2)$. A measurement of two independent singlet structure functions leads to a normalization of the quark and the gluon densities. In another approach one can use measurements of the same singlet structure function at different values of Q^2 to extract the information on $q^S(x, Q^2)$ and $G(x, Q^2)$. This follows from the fact that in general $M_n^S(Q^2)$ and $d\,M_n^S(Q^2)/d\ln Q^2$ will be given by linearly independent combinations of $q_n^S(Q^2)$ and $G_n(Q^2)$.

The Polarized Scattering

In this section we concentrate on the peculiar properties of the first moment of the polarized structure function $g_1^S(x, Q^2)$:

$$M_1^S(Q^2) = \int_0^1 dx\, g_1^S(x, Q^2). \quad (10)$$

The analysis of $M_1^S(Q^2)$ must be different from the general discussion presented above. This follows from the fact that in the OPE expansion of the product of the electroweak currents there is no contribution from a twist-2 spin-1 gluon operator i.e., only the singlet axial current appears: $\mathcal{O}_1^{(q)} \equiv j_5^S$. The first moments of the forward Compton amplitudes $S_1^{(q)}$ and $S_1^{(G)}$ are related to the quark and gluon Green's function of the operator j_5^S. For simplicity we can define the renormalized operator $\mathcal{O}_1^{(q)}|_{\mu^2} \equiv j_5^S|_{\mu^2}$ by a requirement that its Wilson coefficient $c_1^S(\alpha)$ is identically equal to the Wilson coefficient of the NS axial currents $c_1^S(\alpha) = c_1^{NS}(\alpha)$ where the axial currents obey Ward identities. In such a case the quark two-point function of j_5^S gets radiative corrections at the order $\left(\frac{\alpha}{2\pi}\right)^2$. For example at $p^2 = -\mu^2$ it is given by the expression:

$$V_{j_5^S}^{(qq)}(1, \alpha) \equiv v_{j_5^S}^{(qq)}(\alpha) = 1 + v_{j_5^S}^{(2)}\left(\frac{\alpha}{2\pi}\right)^2 + \ldots \quad (11)$$

The two-point gluon Green's function starts at the order $\frac{\alpha}{2\pi}$:

$$V_{j_5^S}^{(qG)}(1, \alpha) \equiv v_{j_5^S}^{(qG)}(\alpha) = k_1 \frac{\alpha}{2\pi} + k_2\left(\frac{\alpha}{2\pi}\right)^2 \ldots \quad (12)$$

Note that the value of k_1 is special: it doesn't depend on the definition of the renormalized operator j_5^S. In particular, for the massless case (which we are presently discussing): $k_1 = -1$, independent of any choice of the renormalization scheme that defines j_5^S.

Using arguments analogous to those of the previous section we can write an expression for the first moment of the polarized singlet structure function:

$$M_1^S(Q^2) = c_1^S\left(\alpha(Q^2)\right) e^{-\int_{t_\Lambda}^{t_Q} dz \gamma_{j_S^S}(\bar{\alpha}(z))} \times \\ \times \left\{ v_{j_S^S}^{(qq)}\left(\alpha(\Lambda_{con}^2)\right) A_1^S + v_{j_S^S}^{(qG)}\left(\alpha(\Lambda_{con}^2)\right) B_1^G \right\}, \quad (13)$$

where the anomalous dimension $\gamma_{j_S^S}(\alpha)$ starts at the order $\left(\frac{\alpha}{2\pi}\right)^2$[3]: $\gamma_{j_S^S}(\alpha) = 2N_f\left(\frac{\alpha}{2\pi}\right)^2 + \mathcal{O}(\alpha^3)$. Now we see the peculiarity of the first moment $M_1^S(Q^2)$ of the singlet polarized structure function $g_1^S(x, Q^2)$: all of its derivatives with respect to Q^2 correspond to the same linear combination of the first moments of the polarized densities A_1^S and B_1^G. Recall that in principle a measurement of a singlet structure function at different values of Q^2 is sufficient to define both the quark and the gluon content of the proton (in general $M_n^S(Q^2)$ and $dM_n^S/d\ln Q^2$ correspond to two linearly independent combinations of the n-th moments of the quark and gluon densities). The first moment of $g_1^S(x, Q^2)$ is an exception because no matter how well its Q^2 variation is known, it is possible to read off only one linear combination of A_1^S and B_1^G. Situation is even more confusing when one thinks about the other polarized structure function $g_2(x, Q^2)$. $g_2(x, Q^2)$ contains a piece that is directly related to $g_1(x, Q^2)$ and a piece $\bar{g}_2(x, Q^2)$ that is independent of $g_1(x, Q^2)$[4]. One could think that $\bar{g}_2(x, Q^2)$ provides the missing information about the first moments of the polarized densities. Unfortunately, $\bar{g}_2(x, Q^2)$ cannot be expressed in terms of the one parton densities[5] and for this reason cannot be used to normalize* A_1^S and B_1^G. One must conclude that the data on the inclusive deep-inelastic leptoproduction does not provide the physical information needed to normalize all of the polarized parton densities. Of course this is consistent with OPE. One knows that the first moments of the polarized structure functions $g_1(x, Q^2)$ are described by N_f objects, that correspond to the proton matrix elements of the diagonal axial currents. On the other hand, the parton model needs $N_f + 1$ objects to normalize all of the first moments of the polarized quark and gluon densities. Below we speculate where the missing information may be coming from.

So far we haven't introduced the running densities. Recall that the analysis is greatly facilitated if one introduces a matrix expression for the given moment of the singlet structure function. In the case of $M_1^S(Q^2)$ the formula must look like this:

$$M_1^S(Q^2) = \left(c_1^S(\alpha(Q^2)), 0\right) \begin{pmatrix} e^{-\int_{t_\Lambda}^{t_Q} dz \gamma_{j_S^S}(\bar{\alpha}(z))} & x \\ y & z \end{pmatrix} \times \\ \times \begin{pmatrix} v_{j_S^S}^{(qq)}\left(\alpha(\Lambda_{con}^2)\right) & v_{j_S^S}^{(qG)}\left(\alpha(\Lambda_{con}^2)\right) \\ v_1^{(Gq)}\left(\alpha(\Lambda_{con}^2)\right) & v_1^{(GG)}\left(\alpha(\Lambda_{con}^2)\right) \end{pmatrix} \begin{pmatrix} A_1^S \\ B_1^G \end{pmatrix}, \quad (14)$$

We have put the "gluon" Wilson coefficient identically to zero ($c_1^G \equiv 0$) to be consistent with OPE. Additionally, we have to make sure that the definition of the other quantities x, y, z and $v_1^{(Gq)}$ and $v_1^{(GG)}$ does not change the relation (13). Note that (14) is equivalent to (13)

---

* Contrast this with the unpolarized scattering, where in principle a measurement of two independent singlet structure functions $F_1(x)$ and $F_2(x)$ can be used to normalize both the quark and gluon densities.

under just one assumption: $x = 0$. This is a very important observation. Suppose we find an operator $\mathcal{O}_1^{(G)}$ which under renormalization mixes with $\mathcal{O}_1^{(q)} = j_5^S$ such that the anomalous dimension matrix equals:

$$\gamma_1^S(\alpha) = \begin{pmatrix} \gamma_{j_5^S}(\alpha) & 0 \\ \gamma_1^{(Gq)}(\alpha) & \gamma_1^{(GG)}(\alpha) \end{pmatrix}. \quad (15)$$

In general a time ordered integral of the matrix $\gamma_1^S(\alpha)$ has a form:

$$\mathrm{T} e^{-\int_0^t dz \gamma_1^S(\bar{\alpha}(z))} = \begin{pmatrix} e^{-\int_0^t dz \gamma_{j_5^S}(\bar{\alpha}(z))} & 0 \\ a & b \end{pmatrix}. \quad (16)$$

This means that the matrix which connects the scales in eqn.(14) can be interpreted as coming from the integration of the RG equations:

$$\begin{pmatrix} e^{-\int_{t_\Lambda}^{t_Q} dz \gamma_{j_5^S}(\bar{\alpha}(z))} & x = 0 \\ y & z \end{pmatrix} = \mathrm{T} e^{-\int_0^{t_Q} dz \gamma_1^S(\bar{\alpha}(z))} \bar{\mathrm{T}} e^{\int_0^{t_\Lambda} dz \gamma_1^S(\bar{\alpha}(z))}. \quad (17)$$

Now eqn.(14) can be cast in a general form discussed in the previous section:

$$M_1^S(Q^2) = \left(c_1^S(\alpha(Q^2)), 0 \right) \mathcal{V}_1 \begin{pmatrix} A_1^S \\ B_1^G \end{pmatrix}, \quad (18)$$

where

$$\mathcal{V}_1 = \mathrm{T} e^{-\int_0^{t_Q} dz \gamma_1^S(\bar{\alpha}(z))} \bar{\mathrm{T}} e^{\int_0^{t_\Lambda} dz \gamma_n^S(\bar{\alpha}(z))} \begin{pmatrix} v_{j_5^S}^{(qq)}(\alpha(\Lambda_{con}^2)) & v_{j_5^S}^{(qG)}(\alpha(\Lambda_{con}^2)) \\ v_1^{(Gq)}(\alpha(\Lambda_{con}^2)) & v_1^{(GG)}(\alpha(\Lambda_{con}^2)) \end{pmatrix}, \quad (19)$$

and $v_1^{(Gq)}$ and $v_1^{(GG)}$ are interpreted as the quarks' and gluons' Green's functions of the gluon operator $\mathcal{O}_1^{(G)}$. The moments of the quark and gluon polarized densities can be now defined in accordance with the general formalism:

$$\begin{pmatrix} \Delta q^S(Q^2) \\ \Delta G(Q^2) \end{pmatrix} = \mathcal{D}_1 \left(\alpha(Q^2) \right) \mathcal{V}_1 \begin{pmatrix} A_1^S \\ B_1^G \end{pmatrix}. \quad (20)$$

In fact the form of the anomalous dimension matrix is further constrained. It is necessary that our approach leads to the results that are consistent with the Altarelli-Parisi analysis. This means that at the lowest order of the perturbative expansion all the entries in the matrix $\gamma_1^S(\alpha)$ are known [6]: $\gamma_1^{(Gq)}(\alpha) = \frac{3}{2} C_F N_f \frac{\alpha}{2\pi} + ...$, where $C_F = 4/3$, and $\gamma_1^{(GG)}(\alpha) = \frac{11C_A - 4T}{6} \frac{\alpha}{2\pi} + ...$, where $C_A = 3, T = \frac{1}{2} N_f$.

Our "creative" interpretation of OPE leads to a treatment of the first moments of the polarized densities that is fully consistent with the rest of the formalism. Namely, we have supplemented the usual OPE expansion with a gluon operator $\mathcal{O}_1^{(G)}$. The operator comes with the Wilson coefficient which is identically equal zero $c_1^G \equiv 0$ and, if the anomalous dimension matrix γ_1^S obeys (15), its presence does not change anything in the analysis of the deep inelastic inclusive leptoproduction. In particular, the presence of the gluon operator does not introduce any inconsistency with the OPE analysis of the inclusive leptoproduction. On the other hand, $\mathcal{O}_1^{(G)}$ is indispensable for a sensible definition of the gluon densities and leads to a consistent picture of the parton model, including a successful reproduction of the Altarelli-Paris equations. The only conditions are that $\mathcal{O}_1^{(G)}$ is a "spin-1" operator, and that it mixes with j_5^S with the anomalous dimension matrix given by eqn.(15). Note that in the lowest order the entries in the anomalous dimension matrix $\gamma_1^S(\alpha)$ have very specific values. Namely, to order $\left(\frac{\alpha}{2\pi}\right)^2$ they obey the relations: $\frac{\alpha}{2\pi}\gamma_1{}^{(GG)} = -\beta$, where β is the beta function of QCD, and $\frac{\alpha}{2\pi}\gamma_1{}^{(Gq)} = \gamma_{j_5^S}$, where $\gamma_{j_5^S}$ the anomalous dimension of the singlet axial current. It means that, at least through order $\left(\frac{\alpha}{2\pi}\right)^2$,:

$$\frac{d}{d\ln\mu^2}\left(j_5^S\big|_{\mu^2} - \frac{\alpha(\mu^2)}{2\pi}\mathcal{O}_1^{(G)}\big|_{\mu^2}\right) = 0, \quad (21)$$

At this point we may be very bold and assume that the "anomaly" relation (21) holds to all orders in $\frac{\alpha}{2\pi}$. This conjecture can be perturbatively verified to a higher and higher order by a study of the Altarelli-Parisi splitting functions. If correct, it suggests that the operator $\mathcal{O}_1^{(G)}$ must have a lot in common with the topological current:

$$K_\mu = \epsilon_{\mu\nu\rho\sigma}\,\mathrm{Tr}\,A^\nu(G^{\rho\sigma} - \frac{2}{3}A^\rho A^\sigma), \quad (22)$$

which precisely obeys eqn.(21). In fact, perturbative calculations do not care about the long distance structure of QCD, which means that, if eqn.(21) holds, in the perturbative calculations operator $\mathcal{O}_1^{(G)}$ might be indistinguishable from K_μ. In other words, all perturbatively calculated quark and gluon Green's functions with once inserted $\mathcal{O}_1^{(G)}$ would behave exactly as the perturbatively calculated Green's functions with once inserted K_μ (have the same momentum, Lorentz and scale dependence). On the other hand, the operator equation $\mathcal{O}_1^{(G)} = K_\mu$ cannot hold. For example the forward proton matrix element of $\mathcal{O}_1^{(G)}$ is obtained in terms of the first moments of the polarized quark and gluon densities. As such it must be a physically measurable, well defined quantity. On the contrary, the forward matrix element of K_μ is neither well defined nor physical. This is as a result of the gauge non-invariance of K_μ under large gauge transformations [7] and it doesn't show up in the perturbative analysis.

The issue of the proper definition of $\mathcal{O}_1^{(G)}$ has grown into a large industry. A discussion of various conjectures is beyond the scope of the present talk. Let us just note that all approaches preserve the "anomaly" equation (21) and this indicate some (probably non-local) relationship between $\mathcal{O}_{(G)}$ and K_μ.

Acknowledgements

The results quoted in this talk have been obtained in the most pleasant collaboration with Lech Mankiewicz. Also some of the discussion presented here can be found in Ref. [8]. This research was supported in part by DOE grant DE-AC02-76ER03064 and NSF grant No. PHY-8714654.

References

1. see for example: A. J. Buras, Rev. Mod. Phys. **52** (1980) 199.

2. G. Altarelli, R. K. Ellis and G. Martinelli, Nucl. Phys. **B143** (1978) 521.

3. J. Kodaira, Nucl. Phys. **B165** (1980) 129.

4. R. L. Jaffe, Comm.Nucl.Part.Phys. **14** (1990) 239.

5. L. Mankiewicz and Z. Ryzak, CAMK preprint CAMK-217/90, Harvard preprint HUTP-90/B011, to appear in Phys. Rev. **D**.

6. G. Altarelli, Phys. Rep. **81** (1982) 1.

7. Z. Ryzak, Harvard preprint HUTP-90/A033.

8. Z. Ryzak and L. Mankiewicz, Harvard preprint HUTP-90/A067.

FLAVOUR SYMMETRY BREAKING AND GENERALIZED GOLDBERGER-TREIMAN RELATIONS

Nils A. Törnqvist
University of Helsinki, Research Institute for High Energy Physics
Siltavuorenpenger 20C, SF00170 Helsinki, Finland

ABSTRACT

We discuss in what way flavour symmetry breaking through nondegenerate light quark masses and through the mixing of π^0, η, η' modifies the Goldberger-Treiman relations for the triplet, octet and in particular, for the generalizations to the flavour singlet channel. We derive simple new algebraic expressions for the $\pi^0 - \eta$ and $\pi^0 - \eta'$ mixing and show that the seemingly large isospin and flavour breaking which result from nondegenerate quark masses are to lowest order cancelled by the $\pi^0 - \eta - \eta'$ mixing.

One of the most celebrated results of current algebra from the fifties is the well known Goldberger-Treiman[1] relation between the axial vector coupling constant and the πN strong coupling constant. The famous EMC experimental results[2] on the polarized structure function of the proton has motivated an interest in the question, if and how this relation can be generalized to the flavour singlet channel.

As was discussed in A. Efremovs talk[3] we recently[4] derived, taking into account the U(1) problem in the singlet channel, such a generalized Goldberger Treiman (GGT) relation which relates the spin fraction carried by the quarks $\Delta\Sigma$ and the $\eta' N N$ coupling constant

$$\Delta\Sigma = \frac{\sqrt{N_f} f_{\eta'} g_{\eta' NN}}{2M_N} , \qquad (1)$$

This relation can be generalized to include a possible direct ghost nucleon coupling g_{QNN}:

$$\Delta\Sigma = \frac{G}{2M_N} = \frac{\sqrt{N_f} f_{\eta'}}{2M_N}(g_{\eta' NN} - \Delta m_{\eta'} g_{QNN}) , \qquad (2)$$

where G is the residue of the ghost pole ($G = \lim_{q^2 \to 0} q^2 \widetilde{G}_2$) in the matrix element of the topological current K_μ, and where $g_{\eta' NN} - \Delta m_{\eta'} g_{QNN} = g_{\eta'_0 NN}$ is composed of the ghost coupling via the η' minus a presumably small term involving the possible direct ghost-nucleon coupling. Here the notation $Q \equiv \partial_\nu \widetilde{K}_\nu$ is used. This relation (2) was derived from the Adler-Bardeen relation

$$\partial_\nu J_\nu^5 = \partial_\nu \widetilde{K}_\nu = N_f \frac{\alpha_s}{2\pi} F^a_{\rho\sigma} \widetilde{F}^a_{\rho\sigma} , \qquad (3)$$

assuming that one can neglect light quark masses and other flavour breaking contributions.

We here relax this assumption of exact $SU3_f$ symmetry following our recent paper[5]. In another recent talk[6] we discussed a preliminary and less general version of these results. In fact, our discussion is rather independent of the exact form and interpretation of the flavour singlet Goldberger-Treiman relation (eq.(1-2)), which still is under dispute in the literature[7-10]. But in in our discussion, we keep our form of the relation as the first order approximation.

When considering corrections from finite quark masses the Adler-Bardeen relation takes the form

$$\partial_\nu J_\nu^5 = \partial_\nu \widetilde{K}_\nu + \sum_i 2m_i \bar{q}_i i\gamma_5 q_i ,\qquad (4)$$

which, together with the well known relations for the triplet and octet, leads to the following relations in the triplet, octet and singlet channels, respectively (for $N_f = 3$ and $q^2 = 0$):

$$2M_N g_A^3 = 2(m_u \nu_u - m_d \nu_d) ,\qquad (5a)$$

$$2M_N g_A^8 = 2(m_u \nu_u + m_d \nu_d - 2m_s \nu_s) ,\qquad (5b)$$

$$2M_N \Delta\Sigma = 2(m_u \nu_u + m_d \nu_d + m_s \nu_s) + G ,\qquad (5c)$$

where $\nu_i = \langle N|\bar{q}_i i\gamma_5 q_i|N\rangle / \bar{N} i\gamma_5 N$, and G is the residue of the ghost pole contribution (see eq.(2)).

The flavour symmetry breaking enters into these equations in two ways:
i) nondegenerate m_q's
ii) $\pi^\circ - \eta - \eta'$ mixing in the coupling constants.

The effect of i) is explicitely seen from eqs.(5a-c) and we can solve for $\Delta\Sigma$ by eliminating the $m_q \nu_q$'s from the r.h.s. of eq.(5c) using the other two. First, one can use eq.(5b) to eliminate ν_s and second, due to the identity

$$3(m_u \nu_u + m_d \nu_d) \equiv 3/2 \left[(m_u + m_d)(\nu_u + \nu_d) + (m_u - m_d)\nu_3\right] ,$$

together with $\nu_3 = \nu_u - \nu_d = \sqrt{2} f_\pi g_{\pi NN}/(m_u + m_d)$ from eq.(5a), neglecting the term $(m_u - m_d)(\nu_u + \nu_d)$ (which leads to higher order isospin corrections), one can reexpress eq.(5c) as

$$\Delta\Sigma = G/2M_N \pm \frac{3}{2} \frac{m_u - m_d}{m_u + m_d} g_A^3 - \frac{1}{2} g_A^8 ,\qquad (6)$$

where the second terms "violates isospin" and where the + or − sign is for proton or neutron respectively. The third term "violates $SU3_f$". Both terms are large, reducing $\Delta\Sigma$ by more than 50%!

But, we must also take into account $\pi^\circ - \eta - \eta'$ mixing, which for eq.(5c) enters into G through the $g_{\eta' NN}$ coupling constant; instead of one η'-pole we have to put the singlet combination of η'-,η-,π°-poles.

$$\frac{g_{\eta' NN}}{m_{\eta'}^2} \longrightarrow \frac{g_{\eta' NN}}{m_{\eta'}^2} cos\theta_3 \pm \frac{g_{\pi NN}}{m_\pi^2} \theta_2 - \frac{g_{\eta NN}}{m_\eta^2} sin\theta_3 ,\qquad (7)$$

where θ_2 and θ_3 are the $\pi^\circ - \eta'$ and $\eta - \eta'$ mixing angles respectively. Inserting conventional values for these angles and for the couplings $g_{\eta NN}$ and $g_{\eta' NN}$, the second "isospin violating" terms of eqs.(6) and (7) cancel each others and similarly the $SU3_f$ violating third terms almost cancel each others. As a result the numerical value of $\Delta\Sigma$ changes very little. As we shall see this cancellation is exact to first order in the limit of large anomaly mass $\Delta m_{\eta'}^2$.

This cancellation of large correction terms looks accidental, but obviously there must be a physical reason for it which we now clarify. Using different techniques other authors[9,11] have reached similar, but less general conclusions. In short we write eqs. (5a,b,c) in a matrix form :

$$2M_N\alpha \begin{pmatrix} g_A^3/\sqrt{2} \\ g_A^8/\sqrt{6} \\ \overline{\Delta\Sigma}/\sqrt{3} \end{pmatrix} = \left[\alpha\mathcal{O} \begin{pmatrix} 2m_u & 0 & 0 \\ 0 & 2m_d & 0 \\ 0 & 0 & 2m_s \end{pmatrix} \mathcal{O}^{-1} + \begin{pmatrix} 0 & 0 & 0 \\ 0 & 0 & 0 \\ 0 & 0 & \Delta m_{\eta'}^2 \end{pmatrix} \right] \begin{pmatrix} \nu_3 \\ \nu_8 \\ \nu_1 \end{pmatrix} \quad (5')$$

where

$$\mathcal{O} = \begin{pmatrix} 1/\sqrt{2} & -1/\sqrt{2} & 0 \\ 1/\sqrt{6} & 1/\sqrt{6} & -2/\sqrt{6} \\ 1/\sqrt{3} & 1/\sqrt{3} & 1/\sqrt{3} \end{pmatrix}, \quad \begin{pmatrix} \nu_3 \\ \nu_8 \\ \nu_1 \end{pmatrix} = \mathcal{O} \begin{pmatrix} \nu_u \\ \nu_d \\ \nu_s \end{pmatrix},$$

and

$$\overline{\Delta\Sigma} = \Delta\Sigma - \left(G - \Delta m_{\eta'}^2 \nu_1/\alpha\right)/2M_N ,$$

and in which $\mathcal{O}$ is the orthogonal matrix which transforms from the ideal frame $(u\bar{u}, d\bar{d}, s\bar{s})$ to the (triplet, octet, singlet) frame. In this construction α can have an arbitrary value, which we shall fix below. Now, one observes that the matrix in the square brackets of eq. (5') is of the same form as the 0^{-+} squared mass matrix with quark mass terms and an anomaly term $(\Delta m_{\eta'}^2)$. If one can choose $\alpha \simeq m_\pi^2/(m_u + m_d)$ this matrix will be equal to the true pseudoscalar mass matrix. As we shall see, then the difference between $\Delta\Sigma$ and $\overline{\Delta\Sigma}$ reduces to the contribution from g_{QNN} only, i.e. to that from the second term in eq.(2).

So, let

$$M_{0^{-+}}^2 = \alpha\mathcal{O} \begin{pmatrix} 2m_u & 0 & 0 \\ 0 & 2m_d & 0 \\ 0 & 0 & 2m_s \end{pmatrix} \mathcal{O}^{-1} + \begin{pmatrix} 0 & 0 & 0 \\ 0 & 0 & 0 \\ 0 & 0 & \Delta m_{\eta'}^2 \end{pmatrix}$$

$$= \alpha \begin{pmatrix} m_u + m_d & (m_u - m_d)/\sqrt{3} & (m_u - m_d)/\sqrt{2/3} \\ (sym.) & (m_u + m_d + 4m_s)/3 & (m_u + m_d - 2m_s)/3\sqrt{2} \\ (sym.) & (sym.) & 2(m_u + m_d + m_s)/3 + \Delta m_{\eta'}^2/\alpha \end{pmatrix}, \quad (8)$$

be the mass matrix. The physical π°, η, η' squared masses are the eigenvalues of this matrix which is diagonalized by another orthogonal matrix, the mixing matrix, Ω :

$$\Omega = \begin{pmatrix} 1 & 0 & 0 \\ 0 & \cos\theta_3 & -\sin\theta_3 \\ 0 & \sin\theta_3 & \cos\theta_3 \end{pmatrix} \begin{pmatrix} 1 & -\theta_1 & \theta_2 \\ \theta_1 & 1 & 0 \\ -\theta_2 & 0 & 1 \end{pmatrix}, \qquad (9)$$

where θ_1 is the $\pi^\circ - \eta$ mixing angle and θ_2, θ_3 are as in eq.(7) above, the $\pi^\circ - \eta'$ and $\eta - \eta'$ mixing angles respectively.

Now Ω diagonalizes the sum of the quark mass term (Q) and the anomaly term (A):

$$\begin{pmatrix} m_{\pi^\circ}^2 & 0 & 0 \\ 0 & m_\eta^2 & 0 \\ 0 & 0 & m_{\eta'}^2 \end{pmatrix} = Q + A, \qquad (10)$$

where

$$Q = 2\alpha\Omega\mathcal{O} \begin{pmatrix} m_u & 0 & 0 \\ 0 & m_d & 0 \\ 0 & 0 & m_s \end{pmatrix} \mathcal{O}^{-1}\Omega^{-1}, \qquad (11)$$

$$A = \Omega \begin{pmatrix} 0 & 0 & 0 \\ 0 & 0 & 0 \\ 0 & 0 & \Delta m_{\eta'}^2 \end{pmatrix} \Omega^{-1}, \qquad (12)$$

but both Q and A have off-diagonal terms of same order as eq.(8). The mixing angles θ_i are determined by the cancellation of $Q_{ij} + A_{ij} = 0$ (for $i \ne j$). To first order in the small isospin mixing angles θ_1 and θ_2 one finds simple algebraic relations for these (denoting $a = \Delta m_{\eta'}^2/\alpha$):

$$\theta_1 = -\theta_2 \frac{2m_s + a}{2\sqrt{2}m_s} = -\theta_2 \frac{m_{\eta'}^2 + m_\eta^2 - 2m_\pi^2}{2\sqrt{2}(m_K^2 - m_\pi^2)} = -0.017, \qquad (13a)$$

$$\theta_2 = \sqrt{\frac{3}{2}} \frac{m_d - m_u}{a} = \sqrt{\frac{3}{2}} \frac{m_{K^+}^2 - m_{K^\circ}^2 + m_{\pi^\circ}^2 - m_{\pi^+}^2}{m_{\eta'}^2 + m_\eta^2 - 2m_K^2} = +0.009, \qquad (13b)$$

$$\theta_3 = \frac{1}{2} tg^{-1} \frac{-4\sqrt{2}m_s}{a + 2m_s} = \frac{1}{2} tg^{-1} \frac{-4\sqrt{2}(m_K^2 - m_\pi^2)}{3m_{\eta'}^2 + 3m_\eta^2 + 2m_\pi^2 - 8m_K^2} = -0.31, \qquad (13c)$$

where we also give the results when expressed in terms of squared pseudoscalar masses and with their numerical values. The latter agree very well with the computer diagonalizations of ref.[12]. Of these, the formula for θ_3 is well known, but θ_1 and θ_2 have not appeared in the literature, although the scheme for the mass matrix is well known (see e.g. ref.[12]). Phenomenological consequences of the isospin mixing have been discussed by many authors[13-18]. These estimates of θ_1 and θ_2 are also in reasonable agreement with a phenomenological analysis using data from the decays $\eta \to 3\pi^\circ$, $\eta' \to 3\pi^\circ$, $\eta' \to \eta 2\pi^\circ$ and $\psi' \to \psi\pi^\circ$ from which it is found, within a model[18], $<\pi^\circ|H|\eta> = -0.0059$ GeV^2 and $<\pi^\circ|H|\eta'> = -0.0055$ GeV^2 with 10% errors. These correspond to $\theta_1 = -0.021$ and $\theta_2 = +0.006$ with the same type of error.

In the limit of large anomaly term ($m_{\eta'} \to \infty$ whereby also $\eta \to \eta_8$ and $m_\eta^2 = (4m_K^2 - m_\pi^2)/3$) the expression for θ_1 reduces to the well known formula of U-spin invariance[19] giving

$$\theta_1 \to \theta_1^{U\text{-}spin} = -\frac{\sqrt{3}}{4}\frac{m_d - m_u}{m_s} = \frac{m_{K^+}^2 - m_{K^\circ}^2 + m_{\pi^\circ}^2 - m_{\pi^+}^2}{\sqrt{3}(m_{\eta_8}^2 - m_\pi^2)} = -0.011 \ . \quad (14)$$

In the same limit, (keeping $m_s \gg m_u, m_d$) $\theta_3 \to -2\sqrt{2}m_s/a \simeq -2\sqrt{2}m_K^2/(3m_{\eta'}^2)$ and one sees that the $\sin\theta_3$ term of eq.(7) also cancels to first order the $\frac{1}{2}g_A^8$ term of eq.(6), i.e., in this limit, the $SU3_f$ part cancels to first order just as the isospin part. This can be seen explicitly from :

$$-\frac{\sqrt{3}}{2M_N}(f_\eta g_{\eta NN})\sin\theta_3 \frac{\Delta m_{\eta'}^2}{m_\eta^2} \to \frac{\sqrt{3}}{2M_N}\left(2M_N g_A^8 \frac{1}{\sqrt{6}}\right) \frac{2\sqrt{2}(m_K^2 - m_\pi^2)}{3m_{\eta'}^2} \frac{\Delta m_{\eta'}^2}{m_\eta^2}$$

$$\simeq g_A^8 \frac{2(m_K^2 - m_\pi^2)}{3m_\eta^2} = g_A^8 \frac{2(m_K^2 - m_\pi^2)}{(4m_K^2 - m_\pi^2)} \simeq \frac{1}{2}g_A^8 \ .$$

Turning back to our original problem and eqs.(5a-c) or (5') we see that these equations are diagonalized by the same Ω :

$$2M_N \begin{pmatrix} g_A^3/\sqrt{2} \\ g_A^8/\sqrt{6} \\ \overline{\Delta\Sigma}/\sqrt{3} \end{pmatrix} = \Omega^{-1} \begin{pmatrix} m_\pi^2 & 0 & 0 \\ 0 & m_\eta^2 & 0 \\ 0 & 0 & m_{\eta'}^2 \end{pmatrix} \frac{1}{\alpha} \begin{pmatrix} \nu_\pi \\ \nu_\eta \\ \nu_{\eta'} \end{pmatrix} , \quad (15)$$

where the ν's on the r.h.s., transformed by the same Ω, are the pseudoscalar densities, which in the pole approximation can be replaced by only one physical pole. In order to obtain the Goldberger-Treiman relation for g_A^3 with mixing it has to satisfy

$$\Omega\frac{1}{\alpha}\begin{pmatrix} \nu_3 \\ \nu_8 \\ \nu_1 \end{pmatrix} = \frac{1}{\alpha}\begin{pmatrix} \nu_\pi \\ \nu_\eta \\ \nu_{\eta'} \end{pmatrix} \simeq \begin{pmatrix} f_\pi g_{\pi NN}/m_\pi^2 \\ f_\eta g_{\eta NN}/m_\eta^2 \\ f_{\eta'}g_{\eta' NN}/m_{\eta'}^2 \end{pmatrix} , \quad (16)$$

where $g_{\pi NN}$ etc. are the physical coupling constants after mixing is taken into account.

Now we can show that in the same one-pole approximation the difference of $\Delta\Sigma$ and $\overline{\Delta\Sigma}$ in the third row of eq.(15) reduces to the contribution from

g_{QNN} only. For this, it is enough to observe that from eq.(16), assuming that $f_\pi = f_\eta = f_{\eta'} = f$, one gets

$$\frac{1}{\alpha}\nu_1 = f\left(\frac{g_{\eta'NN}}{m_{\eta'}^2}\cos\theta_3 - \frac{g_{\eta'NN}}{m_{\eta'}^2}\sin\theta_3 \pm \frac{g_{\pi NN}}{m_\pi^2}\theta_2\right) \ .$$

So this relation which fixes α, also leads to the cancellation of the first term of G (see eq.(2) with the substitution (7)) which appears in the definition of $\overline{\Delta\Sigma}$. Thus the three Goldberger-Treiman relations, with $SU3_f$ breaking taken into account, can be written as

$$\Omega\begin{pmatrix} g_A^3/\sqrt{2} \\ g_A^8/\sqrt{6} \\ \Delta\Sigma/\sqrt{3} + f\Delta m_{\eta'}g_{QNN}/2M_N \end{pmatrix} = \frac{f}{2M_N}\begin{pmatrix} g_{\pi NN} \\ g_{\eta NN} \\ g_{\eta'NN} \end{pmatrix}, \quad (17)$$

where all quantities in the r.h.s. include the $\pi^\circ - \eta - \eta'$ mixing. The transition from the exact $SU3_f$ to broken $SU3_f$ is seen to be a very smooth one, provided Ω is $\simeq 1\!\!1$ as is physically the case.

Note that by separating off the quark and the anomaly terms, one obtains similar cancellations as in eq.(10) for off-diagonal terms of (Q) and (A). Some of these terms are enhanced by large ratios like $m_{\eta'}^2/m_\pi^2$. coming from the ν's.

Now, using the third row of this equation one can obtain

$$\Delta\Sigma = \frac{\sqrt{3}f(g_{\eta'NN}/\cos\theta_3 - \Delta m_{\eta'}g_{QNN})}{2M_N} - \frac{g_A^8}{\sqrt{2}}tg\theta_3 \pm \sqrt{\frac{2}{3}}g_A^3(\theta_2 - \theta_1 tg\theta_3) \ , \quad (18)$$

or up to first order in the mixing angles

$$\Delta\Sigma = \frac{\sqrt{3}f(g_{\eta'NN} - \Delta m_{\eta'}g_{QNN})}{2M_N} - \frac{g_A^8}{\sqrt{2}}\theta_3 \pm \sqrt{\frac{2}{3}}g_A^3\theta_2 \ . \quad (18')$$

Several comments are now in order: Firstly, all large symmetry breaking corrections are cancelled and only a small one proportional to the θ_j's survives. The transition from exact $SU3_f$ to broken $SU3_f$ is a smooth one provided one simultaneously takes into account the nondegenerate m_q and the $\pi^\circ - \eta - \eta'$ mixing. Such cancellations within another context were already discussed long ago[13,14] and recently also for the pion mass[20]. For the pion mass the cancellation can easily be seen in our framework directly from the pseudoscalar mass matrix. Only if one insists in separating, in a pure $SU3_f$ reference frame, a purely singlet gluonic quantity from the quark mass contributions, one does get large cancellations of terms of order $(m_u - m_d)/(m_u + m_d)$ and $\theta_2 m_{\eta'}^2/m_\pi^2$. Physically this means only that the small gluonic admixture in the pion gives a large contribution in the singlet channel because of the smallness of m_π^2 compared to $m_{\eta'}^2$. This is, however, a very natural contribution and the transition to exact $SU3_f$ is a very smooth one, due to the cancellation with quark mass terms. On the other hand, if one neglects the mixing keeping only the nondegenerate m_q's one gets large unphysical isospin and $SU3_f$ breaking terms and one can even get a gluonic contribution of the wrong sign[21].

Secondly, in the limit of $m_u, m_d \ll \Delta m_{\eta'}, m_s$, $\theta_1, \theta_2 \to 0$ but $\theta_3 \neq 0$, eq.(18') reproduces the result of Veneziano[7] if the first order correction from θ_3 is also neglected. Thirdly, there is a difference in $\Delta\Sigma$ for the proton and neutron in eq.(18'). However, it is so small that it can produce the breaking of the Bjorken sum rule, only in the fourth digit.

Acknowledgements. Discussions with my collaborators A. Efremov and J. Soffer are gratefully acknowledged as well as with M. Chaichian, M. Sainio and J. Niskanen.

References
1. M. L. Goldberger, S. B. Treiman, Phys. Rev. **110**, 1178 (1958).
2. J. Ashman et. al. (EMC) Nucl.Phys. **B328**, 1 (1989).
3. A. V. Efremov, These proceedings.
4. A. V. Efremov, J. Soffer, N. Törnqvist, Phys. Rev. Lett. **64**, 1495 (1990) and Marseille preprint CPT-90/P.2402 (unpublished).
5. A. V. Efremov, J. Soffer, N. Törnqvist, Joint Marseille-Helsinki preprint CPT-90/P.2457, HU-TFT-90-96.
6. N.A. Törnqvist, Talk at the 9th Int. Symposium on High Energy Spin Systems, Bonn, Sept. 10-15 1990, Marseille preprint CPT-90/P.2446, to appear in Springer Verlag conference proceedings.
7. G. Veneziano, Mod. Phys. Lett. **A4**, 1605 (1989);
   G. M. Shore, G. Veneziano, Phys. Lett. **B244**, 75 (1990).
8. G. Veneziano, Nucl. Phys. **B159**, 213 (1979).
9. H. Fritzsch, Phys. Lett. **B229**, 122 (1989) and Phys. Lett. **242**, 451 (1990).
10. T. Hatsuda, Nucl. Phys. **B329**, 376 (1990).
11. J. Schechter, V. Soni, A. Subbaraman and H. Weigel, Phys. Rev. **D42**, 2998 (1990).
12. K. Kawabayashi and N. Ohta, Prog. Theor. Phys. **66**, 1789 (1981).
13. D. Gross, S.B. Treiman and F. Wilczek, Phys. Rev. **D19**, 2188 (1979).
14. B.L. Ioffe, Sov. J. Nucl. Phys. **29** (1979) 827; Yad. Fiz. **29**, 1611 (1979).
15. N. Isgur, H. Rubinstein, A. Schwimmer and H. Lipkin, Phys. Lett. **B89**, 79 (1979);
    N. Isgur, Phys. Rev. **D21**, 779 (1980);
    S. Godfrey, N. Isgur, Phys. Rev. **D34**, 899 (1986).
16. T.N. Pham, Phys. Lett. **B134**, 133 (1984).
17. N.A. Törnqvist, Phys. Lett. **B40**, 109 (1972).
18. S.A. Coon, B.H.J. McKellar and M.D. Scadron, Phys. Rev. **D34**, 2784 (1986).
19. S. Okubo and B. Sakita, Phys. Rev. Lett. **11**, 50 (1963).
    R.H. Dalitz and F. von Hippel, Phys. Lett. **10**, 153 (1964).
20. Fayyazuddin and Riazuddin, Phys. Rev. **D42**, 2347 (1990).
21. T.P. Cheng and L.F. Li, Phys. Rev. Lett. **62**, 1441 (1989).

IS THERE A HARD GLUONIC CONTRIBUTION TO THE FIRST MOMENT OF g_1?*

Geoffrey T. Bodwin[†]
High Energy Physics Division, Argonne National Lab, Argonne, IL 60439

Jianwei Qiu[‡]
Institute for Theoretical Physics, SUNY, Stony Brook, NY 11794

ABSTRACT

We show that the size of the hard gluonic contribution to the first moment of the proton's spin-dependent structure function g_1 is entirely a matter of the convention used in defining the quark distributions. If the UV regulator for the spin-dependent quark distributions respects the gauge invariance of Green's functions (allows shifts of loop momenta) and respects the analyticity structure of the unregulated distributions, then the hard gluonic contribution to the first moment of g_1 vanishes. This is the case, for example, in dimensional regularization. By relaxing the requirement that the regulator allow shifts of loop momenta, we are able to obtain a nonvanishing hard gluonic contribution to the first moment of g_1. However, the first moments of the resulting quark distributions correspond to matrix elements that are either gauge variant or involve nonlocal operators and, hence, have no analogue in the standard operator-product expansion.

THE "SPIN CRISIS" AND A PROPOSED EXPLANATION

The hadronic part of the deep-inelastic scattering probability $W_{\mu\nu}$ is given by the forward matrix element in the proton state of two factors of the electromagnetic current:

$$W_{\mu\nu} = \sum_{\mathcal{X}} \langle \mathcal{X}|j_\nu|P,S\rangle^\star \langle \mathcal{X}|j_\mu|P,S\rangle \, 2\pi\delta(M_{\mathcal{X}}^2 - (P+q)^2), \tag{1}$$

where $j_\mu = \bar{\psi}\gamma_\mu\psi$ is the electromagnetic current, ψ is the quark field, q is the four-momentum of the virtual photon, and P and S are the momentum and spin four-vectors of the proton, respectively. The spin-dependent part $\Delta W_{\mu\nu}$ can be decomposed into two form factors g_1 and g_2 as follows:

$$\Delta W_{\mu\nu} = \frac{4\pi i}{P\cdot q}\epsilon_{\mu\nu\lambda\sigma}q^\lambda M_P \left[S^\sigma g_1(x,Q^2) + (S^\sigma - \frac{S\cdot q}{P\cdot q}P^\sigma)g_2(x,Q^2)\right], \tag{2}$$

where $Q^2 = -q^2$ and $x = Q^2/(2P\cdot q)$. For longitudinally polarized protons, only $g_1(x)$ contributes to $\Delta W_{\mu\nu}$ in the scaling limit. The EMC collaboration[1] has measured $g_1(x)$ at $\langle Q^2 \rangle = 10.7$ GeV^2.

In the naive parton model, the first moment of $g_1(x)$ is simply related to the first

---

*Talk presented by G. Bodwin.
[†]Work supported by the U.S. Department of Energy, Division of High Energy Physics, Contract W-31-109-ENG-38.
[‡]SSC Fellow. Work supported in part by the National Science Foundation under grant No. PHY-89-08495.

moments of the spin-dependent quark distributions:

$$\int_0^1 dx\, g_1(x) = 1/2(4/9\Delta u + 1/9\Delta d + 1/9\Delta s), \tag{3}$$

where $\Delta p = \int_0^1 dx\, \Delta p(x)$. The spin-dependent parton distribution $\Delta p(x)$ is defined as the difference between the probabilities to find the parton with spin parallel to the proton or antiparallel to the proton at longitudinal momentum fraction x. Gluonic radiative corrections of $O(\alpha_S)$ introduce a factor $(1 - \alpha_S/\pi)$ on the RHS of (3).

Usually, the spin-dependent quark distributions Δq_i are assumed to be related to the expectation value of the axial-vector current:

$$2M_P S^\mu \Delta q_i = \langle P, S | j_{5i}^\mu | P, S \rangle, \tag{4}$$

where $j_{5i}^\mu = \overline{\psi}_i \gamma^\mu \gamma_5 \psi_i$. Using (4), one can obtain a relationship between $\Delta u - \Delta d$ and g_A/g_V, which is known as the Bjorken sum rule,[2] and a relationship between the combination $\Delta u + \Delta d - 2\Delta s$ and the F/D ratio[3] from hyperon decay. By making use of these relationships and the measured value of the first moment of g_1, one can determine the value of the singlet combination $\Delta\Sigma = \Delta u + \Delta d + \Delta s$. The result is that $\Delta\Sigma$ differs significantly from unity and is consistent with zero, which seems to imply that quarks carry very little of the proton's total spin. On the basis of the static quark model, one would expect $\Delta\Sigma$ to be of order unity. This apparent conflict between the EMC measurement and intuition from the static quark model has come to be known as the "spin crisis."

Efremov and Teryaev[4] (ET); Altarelli and Ross[5] (AR); and Carlitz, Collins, and Mueller[6] (CCM) have suggested that the first moment of g_1 contains, in addition to the usual quark contributions, a contribution proportional to the first moment of spin-dependent gluon distribution $\Delta g(x)$. ET, AR, and CCM conjecture that this gluonic contribution could bring the value deduced for $\Delta\Sigma$ into agreement with intuition from the static quark model. Since a gluon has no electric charge, it first contributes to deep-inelastic scattering at one-loop order ($O(\alpha_S)$). However, to leading order, the combination $\alpha_S(Q^2)\Delta g(Q^2)$ does not evolve with Q^2, so $\alpha_S \Delta g$ could be large even at large Q^2.

IDENTIFYING THE HARD GLUONIC CONTRIBUTION: GENERALITIES

Fig. 1 depicts one of the $O(\alpha_S)$ diagrams that yield a gluonic contribution to the deep-inelastic scattering cross section. In this order there are, in addition, three diagrams obtained by permuting the photon and gluon attachments to the quark line and a set diagrams obtained by reversing the arrows on the quark lines (quark goes to antiquark). We work in a frame in which the proton is moving in the plus z direction and the virtual photon has no transverse components of momentum.

A given Feynman diagram does not, in general, have a single interpretation in terms of partonic distributions and cross sections. Rather, the interpretation depends on the values of loop momenta under consideration. In order to obtain a contribution from the diagram of Fig. 1 that is proportional to $\Delta g(x)$, we focus on momenta p such that the gluon is nearly on its mass shell.

Even in this restricted region of momentum space, the diagram has more than one interpretation. When k_T is large compared with any hadronic scale, then the q and $\bar{q}$ produced by the gluon are at large angles. The quark propagator is far off the mass shell, and, consequently, asymptotic freedom allows one to compute the box subdiagram perturbatively. This contribution has the following interpretation: the

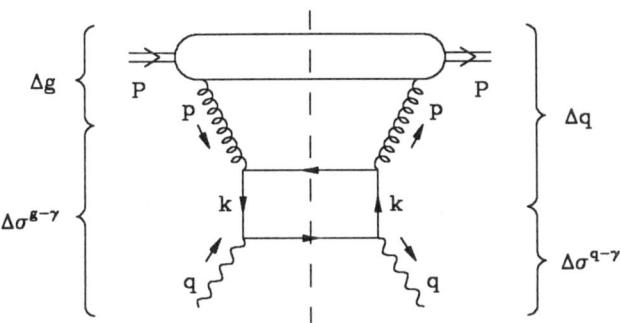

Fig. 1. An order α_S process that yields a gluonic contribution to deep-inelastic scattering.

proton emits a nearly-collinear, nearly-on-shell gluon with spin-dependent probability $\Delta g(x)$; the gluon then undergoes a hard interaction with the virtual photon, via the quark box subdiagram, with spin-dependent probability $\Delta\sigma^{g-\gamma}(x)$. Hence, we have a contribution to $\Delta\sigma^{g-\gamma}(x) \otimes \Delta g(x)$, as indicated by the left-hand set of braces in Fig. 1. The symbol $\otimes$ denotes the convolution $B(x) \otimes C(x) = \int_x^1 (dy/y) \, B(x/y) C(y)$, which has the property that the nth moment of the convolution is the product of the nth moments: $(B \otimes C)_n = \int_0^1 dx \, x^{n-1} B(x) \otimes C(x) = B_n C_n$. We call $\Delta\sigma^{g-\gamma}(x) \otimes \Delta g(x)$ the hard gluonic contribution to $g_1(x)$, and we call $\Delta\sigma^{g-\gamma}(x)$ the hard gluonic coefficient.

When k_T is small, the q and $\bar{q}$ are nearly collinear to each other and, hence, to the proton. The quark propagator is near the mass shell. In fact, when k_T is zero, the quark propagator becomes singular. In this region, because k^2 is small, the box diagram cannot be calculated perturbatively. Rather, one must absorb such soft contributions, including the collinear singularity, into the definition of $\Delta q(x)$. That is, the diagram of Fig. 1 must be given the following interpretation: the proton emits a nearly-collinear, nearly-on-shell quark with spin-dependent probability $\Delta q(x)$; the quark then undergoes a hard interaction with the virtual photon with spin-dependent probability $\Delta\sigma^{q-\gamma}(x)$. That is, we have a contribution to $\Delta\sigma^{q-\gamma}(x) \otimes \Delta q(x)$, as indicated by the right-hand set of braces in Fig. 1.

The central issue in determining the size of the hard gluonic coefficient $\Delta\sigma^{g-\gamma}(x)$ is the partitioning of the complete proton–virtual-photon spin-dependent cross section $\Delta\sigma^{p-\gamma}(x)$ into a quark piece plus a gluon piece:

$$\Delta\sigma^{p-\gamma}(x) = \sum_i \Delta\sigma_i^{q-\gamma}(x) \otimes \Delta q_i(x) + \Delta\sigma^{g-\gamma}(x) \otimes \Delta g(x). \qquad (5)$$

Of course, $\Delta\sigma^{p-\gamma}(x)$, being a physical (measured) quantity, is independent of the partitioning. The precise method of partitioning the contributions on the RHS of (5) is known as the factorization scheme. A change of factorization scheme merely shifts contributions from $\Delta\sigma^{g-\gamma}(x)$ to $\Delta q(x)$ in such a way that $\Delta\sigma^{p-\gamma}(x)$ is unchanged.

IDENTIFYING THE HARD GLUONIC CONTRIBUTION: SPECIFICS

Now let us define the hard gluonic contribution more precisely. First we define $\Delta\sigma^{q-\gamma}(x) \otimes \Delta q(x)$. Then we subtract this quantity from the diagram of Fig. 1 plus its crossed-box and reversed-arrow partners. The remainder is the hard gluonic contribution $\Delta\sigma^{g-\gamma}(x) \otimes \Delta g(x)$. In order to simplify the discussion, we specialize to the light-cone gauge $n \cdot A = 0$, where n is a unit vector in the minus light-cone direction. We have given a manifestly gauge-invariant discussion of these issues previously.[7]

In order to identify $\Delta\sigma^{q-\gamma}(x) \otimes \Delta q(x)$, we first approximate the box and crossed-box subdiagrams by expressions that are correct when the q and $\bar{q}$ are collinear: we drop k_T in the q-γ vertices and in the lower final-state–mass-shell δ function. Then, the k_T integration becomes UV divergent, so we must impose a regulator to ensure that $k_T^2 \lesssim \mu_{fact}^2$, where μ_{fact}^2 is the "factorization scale", with $\mu_{fact}^2 \gg \Lambda_{QCD}^2$. The choice of regulator specifies the factorization convention. In light-cone gauge, the crossed-box subdiagrams vanish in the collinear approximation, so only the box subdiagram contributes to $\Delta\sigma^{q-\gamma}(x) \otimes \Delta q(x)$. Owing to the collinear approximation, this contribution factors, as shown in Fig. 2, into piece corresponding to $\Delta q(x)$, which contains a cut-triangle subdiagram with a cut vertex[8] $\gamma^+\gamma^5\delta[x-(k^+/P^+)]$, and a piece corresponding to $\sigma^{q-\gamma}(x) = e_q^2\delta[x - (Q^2/2P \cdot q)]$.

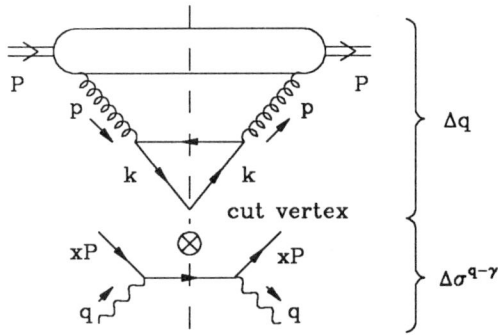

Fig. 2. The factored form that arises from applying the collinear approximation to the diagram of Fig. 1.

It was shown in Ref. 7 that, if one chooses conventional UV regulators for $\Delta q(x)$, such as $\overline{MS}$ or Pauli-Villars regulators, the first moment of the hard gluonic contribution vanishes because the first moment of the hard coefficient vanishes: $\Delta\sigma^{g-\gamma} \equiv \int_0^1 dx\,\Delta\sigma^{g-\gamma}(x) = 0$.

GENERAL THEOREM ON THE VANISHING OF $\Delta\sigma^{g-\gamma}$

The vanishing of the first moment of the hard gluonic coefficient is actually quite general. In Ref. 7 it was shown that, if the UV regulator for Δq respects the analyticity of the unregulated graph (parton-level analyticity) and respects the gauge invariance of Green's functions (allows shifts of loop momenta), then Δq is given by a gauge-invariant matrix element of j_5^μ and $\Delta\sigma^{g-\gamma}$ vanishes. The main features of the proof are as follows. The analyticity condition allows one to use contour methods to convert the cut-triangle subgraph to an uncut triangle. Then the Δq_i are given by matrix elements of the axial-vector current according to (4). Gauge invariance of the regulator implies that the value of the uncut triangle is uniquely determined.[9] Thus, Δq and, implicitly, $\Delta\sigma^{g-\gamma}$, are independent of the choice of regulator, provided that the regulator satisfies the unitarity and gauge-invariance requirements. But we already know at least one such regulator, namely $\overline{MS}$, for which $\Delta\sigma^{g-\gamma}$ vanishes. Hence, $\Delta\sigma^{g-\gamma}$ vanishes for all such regulators.

SIDESTEPPING THE THEOREM

It is possible to sidestep the theorem on the vanishing of $\Delta\sigma^{g-\gamma}$ by relaxing the requirement that the UV regulator for $\Delta q(x)$ respect shifts of loop momenta. Specifically, one can regulate with a direct cutoff on the k_T integration:

$$\Delta q^{cutoff}(x, \mu^2_{fact}) = \int_0^{\mu^2_{fact}} dk_T^2 \, \Delta\mathcal{P}(x, k_T) \neq \Delta q^{\overline{MS}}(x, \mu^2_{fact}). \qquad (6)$$

Here, $\Delta\mathcal{P}(x, k_T)$ is the spin-, x-, and k_T-dependent distribution to find a quark in the proton. One can measure $\Delta\mathcal{P}(x, k_T)$ by measuring the k_T of q and $\bar{q}$ jets accompanying deep-inelastic scattering. Indeed, CCM have suggested such a measurement as a means for identifying the hard gluonic contribution. The quantity $\Delta\mathcal{P}(x, k_T)$ also appears in the expression for the k_T distribution of lepton pairs in spin-dependent Drell-Yan production. With the k_T-cutoff definition of $\Delta q(x)$, the first moment of the hard gluonic coefficient is nonvanishing: $\int_0^1 dx \, \Delta\sigma^{g-\gamma}(x) = -\alpha_S N_F/(2\pi)$, as advocated by ET, AR, and CCM. The k_T-cutoff regulator looks Lorentz variant because it singles out the transverse plane. However, P and q already define a transverse plane, so the cross section can be written in terms of invariants. This definition amounts to defining $\Delta q(x)$ in terms of the triangle graph, but with the "anomaly part" omitted.

RELATIONSHIP TO THE OPERATOR-PRODUCT EXPANSION

Since the k_T-cutoff regulator does not allow shifts of loop momenta, it does not respect the gauge invariance of Green's functions. Consequently, the singlet quark distributions are gauge-variant matrix elements of j_5^μ:

$$2M_P S^+ \sum_i \Delta q_i^{cutoff} = \langle P, S| \sum_i j_{5i}^+ |P, S\rangle_{cutoff-l.c.g.}. \qquad (7)$$

Here, the subscript on the matrix element indicates that the current is to be regulated using the k_T-cutoff method and that the matrix element is to be evaluated in the light-cone gauge. If we express the RHS of (7) in terms of a dimensionally-regulated (gauge-invariant) matrix element of j_{5i}^+, then a gauge-variant gluon operator appears:

$$\langle P, S| \sum_i j_{5i}^+ |P, S\rangle_{cutoff-l.c.g.} = \langle P, S| \sum_i j_{5i}^+ |P, S\rangle_{\overline{MS}} - \langle P, S|K^+|P, S\rangle_{l.c.g.}, \qquad (8)$$

where $K^+ = [\alpha_S N_F/(2\pi)]\epsilon^{+\mu\nu\rho} A_\mu^a \partial_\nu A_\rho^a$ is the winding-number current. The matrix element of K^+ is proportional to Δg.

In the manifestly gauge-invariant formalism for Δq presented in Ref. 7, gauge invariance is maintained through the appearance nonlocal path integrals of the gauge field. These path integrals lead to matrix elements of a nonlocal gluonic operator:

$$2M_P S^+ \sum_i \Delta q_i^{cutoff} = \langle P, S| \sum_i j_{5i}^+ - \tilde{K}^+ + K^+ |P, S\rangle_{cutoff}, \qquad (9)$$

where $\tilde{K}^+ = [\alpha_S N_F/(4\pi)] \text{Tr}\, \epsilon^{+\mu\nu\rho} F_{\mu\sigma}(D\cdot n)^{-1} F_{\nu\rho} n^\sigma$. In the light-cone gauge, matrix elements of $\tilde{K}^+$ reduce to matrix elements of K^+. The inverse derivative is defined by

$(\partial_y)^{-1} f(y) = \int_{-\infty}^{y} dy' \, f(y')$, and the inverse covariant derivative $(D)^{-1}$ is defined by its Taylor-series expansion in powers of the gauge field. Factors of the inverse derivative correspond to "eikonal" propagators that arise from the Feynman rules for the path integral. Comparing (9) with (7) and (8), we see that the cutoff matrix element of j_{5i}^+ can be expressed in terms a dimensionally-regulated matrix element of j_{5i}^+ and a matrix element of the nonlocal gluon operator $\tilde{K}^+$:

$$\langle P, S | \sum_i j_{5i}^+ | P, S \rangle_{cutoff-l.c.g.} = \langle P, S | \sum_i j_{5i}^+ | P, S \rangle_{\overline{MS}} - \langle P, S | \tilde{K}^+ | P, S \rangle. \tag{10}$$

We conclude that the matrix element corresponding to Δq^{cutoff} is either gauge variant or involves a nonlocal operator. Hence, it does not correspond to the any of the matrix elements that appear in a conventional operator-product expansion.[11]

WHY USE THE k_T-CUTOFF REGULATOR TO DEFINE Δq?

The k_T-cutoff method for defining the quark distributions seems somewhat awkward, given that Δq^{cutoff} does not correspond to a gauge-invariant matrix element of a local operator. In fact, if one wishes to preserve the Bjorken sum rule and the standard relationship of the quark distributions to hyperon decay, it is necessary to use a more conventional UV regulator for the flavor-nonsinglet combinations of the quark distributions. On the other hand, $\Delta q^{cutoff}(x, \mu_{fact}^2)$ is directly related to a measurable quantity, namely, $\Delta \mathcal{P}(x, k_T)$. Perhaps a more important point is that, unlike conventional regulators, the k_T-cutoff method respects quark helicity conservation. For example, in dimensional regularization $\{\gamma_5, \gamma_\mu\} \neq 0$ for the higher-dimensional γ_μ's that appear in quark propagators in loops. Thus, the combination of a gluon-quark vertex and a quark propagator can change the Dirac matrices $\frac{1}{2}(1 \pm \gamma_5)$, which correspond in four dimensions to helicity projectors, to $\frac{1}{2}(1 \mp \gamma_5)$. The k_T-cutoff method produces no such helicity flip and, hence, may lead to a definition of Δq that is closer to intuition about the quark's spin.

CONCLUSIONS AND DISCUSSION

We have seen that, if $\Delta q(x)$ is defined using a UV regulator that respects the optical theorem at the parton level and respects gauge invariance of Green's functions (allows shifts of loop momenta), then the hard gluonic contribution to the first moment of $g_1(x)$ vanishes. By relaxing the requirement that the UV regulator allow shifts, one can obtain a nonvanishing hard gluonic contribution to the first moment of $g_1(x)$, as advocated by ET, AR, and CCM. However, in this case, Δq is no longer given by a gauge-invariant matrix element of j_5^μ: either the matrix element is gauge variant, or the operator is nonlocal.

The presence or absence of a hard gluonic contribution to $\int_0^1 dx \, g_1(x)$ is purely a matter of the factorization convention chosen in defining $\Delta q(x)$ and, implicitly, $\Delta \sigma^{g-\gamma}(x)$. The choice of factorization scheme has no effect on predictions for high-energy cross sections: it merely shifts contributions from $\Delta \sigma^{g-\gamma}(x)$ to $\Delta q(x)$ in (5). Of course, one must take care to apply a given factorization scheme consistently for all p-QCD processes and sum rules. Although $\Delta \sigma^{g-\gamma}(x)$ is arbitrary, we wish to emphasize that $\Delta g(x)$ is well-defined at this order in α_S and measurable.

The original motivation for the suggestion that there is a hard gluonic contribution to the first moment of $g_1(x)$ was the desire to bring $\Delta \Sigma$ into agreement with intuition from the static quark model. In fact, unlike dimensional regulators, the k_T-

cutoff UV regulator that we have discussed does not flip quark helicity. Thus, the k_T-cutoff–regulated Δq might correspond more closely with intuition than does the $\overline{MS}$-regulated Δq. On the other hand, if one could carry out a renormalization-group integration from the soft scale of the constituent-quark model to the hard scale of deep-inelastic scattering, then it is clear that one would encounter nonperturbative helicity-flipping phenomena associated with the breaking of chiral symmetry. Thus, it seems likely that the quark-helicity information contained in the static quark model would be lost anyway.

Suppose one takes the point of view that, for reasons not as yet understood, the nonrelativistic quark model is approximately correct at deep-inelastic scattering scales. In a nonrelativistic bound state, it seems likely that Δg would vanish as v/c. In positronium, for example, the spin-dependent photon distribution is suppressed by at least one power of v/c (although the spin-independent distribution is not). Hence, in the static limit, it seems that the various definitions of Δq would all lead to a vanishing hard gluonic contribution to the first moment of g_1.

So far, there is no reason to believe that one definition of Δq is any closer to intuition from the static quark model than another. Measurement of Δg might shed some light on this issue. Lattice simulations[10] already suggest that Δg is too small for the presence of a hard gluonic contribution to "explain" the spin crisis.

Of course, the spins of the quarks and gluons are not the whole story: there is also orbital angular momentum. Unfortunately, there is no leading-twist operator that corresponds to orbital angular momentum.[11] The twist expansion is essentially an expansion in powers of the parton transverse momentum. Since the quark angular momentum operator $\overline{\psi}(\vec{r} \times \vec{p})\psi$ involves the operator $\vec{r}$, it cannot be written in terms of a finite number of powers of the transverse momentum. In this sense, the quark angular momentum is actually an infinite-twist quantity. Consequently, there is no experimental test in hard scattering of the sum rule $(1/2)\sum_i \Delta q_i + \Delta g + \langle L_z \rangle = 1/2$. It is clear, then, that the EMC result does not represent a fundamental disagreement between theory and experiment: there is a spin crisis only if one believes that the static quark model has something to do with spin phenomena at deep-inelastic scales.

We wish to thank S.J. Brodsky, R.D. Carlitz, J.C. Collins, G.P. Lepage, and A.H. Mueller for helpful discussions.

REFERENCES

1. J. Ashman et al., Phys. Lett. 206B, 364 (1988).
2. J.D. Bjorken, Phys. Rev. D1, 1376 (1970).
3. M. Bourquin et al., Z. Phys. C21, 27 (1983).
4. A.V. Efremov and O.V. Teryaev, JINR Dubna preprint E2-88-287 (1988).
5. G. Altarelli and G.G. Ross, Phys. Lett. 212B, 391 (1988).
6. R.D. Carlitz, J.C. Collins and A.H. Mueller, Phys. Lett. 214B, 229 (1988); ANL-HEP-CP-89-69, to be published in the proceedings of the *Rencontre de Moriond*, Les Arcs, France, March 12-17, 1989.
7. G.T. Bodwin and J.-W. Qiu, Phys. Rev. D41, 2755 (1990).
8. A.H. Mueller, Phys. Rev. D18, 3705 (1978); S. Gupta and A.H. Mueller, Phys. Rev. D20, 118 (1979).
9. S.L. Adler and W. Bardeen, Phys. Rev. 182, 1517 (1969).
10. J.E. Mandula, Phys. Rev. Lett. 65, 1403 (1990); in these proceedings.
11. R.L. Jaffe and A. Manohar, Nucl. Phys. B337, 509 (1990).

HIGH-TWIST CONTRIBUTIONS TO $g_2(x, Q^2)$

Xiangdong Ji
Center for Theoretical Physics
Laboratory for Nuclear Science and Department of Physics
Massachusetts Institute of Technology
Cambridge, Massachusetts 02139

ABSTRACT

Based on general arguments and the explicit bag-model calculation, we show that the twist-3 contribution to the nucleon's transverse spin-dependent structure function $g_2(x)$ is significant and its measurement allows us a clean study of high-twist effects.

$g_2(x)$ AND OPERATOR-PRODUCT EXPANSION

We begin with the definition of the nucleon's spin-dependent structure function $g_2(x)$. Then, we summarize the results from the operator-product-expansion (OPE) analysis.

In deep-inelastic electron-nucleon scattering, we measure the nucleon's hadron tensor,

$$W_{\mu\nu}(q, P, S) = \frac{1}{4\pi} \int d^4\xi \, e^{iq\cdot\xi} \langle PS \,|[J_\mu(\xi), J_\nu(0)]|\, PS \rangle \tag{1}$$

where q is the virtual photon four momentum, P and S are the target four-momentum and spin, respectively. Spin-dependent effects are related to the antisymmetric part of the hadron tensor $W^A_{\mu\nu}$. By Lorentz invariance and gauge invariance $W^A_{\mu\nu}$ can be constructed from two scalar functions $g_1(x, Q^2)$ and $g_2(x, Q^2)$ for a spin-half target,

$$W^A_{\mu\nu} = i\epsilon_{\mu\nu\lambda\sigma} \frac{q^\lambda}{\nu} \left[g_1(x, Q^2) S^\sigma + g_2(x, Q^2) \left(S^\sigma - P^\sigma \frac{q \cdot S}{\nu} \right) \right] \tag{2}$$

where $Q^2 = -q^2$ and $\nu = P \cdot q$. In the Bjorken limit in QCD, both $g_1(x, Q^2)$ and $g_2(x, Q^2)$ scale to $g_1(x)$ and $g_2(x)$ modulo logarithms.

Introduce twist-2 and twist-3 operators, $O_{2,i}^{\sigma\mu_1\cdots\mu_n}$ and $O_{3,i}^{\sigma\mu_1\cdots\mu_n}$, and define their nucleon forward matrix elements,

$$\langle PS | O_{2,i}^{\sigma\mu_1\mu_2\cdots\mu_n} | PS \rangle = 2a_i^n \mathcal{S}_1 \{ S^\sigma P^{\mu_1} P^{\mu_2} \cdots P^{\mu_n} \} - (\text{Traces}) \tag{3}$$

where $\mathcal{S}_1$ symmetrizes all indices, and

$$\langle PS|O_{3,i}^{\sigma\mu_1\mu_2\cdots\mu_n}|PS\rangle = 2d_i^n \mathcal{S}\mathcal{A}\{S^\sigma P^{\mu_1}P^{\mu_2}\ldots P^{\mu_n}\} - (\text{Traces}) \quad (4)$$

where $\mathcal{A}$ antisymmetrizes σ and μ_1 and $\mathcal{S}$ symmetrizes $\mu_1, \mu_2, \ldots, \mu_n$. The scalar matrix elements a_i^n and d_i^n depend on the nucleon structure and the renormalization scale μ^2 at which the operators are defined. The operator-product expansion and dispersion relations provide us an infinite set of sum rules for the moments of the structure functions,

$$\int_0^1 x^n g_1(x, Q^2)dx = \frac{1}{2}\sum_i a_i^n(\mu^2) F_{2,i}^n(Q^2, \mu^2) \quad n = 0, 2, 4\ldots \quad (5)$$

$$\int_0^1 x^n g_2(x, Q^2)dx = -\frac{n}{2(n+1)}\left[\sum_i a_i^n(\mu^2) F_{2,i}^n(Q^2, \mu^2) - \sum_i d_i^n(\mu^2) F_{3,i}^n(Q^2, \mu^2)\right] \quad n = 2, 4, \ldots \quad (6)$$

Eqs. (5) and (6) show that $g_1(x, Q^2)$ receives contribution from twist-2 operators alone, whereas $g_2(x, Q^2)$ receives contributions from both twist-2 and twist-3 operators. The twist-2 part of the $g_2(x, Q^2)$ can be constructed from $g_1(x, Q^2)$,

$$g_2(x, Q^2)^{WW} = -g_1(x, Q^2) + \int_0^1 \frac{g_1(y, Q^2)}{y} dy. \quad (7)$$

In general, we write,

$$g_2(x, Q^2) = g_2(x, Q^2)^{WW} + \bar{g}_2(x, Q^2) \quad (8)$$

where $\bar{g}_2(x, Q^2)$ represents the twist-3 contributions only,

$$\int_0^1 x^n \bar{g}_2(x, Q^2)dx = \frac{n}{2(n+1)}\sum_i d_i^n(\mu^2) F_{3,i}^n(Q^2). \quad n = 2, 4, \ldots \quad (9)$$

The number of twist-3 operators is an increasing function of n. The following two sets of operators are all twist-3,

$$R_\ell^{\sigma\mu_1\cdots\mu_n} = \frac{i^{n-3}}{4} g\mathcal{S}\{\bar{\psi}(0)D^{\mu_1}\ldots D^{\mu_{\ell-1}} G^{\sigma\mu_\ell} D^{\mu_{\ell+1}}\ldots D^{\mu_{n-1}}\gamma^{\mu_n}\gamma^5\psi(0)\} \quad (10)$$

and

$$S_\ell^{\sigma\mu_1\cdots\mu_n} = \frac{i^{n-2}}{4} g\mathcal{S}\{\bar{\psi}(0)D^{\mu_1}\ldots D^{\mu_{\ell-1}} \tilde{G}^{\sigma\mu_\ell} D^{\mu_{\ell+1}}\ldots D^{\mu_{n-1}}\gamma^{\mu_n}\psi(0)\} \quad (11)$$

However, not all of them enter the operator-product expansion independently. The current product is even under the charge conjugation, whereas operators R_ℓ and S_ℓ are not charge-conjugation eigenstates. It is easy to show that the combinations $R_\ell - R_{n-\ell}$ and $S_\ell + S_{n-\ell}$ are charge-conjugation even. The former are $n/2 - 1$ in number and the latter are $n/2$, giving total of $n - 1$ operators.

The twist-3 operators are direct manifestation of quark-gluon interaction. The main effect of the gluons inside of the hadron is, of course, confining quarks. Therefore, the twist-3 contributions should be large in any realistic nucleon model with confinement. Their effect on $g_2(x, Q^2)$ cannot a priori be neglected in comparison to the twist-2 operators.

IMPULSE APPROXIMATION AND BAG MODEL

To calculate $g_1(x, Q^2)$ and $g_2(x, Q^2)$ we take the Bjorken limit of Eq. (1) and use the light-cone expansion

$$[J_\mu(\xi), J_\nu(0)] = \bar\psi(\xi)\mathcal{Q}^2\gamma_\mu S(\xi)\gamma_\nu\psi(0) - \bar\psi(0)\mathcal{Q}^2\gamma_\nu S(-\xi)\gamma_\mu\psi(\xi) + \ldots \quad (12)$$

where $\mathcal{Q}$ is the quark charge matrix and the ellipses denote terms less singular on the light-cone. $S(\xi)$ is the free field causal function

$$S(\xi) = \{\psi(\xi), \bar\psi(0)\} = (-i\,\partial\!\!\!/ + m)\Delta(\xi^2, m) = -\frac{\partial\!\!\!/}{2\pi}\delta(\xi^2)\epsilon(\xi^0) + \ldots \quad (13)$$

Substituting for S, performing some γ-matrix algebra and isolating terms antisymmetric in $\mu \leftrightarrow \nu$, we find,

$$W_{\mu\nu}^A = i\epsilon_{\mu\nu\lambda\sigma}q^\lambda\left(-\frac{i}{8\pi}\right)\int d^4\xi\, e^{iq\cdot\xi}\delta(\xi^2)\epsilon(\xi^0)$$
$$\langle PS|\bar\psi(\xi)\mathcal{Q}^2\gamma^\sigma\gamma_5\psi(0) + \bar\psi(0)\mathcal{Q}^2\gamma^\sigma\gamma_5\psi(\xi)|PS\rangle \quad . \quad (14)$$

In Eq. (14) we have ignored terms which vanish as $Q^2 \to \infty$. This result could equally well be derived from the OPE since all the coefficient functions are simple if radiative corrections are ignored. In particular, all twist-3 operators can be summed to give a simple operator,

$$O_3^{\sigma\mu_1\mu_2\cdots\mu_n} = i^n \mathcal{S}\mathcal{A}\{\bar\psi(0)\mathcal{Q}^2\gamma^\sigma\gamma^5 D^{\mu_1}D^{\mu_2}\ldots D^{\mu_n}\psi(0)\} - (\text{Traces}) \quad (15)$$

Combining it with the twist-2 operator, we immediately arrive Eq. (14).

In order to estimate the importance of the twist-3 operator on $g_2(x)$, which determines how interesting $g_2(x)$ is, we evaluate $g_2(x)$ in the bag model. The bag boundary simulates the confinement effects which arise from quark-gluon interactions. The twist-3 effects are therefore due to the bag boundary. As a result, measurement of $g_2(x)$ will provide non-trivial constraints on the nucleon models. We do not expect this calculation to provide a quantitatively accurate prediction for $\bar{g}_2(x)$. Instead, we view this as a toy model or cartoon, lacking

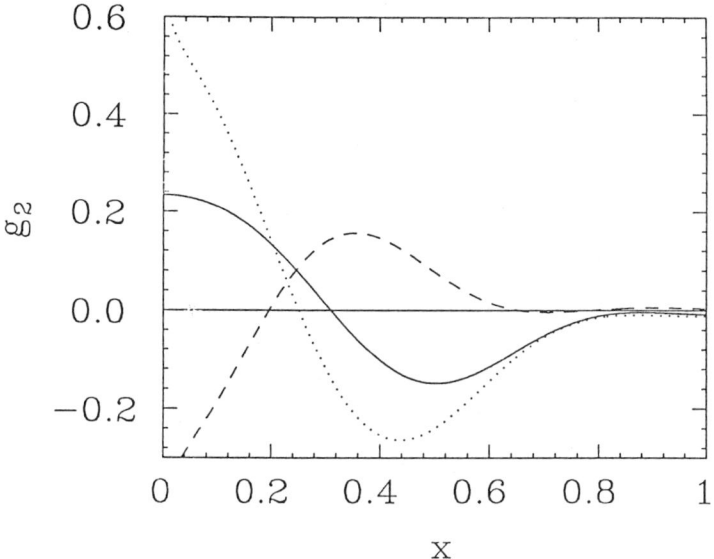

Fig. 1: $g_2(x)$ *in the bag model. The dotted curve represents the twist-2 contribution, the dashed curve represents the twist-3 contribution, and the solid line represents the sum.*

Regge behavior and the proper $x \to 1$ limit, but giving a rough estimate of the size of $\bar{g}_2(x)$. Some of the recent developments in the calculation of bag structure functions could give an improved estimate of $\bar{g}_2(x)$.

To calculate structure functions in a bag, we choose the nucleon rest frame where $P_\mu = (M, 0, 0, 0)$. To calculate $g_1(x)$ we polarize the nucleon in the z-direction (defined to be the opposite direction of the virtual photon 3-momentum), then $S_\mu^z = (0, 0, 0, M)$. Substituting S and P into the general decomposition for $W_{\mu\nu}^A$ in Eq. (2) and the impulse approximation limit, Eq. (14), we identify,

$$g_1(x) = \frac{1}{8\pi} \int d\xi_- e^{iq_+\xi_-} \langle PS^z | \bar{\psi}(\xi_-) Q^2 \gamma^+ \gamma^5 \psi(0) + \bar{\psi}(0) Q^2 \gamma^+ \gamma^5 \psi(\xi_-) | PS^z \rangle \tag{16}$$

where $q_+ = -Mx/\sqrt{2}$. Next consider the nucleon to be polarized perpendicular to the photon momentum, say in the x-direction, then $S_\mu^x = (0, M, 0, 0)$. In this case, we find the transverse structure function,

$$g_1(x) + g_2(x) = \frac{1}{8\pi\sqrt{2}} \int d\xi_- e^{iq_+\xi_-} \langle PS^x | \bar{\psi}(\xi_-) Q^2 \gamma^1 \gamma^5 \psi(0) + \bar{\psi}(0) Q^2 \gamma^1 \gamma^5 \psi(\xi_-) | PS^x \rangle. \tag{17}$$

Now, the structure function $g_2(x)$ can obtained by subtracting out known $g_1(x)$. It is shown on Fig. 1 in solid curve.

The $g_2(x)$ calculated contains contributions from both twist-2 and twist-3 operators. The twist-2 part can be calculated from Eq. (7) once $g_1(x)$ is known and is shown in dotted curve in Fig. 1. The twist-3 part is, therefore, just the difference between the solid and the dotted curve. We show it with dashed curve in the same figure. The result clearly shows that, at least in the bag model, the twist-3 contribution to $g_2(x)$ is not small. It is interesting, however, that the twist-3 and twist-2 contributions are destructive and the net $g_2(x)$ bears the sign of twist-2 part.

The twist-3 contribution can also be calculated from a set of operators which explicitly depends on the interactions. Using the equation of motion for quarks in the bag,

$$i \not{\partial} \psi(\xi) = \delta(\xi - R)\psi(\xi), \qquad (18)$$

we transform the twist-3 operator in Eq. (15) into,

$$O_3^{\sigma\mu_1\mu_2\cdots\mu_n} = \frac{i^n}{2} S[\bar{\psi}(0)\sigma^{\sigma\mu_1}\gamma^5 \partial^{\mu_2} \ldots \partial^{\mu_n} \delta_S \psi(0) + \bar{\psi}(0)\delta_S \sigma^{\sigma\mu_1}\gamma^5 \partial^{\mu_2} \ldots \partial^{\mu_n} \psi(0)] \qquad (19)$$

where we used the shorthand $\delta_S = \delta(\xi - R)$ and $\sigma_{\sigma\mu_1} = \frac{i}{2}[\gamma_\sigma, \gamma_{\mu_1}]$. If there were no bag boundary, this operator is identically zero. The result from Eq. (19) agrees with the dashed curve on the Fig. 1. This confirms the physical interpretation of $\bar{g}_2(x)$ and is a check on the algebra.

COMPARISON WITH PARTON MODEL

The nucleon's deep inelastic structure functions are usually interpreted in terms of Feynman's parton model. For the unpolarized structure functions $F_1(x)$ and $F_2(x)$, the naive parton picture, in which the on-shell partons move co-linearly with the nucleon in the infinite momentum frame, was shown being equivalent to the impulse approximation or the OPE in the simplest version. However, this picture breaks down for $g_2(x)$ because of contributions from twist-3 operators. In fact, there are variety of parton model predictions for $g_2(x)$ in the literature and most of them are incorrect. Apparently the essential ingredient missing in the naive parton model is the off-shellness, or initial state interaction. The quarks inside of the nucleon are off-shell because of confinement. The large off-shellness of the quark in the bag is directly responsible for the large twist-3 effect shown in Fig. 1.

ACKNOWLEDGEMENT

This work was done in collaboration with R. Jaffe and is in part supported by DOE contract DE-AC02-76ER03069.

REFERENCES

1. R. Jaffe and X. Ji, MIT CTP preprint 1848, to be published in Phys. Rev. D. This paper contains a full list of references relevant to this talk.

The Asymptotic S-Matrix, Mass-Shell Anomalies and Observables*

H.F. CONTOPANAGOS
University of Michigan, Ann Arbor, MI 48109, USA
and
M.B. EINHORN
*Institute for Theoretical Physics, University of California
Santa Barbara, CA 93106, USA
&
University of Michigan, Ann Arbor, MI 48109, USA*[†]

Abstract

Asymptotic S-Matrix techniques are applied to the identification of a Standard Model background to polarized deep-inelastic scattering experiments at HERA in search of Right-handed Charged Currents. It is also pointed out that the Asymptotic S-Matrix produces finite observables free of Mass-Shell Anomalies.

1. Introduction

Experiments using polarized beams are important for testing various aspects of the Standard Model (SM). One of these aspects is the existence of left-handed charged currents, but the absence of right-handed charged currents. One of the goals of experimentation at HERA will be to test this hypothesis by means of Deep Inelastic Scattering (DIS) experiments using a right-handed polarized electron beam.[1][2] If such an experiment yields a signal of the type $e_R^- + p \to X$ + missing energy, one may be led to the conclusion that the signal is due to a right-handed coupling of the form $g_R \bar{\Psi}_e \gamma_\mu (1+\gamma_5) \Psi_\nu W_R^{(-)\mu} + h.c.$ In what follows we shall show that this connection should not be immediately made, since the SM left-handed charged current provides a finite background under the given experimental conditions at HERA.

2. Right-handed Charged Current

Before taking up the discussion of the background, let us first record the cross section for a right-handed coupling between a neutrino, electron and charged vector boson W_R. In what follows we shall assume that the corresponding coupling constant is the same as the Standard Model coupling constant for the left-handed coupling. We shall denote the unknown mass of the new vector boson by M_R. The neutrino is assumed massless. Then the dominant process is shown in Fig.

---

* Talk presented by H. F. Contopanagos.
† *Permanent address.*

1a, where kinematic notation is also introduced. The standard kinematic variables are $s = (p+l)^2$, $q^2 = (l-l')^2 \equiv -Q^2$, $x \equiv Q^2/2p \cdot q$. The differential cross section for a right-handed helicity electron to undergo DIS producing a right-handed neutrino via the exchange of W_R is

$$\frac{d\sigma_R^R}{dx} = \frac{\pi \alpha_2^2}{2M_R^2}\left[\left(1+\frac{1}{x\zeta_R}\right)^{-1}U(x)+\left(1+\frac{2}{x\zeta_R}-\frac{2}{x\zeta_R}(1+\frac{1}{x\zeta_R})\ln(1+x\zeta_R)\bar{D}(x)\right)\right] \quad (1)$$

In the above, $U(\bar{D})$ denotes the sum of contributions from quarks (antiquarks) of charge $+2/3$ $(+1/3)$, where the Q^2-evolution has been ignored.[*] We have also used the notation $\alpha_2 \equiv g_2^2/4\pi$, with $g_2 = e/\sin\theta_w$, and $\zeta_R = s/M_R^2$. The above formula shows that an upper limit on σ_R^R may be interpreted as a lower limit on M_R. Estimates have suggested[3] that, in the case of a light, Dirac ν_{eR} such as the one we are using here, experiments at HERA might be sensitive to M_R as large as 300-500 GeV.

3. The HERA Background

This background consists of the process[4] $e_R^- + p \rightarrow \nu_{eL} + \gamma + X$. In other words, a right-handed helicity electron flips its helicity, while radiating an unobservable forward-going photon, and couples to a SM left-handed charged current, see Fig. 1b. The collinear photon goes unobservably down the beam pipe within a forward angle whose size is restricted by the HERA luminosity monitors.

One might think that the cross-section resulting from the amplitude in Fig. 1b is heavily suppressed, not so much due to the extra coupling α relative to the non-radiative process, but primarily due to the helicity-flip factor $\chi_e^2 \equiv (m_e/E)^2 \simeq 4.0 \times 10^{-10}$ (the electron energy E at HERA is $\simeq 26$ GeV). However, upon integration over the forward direction, a cancellation of the helicity-flip factor occurs, due to the collinear singularity of the electron propagator.[5] With the kinematics shown in Fig. 1b, the cross-section can be written

$$d\sigma_L^{R,\gamma} = \frac{\alpha\alpha_2^2}{16\pi^2}\int d^4q \int \frac{d\Omega_\gamma k_0^2}{k \cdot (l-q)}\frac{m_e^2}{(k\cdot l)^2}\left(\frac{p\cdot k}{p\cdot l}\right)^2 \frac{1}{y'}\left[xy'^2 G_1 + (1-y')G_2 + xy'(1-\frac{y'}{2})G_3\right] \quad (2)$$

where $y' = p\cdot q/p\cdot(l-k)$. Integrating over the photon direction inside a forward cone around l, defined by the angular resolution Δ_θ^f of the final state in the actual experimental situation, we have:

$$\frac{k_0^2 m_e^2}{4\pi}\int\frac{d\Omega_\gamma}{(k.l)^2} \simeq \chi_e^2 \int_0^{\Delta_\theta^f}\frac{d\theta^2}{(\theta^2+\chi_e^2)^2} = \chi_e^2\left[\frac{1}{\chi_e^2}-\frac{1}{\chi_e^2+(\Delta_\theta^f)^2}\right] \quad (3)$$

At HERA $\Delta_\theta^f \simeq 10^{-3}$, $\chi_e \simeq 2.0 \times 10^{-5}$. Therefore the above factor is approx-

---

[*] In other words the usual structure functions are given in terms of quark distributions by $G_2 = 2xG_1 = 2x(U(x) + \bar{D}(x))$, $G_3 = 2(U(x) - \bar{D}(x))$.

imately equal to 1!† Inputing the quark-parton model distributions, as before, we obtain:

$$\frac{d\sigma_L^{R,\gamma}}{dx} = \frac{\alpha\alpha_2^2}{4xs}\left\{\left[\frac{x\zeta_L}{2}+1-\left(1+\frac{1}{x\zeta_L}\right)\ln(1+x\zeta_L)\right]U(x)\right.$$
$$\left.+\left[\frac{x\zeta_L}{2}-6+4\left(1+\frac{1}{x\zeta_L}\right)\ln(1+x\zeta_L)+2\left(1-\frac{1}{x\zeta_L}\right)\text{Li}_2(x\zeta_L)\right]\bar{D}(x)\right\} \quad (4)$$

where Li$_2$ is the dilogarithm and $\zeta_L \equiv s/M_W^2$. This cross-section corresponds to a considerable background, as can be seen in Fig. 2.

4. Theoretical problems and the Asymptotic S-Matrix

The fact that the above process, calculated at an electron energy very high relative to m_e, gives a non-zero contribution identical to the massless limit of the process, makes the prediction suspicious.[5] Lee and Nauenberg's conclusion that the helicity-flip cross-section survives in the limit $m_e \to 0$ is profoundly troublesome since, if taken seriously via such arguments as the calculation of section 3, it would correspond to a mass-singularity-induced anomaly (mass-shell anomaly). In other words, a Lagrangian (such as $\mathcal{L}_{QED}$) that is chirally invariant in the massless limit produces chiral symmetry breaking effects, such as the non-decoupling of the electron-helicity states for certain processes, in the massless limit. For a complete review of the problem, the associated difficulties having to do with different regularization schemes of the mass singularities, and its resolution, see Ref.[6] and references contained therein.

Suffice it here to say that the important physical ingredient that becomes relevant in calculating the high-energy limit of certain processes in perturbation theory is the physical degeneracy of free-particle states with different particle content, within the experimental resolutions of the actual *physical* process (experiment). Consider, as an example, the Hamiltonian of Quantum Electrodynamics, $H_{QED} = H_0 + V$, where H_0 is the free-particle Hamiltonian and V the interaction Hamiltonian. Because of the masslessness of the photon, the asymptotic behaviour of scattering states $e^{-iHt}|\psi\rangle$ do not approach free-particle states $e^{-iH_0t}|\psi_0\rangle$ in the limit $t \to \pm\infty$. The long-range tale of the Coulomb potential survives in the remote past and far future, making the one-electron Fock state surrounded by a soft photon cloud. On the other hand, this same masslessness of the photon, allows the soft cloud to be unobservable within the detector energy resolution i.e., degenerate with the one-particle state. Omitting this effect from the usual Feynman-Dyson S-matrix (S_{FD}) i.e., assuming the asymptotic Hamiltonian of QED is H_0, gives rise to IR divergencies *and* doesn't describe the actual physical degeneracy. In general, omitting an asymptotic interaction from the

---

† Notice that we would recover the same result even in the *massless* limit $\chi_e \to 0$, i.e., for an exactly massless electron, after the forward integration.

asymptotic Hamiltonian of a massless gauge theory produces mass-singularities, mass-shell anomalies, and disregards physical degeneracy. The high-energy limit of a massive theory, calculated via the resulting S_{FD}, resurrects these problematic (but complementary) features.

One can properly account for the asymptotic properties of these theories, by choosing an interacting asymptotic Hamiltonian:

$$H_0 \to H_A(\Delta) = H_0 + V_A(\Delta) \tag{5}$$

The physical meaning of Δ will be explained later in this section. The corresponding Asymptotic S-matrix is

$$S_A = \Omega^{(-)\dagger}_{H,H_A} \Omega^{(+)}_{H,H_A} = \Omega^{(-)}_{H_A,H_0} S_{FD} \Omega^{(+)\dagger}_{H_A,H_0} \tag{6}$$

with the Møller wave operators defined as $\Omega^{(\mp)}_{H,H_A} \equiv \lim_{t \to \pm\infty} e^{iHt} e^{-iH_A t} = \Omega^{(\mp)}_{H,H_0} \Omega^{(\mp)\dagger}_{H_A,H_0}$. Transforming all operators in the interaction picture we may write:

$$\Omega^{(\pm)}_{H_A,H_0} = Texp[-i \int_{\mp\infty}^{0} dt V_A(t)] \ , \quad S_{FD} = Texp[-i \int_{-\infty}^{+\infty} dt V(t)] \tag{7}$$

In this picture a perturbative evaluation of S_A is straightforward. Suppose we are interested in QED radiative corrections to a basic process that occurs in a theory with an interaction Hamiltonian $V(t)$. One can write $V(t) = V^{(QED)}(t) + V^{(J)}(t)$, where $V^{(J)}(t)$ is the rest of the Hamiltonian, giving rise to the non-radiative process (such as the one in section 2). Then:

$$V^{(QED)}(t) = e \int d^3x : \bar{\Psi}^{(e)} \gamma_\mu \Psi^{(e)} : A^\mu = e \int \widehat{d^3k_1 d^3k_2} \sum_{l=1}^{8} h_l(\mathbf{k}_1, \mathbf{k}_2, \mathbf{k}_3) exp[-i(S\omega)^l t] \tag{8}$$

In the above expression S is a 8×3 sign matrix and ω stands for the energies of the particles in each vertex h_l. We can now define the asymptotic interaction Hamiltonian as

$$V^{(QED)}_A(\Delta;t) = e \int \widehat{d^3k_1 d^3k_2} \sum_{l=1}^{8} h_l(\mathbf{k}_1, \mathbf{k}_2, \mathbf{k}_3) \Theta(\Delta - |(S\omega)^l|) exp[-i(S\omega)^l t] \tag{9}$$

We see that Δ corresponds to the experimental regions of phase space characterizing a certain physical process within which the energies $(S\omega)^l$ of the particles at the corresponding vertex l are indistinguishable (degenerate). More precisely, one may write:

$$\Delta = \cup \Delta_\alpha \ , \quad \Delta_\alpha \in \{\delta_\theta^{in}, \Delta_\theta^f, \delta_E \ ...\}$$

where δ_θ^{in} = the beam angular resolution, Δ_θ^f = the final-state angular resolution, $\delta_E = \Delta E/E$, $\Delta E = max\{\Delta E^{in}, \Delta E^f\}$ =the energy resolution. It is obvious that

$$V_A^{(QED)} \begin{cases} \neq 0, & \text{if } \exists\, \alpha, i : \chi_i \equiv m_i/\omega_i < \Delta_\alpha \\ = 0, & \text{otherwise.} \end{cases}$$

This construction *defines* the high-energy limit of a theory and automatically includes the massless limit (massless theory). A perturbative evaluation of the asymptotic wave operators may be obtained from the following formula:

$$\Omega_\pm^{(n)}(\Delta) = \sum_{l_1 l_2 \ldots l_n} e^n V_{l_n} V_{l_{n-1}} \ldots V_{l_1} \frac{1}{(S\omega)^{l_1} \pm i\epsilon} \frac{1}{(S\omega)^{l_1} + (S\omega)^{l_2} \pm 2i\epsilon} \cdots \frac{1}{(S\omega)^{l_1} + \ldots (S\omega)^{l_n} \pm ni\epsilon} \tag{10}$$

with $V_{l_j} \equiv \int \widehat{d^3 k_{1_j} d^3 k_{2_j}} h_{l_j}(\mathbf{k}_{1_j}, \mathbf{k}_{2_j}, \mathbf{k}_{3_j}) \Theta_\Delta$.

5. HERA background revisited

Let us calculate again the radiative process of section 3, using the correct matrix S_A. Our choice of initial and final states will be:

$$|i\rangle = |e(1; R)\rangle \, , \quad |f\rangle = |\nu_e(1'; L)\gamma(\mathbf{k}; \lambda)\rangle$$

Considering the hadronic part of the process as an external source of W-bosons we may write:

$$[S_A^{(1)}]_{fi} = \langle f| S_{FD}^{(0)} \Omega_+^{(1)\dagger} |i\rangle + \langle f| S_{FD}^{(1)} |i\rangle \tag{11}$$

The first term is the extra contribution coming from the degeneracy of the initial state within the experimental angular resolution of the electron beam. Indeed one may show

$$\Omega_+^{(1)\dagger} |i\rangle = e \sum_{\alpha \lambda'} \int \widehat{d^3 k'} |\gamma(\mathbf{k}'; \lambda') e(1 - \mathbf{k}'; \alpha)\rangle \bar{U}_\alpha^{(e)}(1 - \mathbf{k}') \displaystyle{\not{\!\epsilon}}_{\lambda'}^*(\mathbf{k}') U_i^{(e)}(1) \frac{\Theta(\Delta - |\nu'|)}{2\omega(1 - \mathbf{k}', m_e)\nu'} \tag{12}$$

In the above $\nu' = \omega(1 - \mathbf{k}', m_e) + \omega(\mathbf{k}', m_\gamma) - \omega(1, m_e)$. Notice how the asymptotic wave operator transforms a one-particle state into an electron-photon state degenerate with that one. The second term in Eq. (11) can be calculated from the Feynman rules. The S_A-matrix element is shown in Fig. 3. Looking at the collinear phase space where the final photon is almost parellel to the initial electron, we deduce the following singularity structure for the cross-section:

$$d\sigma_L^{R,\gamma} \sim \alpha \chi_e^2 \left\{ \frac{1}{\chi_e^2 + (\delta_\theta^{in})^2} - \frac{1}{\chi_e^2 + (\Delta_\theta^f)^2} \right\} \tag{13}$$

i) Massless limit:
Obviously $\lim_{\chi_e^2 \to 0} d\sigma_L^{R,\gamma} = 0$. Therefore there are no mass-shell anomalies in the massless limit, if one computes the physically relevant S_A-matrix elements.
ii) High-energy limit:

In eq. (13) we shall have to input the experimental values of the physical resolutions. At HERA, $\chi_e \simeq 2.0 \times 10^{-5}$, $\Delta_\theta^f \simeq 10^{-3}$. For the inital (beam) angular resolution an adequate estimate can be given by a lower limit to the ability of the accelerator to distinguish a single-electron from a collinear electron-photon state going through the interaction region. Hence an upper limit is $\delta_\theta^{in} < \sigma/d^*$ where σ is the transverse radius of the beam at the interaction region and d the drift distance of the electrons from the final focus to the interaction region. At HERA, $\sigma/d \simeq 0.07\text{mm}/5.5\text{m} \simeq 1.2 \times 10^{-5}$.[†] Notice that in all cases we may write

$$\chi_e^2 + (\Delta_\theta^f)^2 \simeq (\Delta_\theta^f)^2 \;, \quad \chi_e^2 + (\delta_\theta^{in})^2 \simeq \chi_e^2 \tag{14}$$

Therefore $\lim_{exp} d\sigma_L^{R,\gamma} \sim \alpha$, and the prediction of the background made in section 3 is approximately correct. However, survival of this background is not connected to mass-shell anomalies any more since, as we showed, the massless limit of the process is indeed smooth and equal to zero.

6. Conclusion

Perturbative calculations using S_{FD} in massless gauge theories or in massive theories when the high-energy limit of a physical process is sought, are plagued by mass singularities, mass-shell anomalies, and are based on matrix elements that do not account for the physical degeneracy occuring in these cases. S_A, on the other hand, incorporates in its definition the notion of physical degeneracy and is characterized by finite and non-anomalous matrix elements. Hence it allows reliable perturbative calculations in massless gauge theories or when high-energy-limit processes are considered. Processes that involve *exactly* massless particles, and that are anomalous when calculated via S_{FD}, turn out to have a smooth massless limit equal to zero when calculated via S_A. This is due to the fact that the experimental resolutions introduced by the transformation of Fock states into coherent states cut-off the mass singularities in that case. For this reason, a discussion analogous to the helicity-flip process shows that there are no longitudinal massless photons in massless QED, contrary to some recent claims.[7] Processes that involve massive particles, but whose energy is much larger than their mass, may be calculated via S_A as well. These processes, if anomalous in the S_{FD}-approach, have a smooth massless limit equal to zero in the S_A-approach, but their high-energy (experimental) limit may be non- zero, if the experimental resolutions happen to be smaller than the corresponding mass-parameters. This is the criterion defining when the high-energy limit of a process corresponds to a massive or a massless theory. Through this approach, a SM background is identified and will have to be taken into account for the polarized DIS experiments designed at HERA.

---

[*] Actually we suspect that in reality $\delta_\theta^{in} \ll \sigma/d$.
[†] By comparison, at SLC $\sigma/d \simeq 10^{-6}$.

Acknowledgements

We are grateful to D.N. Williams, D.R.T. Jones, A. Mueller, G. Sterman, and A. Wightman for many helpful discussions at various stages of this work.

Figure Captions

Fig. 1
a. Deep-inelastic scattering via a hypothetical W_R exchange.
b. Radiative deep-inelastic scattering via a Standard Model W exchange.

Fig. 2
The ratio of the signal over the background as a function of M_R.

Fig. 3
a. The two-particle state contribution to the S_A-matrix element.
b. The one-particle state contribution, corresponding to the S_{FD}-matrix element.

REFERENCES

1. Proceedings of the Workshop *Experimentation at HERA*, NIKHEF, Amsterdam, June 9-11, 1983, DESY HERA 83/20, Oct. 1983.

2. *Proceedings of the HERA Workshop*, Hamburg, Germany, Oct. 12-14, 1987 (R.D.Peccei, ed.) Hamburg: Deutsches Elektronen Synchrotron, 1988, 2 vols.

3. G. Wolf, Lectures given at the *Advanced Study Institute on Techniques and Concepts of High Energy Physics*, 1986, St. Croix, DESY 86-089, August 1986.

4. H.F. Contopanagos and M.B. Einhorn, UM-TH-89-09 (July, 1989).

5. T.D. Lee and M. Nauenberg, Phys. Rev. 133 (1964) B1549.

6. H.F. Contopanagos, Nucl. Phys. B343 (1990) 571.
   H.F. Contopanagos and M.B. Einhorn, UM-TH-90-11; NSF-ITP-90-165i.
   H.F. Contopanagos and M.B. Einhorn, UM-TH-90-12; NSF-ITP-90-166i.

7. A.S. Gorsky, B.L. Ioffe, and A.Yu. Khodjamirian, Phys. Lett. 227 (1989) 474.

304 The Asymptotic S Matrix

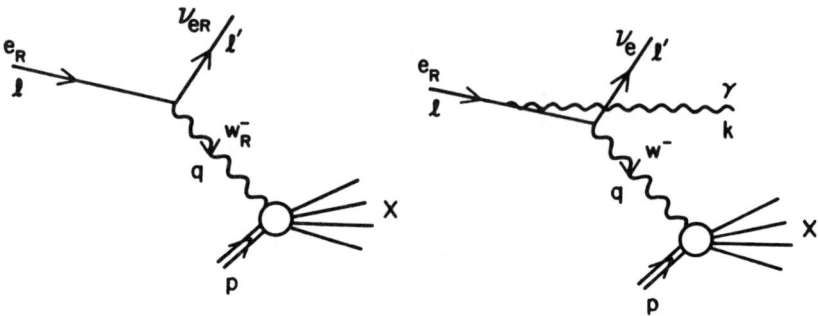

Fig. 1 a,b

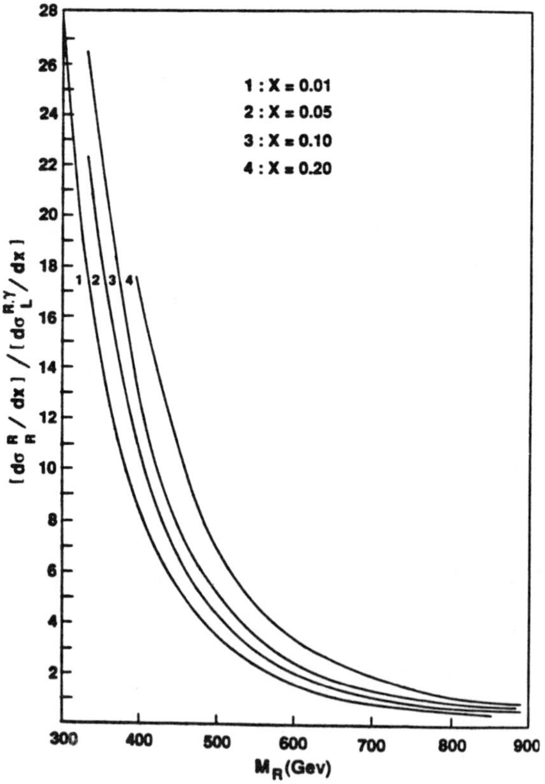

Fig. 2

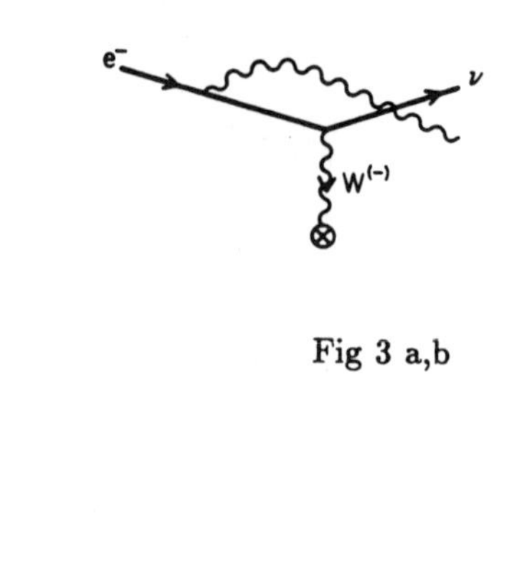

Fig 3 a,b

MEASUREMENT OF THE SPIN-DEPENDENT STRUCTURE FUNCTIONS OF THE PROTON AND NEUTRON AT HERA

R.G. Milner

Massachussets Institute of Technology, Cambridge, MA 02139 USA

ABSTRACT

HERMES is an experiment under preparation by an international collaboration of physicists from the United States, Germany, Canada, Italy and the Soviet Union to measure the deep inelastic spin-dependent structure functions of the nucleon at DESY, Hamburg, Germany. The structure functions of the proton and neutron will be determined by a completely new technique using polarized internal gas targets of hydrogen, deuterium, and ^{3}He and the longitudinally polarized 35 GeV electron beam of the HERA electron storage ring. Measurement of the deep inelastic spin-structure of both isospin states of the nucleon at the same kinematics and using the same apparatus will provide a high precision test of the Bjorken sum rule.

I. INTRODUCTION

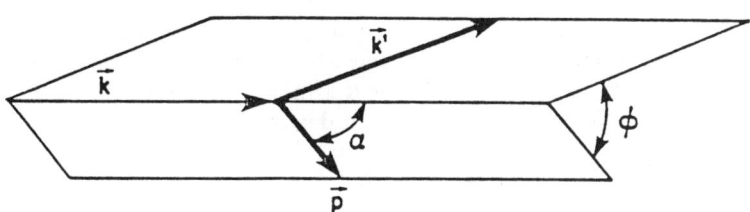

Figure 1. Definition of angles α and ϕ.

The information about the internal spin structure of the nucleons is contained in the structure functions $g_1^{p,n}(x)$ and $g_2^{p,n}(x)$ which can be measured in deep inelastic scattering of longitudinally polarized charged leptons from a polarized nucleon target. As indicated in fig. 1, we define α as the angle between the beam momentum vector $\vec{k}$ and the target polarization vector $\vec{p}$ and ϕ as the angle between the polarization plane formed by $\vec{k}$ and $\vec{p}$ and the lepton scattering plane formed by $\vec{k}$ and by the scattered lepton momentum vector $\vec{k}'$. The spin-dependent part of the deep inelastic cross section is then given by the difference of cross sections for two opposite target polarizations[1]

$$\frac{d^3(\sigma(\alpha) - \sigma(\alpha + \pi))}{dx\,dy\,d\phi} = \frac{e^4}{4\pi^2 Q^2}\left[\cos\alpha\left(\left[1 - \frac{y}{2} - \frac{y^2}{4}\gamma^2\right]g_1(x,Q^2) - \frac{y}{2}\gamma^2 g_2(x,Q^2)\right)\right.$$

$$-\sin\alpha\cos\phi\sqrt{\gamma^2(1-y-\frac{y^2}{4}\gamma^2)}\left(\frac{y}{2}g_1(x,Q^2)+g_2(x,Q^2)\right)\Bigg] \tag{1}$$

The kinematics are described by $\nu = E - E'$, the energy transfer by the virtual photon to the nucleon in the deep inelastic scattering process, and by the negative square of the invariant mass Q^2 of the virtual photon. The Bjorken scaling variables are defined as $x = Q^2/2M\nu$ and $y = \nu/E$. Here E and E' are the energies of the incoming and the scattered lepton and M is the nucleon mass. In the infinite momentum frame, x can be interpreted as the fraction of the nucleon momentum carried by the struck quark. The quantity γ is defined as $\gamma = \sqrt{Q^2}/\nu$. The two structure functions $g_1^{p,n}(x,Q^2)$ and $g_2^{p,n}(x,Q^2)$ can be separated by performing two measurements with different orientations α of the target polarization vector $\vec{p}$. For instance by polarizing the target longitudinally ($\alpha = 0$), one measures the longitudinal asymmetry $A_\|$ and by polarizing the target perpendicular to the beam direction ($\alpha = 90°$), the transverse asymmetry $A_\perp$ is obtained.

Experimentally the situation at present is very unsatisfactory. Only two measurements of $g_1^p(x,Q^2)$ have been performed (with limited statistical and systematic accuracy) for the proton [2,3,4,5]; no data at all exist for $g_1^n(x,Q^2)$ and the structure function $g_2(x,Q^2)$ is completely unexplored. In the case of the deuteron, which is a spin-1 target, there are additional leading twist spin-dependent structure functions beyond spin-$\frac{1}{2}$, namely $b_1(x)$ [6] and $\Delta(x)$ [7]. HERMES will provide the first experimental data on these quantities. Fig. 2 shows the presently available experimental information for $g_1^p(x)$ obtained by the EMC[2,3] and SLAC [4,5] experiments in deep inelastic scattering from frozen ammonia and butanol targets, respectively. The kinematic range extends from about $x = 0.7$ down to $x = 0.015$, but below $x \approx 0.1$ the x dependence is only poorly determined with error bars of 100% to 200%. Nonetheless, it is the data at small x that have sparked the recent interest in spin-dependent structure functions.

From the combined EMC and SLAC data the integral of $g_1^p(x)$, $I_1^p = \int_0^1 g_1^p(x)\,dx$, has been evaluated

$$I_1^p = 0.126 \pm 0.010 \pm 0.015, \tag{2}$$

This deviates by more than two standard deviations from the theoretical prediction of the Ellis-Jaffe Sum Rule[8] which including QCD corrections gives a value of $I_1^p = 0.189\pm0.005$ (or according to a recent analysis[9] $I_1^p = 0.175\pm0.018$).

The Bjorken Sum Rule relates the integral over x of the difference of $g_1^p(x)$ and $g_1^n(x)$ to the ratio of the axial vector to vector coupling constants measured in Gamow-Teller nucleon beta decay, g_A/g_V. In the scaling limit and taking into account QCD radiative corrections[10,11] it can be written

$$\int_0^1 [g_1^p(x) - g_1^n(x)]\,dx = \frac{1}{6}\frac{g_A}{g_V}(1 - \frac{\alpha_s}{\pi}). \tag{3}$$

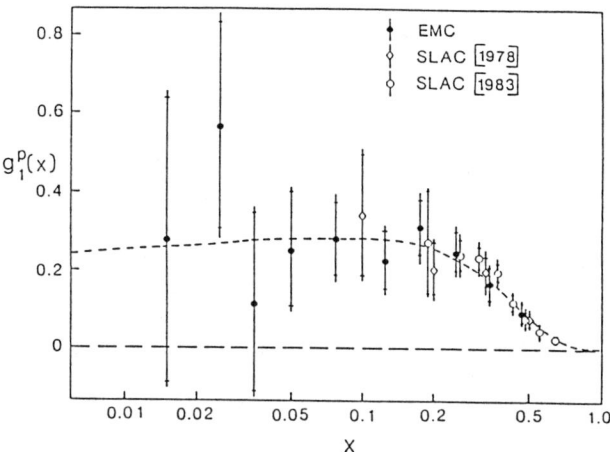

Figure 2. The world's data on $g_1^p(x)$.

Assuming the validity of the Bjorken Sum Rule and a value of 0.191 ± 0.003 for $Q^2 = 10.7$ GeV^2, one concludes that

$$I_1^n = -0.065 \pm 0.010 \pm 0.015 \quad (4)$$

and hence that the spin dependent structure function $g_1^n(x)$ is much larger than hitherto assumed[12] and is negative over a wide range of x.

Using arguments based on SU(3) flavor symmetry [13,14,15], one can calculate the mean z component of the spin carried by each quark flavor in a proton with spin $s_z = 1/2$. The surprising result of this analysis is that little of the proton spin originates from the spin of the quarks

$$<s_z>_{quarks} = 0.060 \pm 0.047 \pm 0.069, \quad (5)$$

and that the strange sea has a substantial polarization opposite to the spin of the parent proton

$$<s_z>_{s-quarks} = -0.095 \pm 0.016 \pm 0.023. \quad (6)$$

These conclusions are supported by the results of a low energy elastic neutrino nucleon scattering experiment[16]. The above results have led to an extensive discussion of the internal spin structure of the nucleon and how quark and gluon spin and orbital angular momentum contribute to the nucleon spin. It is clear that new experimental data are required.

II. HERMES

HERMES will use a new technique to measure the asymmetries in deep inelastic scattering of polarized electrons from polarized protons and neutrons[17]. The experiment will employ internal polarized atomic gas targets of thickness $\approx 10^{14}$cm^{-2} and polarization 50% to 80% placed in the 60 mA circulating longitudinally polarized electron beam of the HERA storage ring, at DESY, Hamburg, Germany. HERA is a new storage ring, 6.3 km in circumference, which will provide collisions between 35 GeV electrons and 820 GeV protons, as shown in Figure 3. The electron ring has been commissioned at 27 GeV and it is planned to operate in excess of 30 GeV in 1991. The electron ring operates at 500 MHz with 220 bunch buckets and each bunch will contain 4×10^{10} electrons. The bunch is 27 ps in length and the spacing between bunches is 96 ns. The electron and proton beams will be separated by about 80 cm at the location of our experiment.

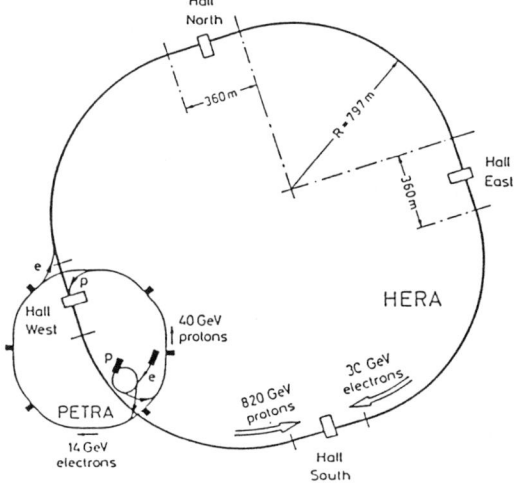

Figure 3. Layout of the HERA storage rings.

To study the proton, a polarized atomic hydrogen target will be used. For the neutron, polarized deuterium and ^{3}He targets will be used. Since the deuteron is an isoscalar target we will measure the sum of proton and neutron asymmetries. The neutron asymmetry can be deduced by subtracting the proton asymmetry from the deuteron asymmetry. In the case of ^{3}He, to a good approximation the two protons in this nucleus have opposite spins, and so the asymmetry is due to the neutron alone. Thus, the deep inelastic neutron asymmetry can be measured in two independent ways. This intuitive picture is supported by recent measurements from TRIUMF[30] and Bates[31] which are in good agreement with a calculation using 2-body and 3-body nucleon momentum distributions[18]. These measurements become feasible because of both the development of a new generation of polarized hydrogen, deuterium, and ^{3}He internal targets and because of the availability of longitudinally polarized high energy electrons in the HERA electron storage ring. The electrons will be polarized transversely by the Sokolov-Ternov effect[19] and their spins rotated into the longitudinal direction at the interaction regions by magnets. Recently transverse electron polarization has been observed at 29 GeV at TRISTAN, KEL, Japan and at 45 GeV at LEP, CERN (see the talks in this proceedings). Measurements of the transverse polarization in HERA will commence in 1991. HERMES will take data with a luminosity of $\approx 10^{32} \text{cm}^{-2} \text{s}^{-1}$.

III. POLARIZED INTERNAL GAS TARGETS

HERMES will employ polarized internal gas targets. The basic scheme is to direct a flux of polarized atoms into a thin-walled storage cell[21,22], which has openings for entry of the atoms and for entry and exit of the circulating beam. With a flux of 10^{17} polarized atoms per second the target thickness for a typical storage cell would be about 10^{14} atoms/cm^2. The polarized atoms typically spend about 5 ms in the cell and suffer about 500 bounces on the wall of the storage cell. For the hydrogen and deuterium atoms, this represents a strong depolarization mechanism and the cell wall must be coated with materials such as drifilm or teflon. For the polarized ^{3}He atom, because of its closed atomic structure, storage times of hundreds of seconds have been achieved[26] and hence it is not necessary to coat the cell. The optimal coating for hydrogen and deuterium has been studied in great detail by the Argonne and Madison groups in our collaboration. In the case of the hydrogen and deuterium targets, a holding field of order several kgauss is necessary to prevent depolarization by the intense magnetic field of the circulating beam[20]. Again, for the polarized ^{3}He atoms, this is not necessary.

To generate the flux of polarized hydrogen and deuterium atoms we will employ an atomic beam source. An electron polarized beam of atoms is generated by passing a flux of atoms from a nozzle through a Stern-Gerlach magnet. This is followed by a region where rf transitions are induced between hyperfine states of the atoms in order to select different nuclear polarization states and to provide rapid spin reversal. A source of this kind for an internal target in an antiproton ring is under construction by the FILTEX collaboration[23,24] and it will be used also for the HERA measurement. It will deliver 10^{17} atoms/sec in a single substate of hydrogen.

At Novosibirsk an Argonne-Novosibirsk collaboration has carried out the first measurements with a polarized internal target of deuterium in the VEPP-3 electron storage ring[25,27]. A drifilm coated storage cell fed by an atomic beam source has been successfully operated over a six month period. The storage cell was found to increase the target thickness as anticipated and no loss in polarization was observed. At present, a second stage internal target is being installed, where the additional capability to open and close the cell has been added.

In addition, at Argonne National Laboratory polarized hydrogen and deuterium gas targets have been under development using the technique of optical pumping spin-exchange. The Argonne group has achieved a polarization of 30% at a flux rate of 2.5×10^{17} polarized atoms per second[25]. Work is in progress to achieve a polarization of at least 50% at a flux rate of 4×10^{17} polarized atoms per second[25]. The target polarization will be monitored by extracting a beam from the storage cell and analyzing it by rf spectroscopy.

At Caltech a polarized ^{3}He external target of thickness 10^{19} atoms/cm^2 has been developed[26]. The technique used is polarization of the 2^3S metastable state via laser optical pumping, followed by polarization of the ground state through metastability exchange. The first measurement of spin-dependent electron scattering from ^{3}He was carried out using this target by a collaboration from Caltech and MIT at the Bates Linear Accelerator Center this year[31]. The measured asymmetry at the quasielastic peak was in good agreement with the picture of the polarized ^{3}He nucleus as an effective polarized neutron. An internal target of polarized ^{3}He atoms is under construction at MIT-Bates. It is anticipated that the first measurements with this target will take place at the Indiana University Cyclotron Facility in 1991.

IV. HERMES ELECTRON SPECTROMETER

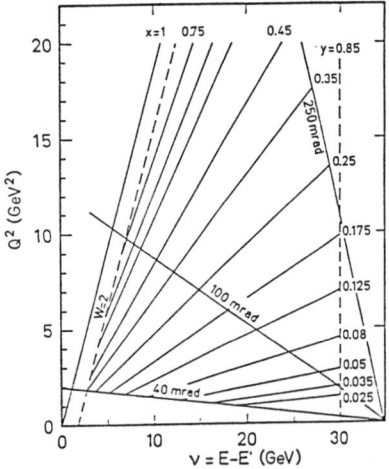

Figure 4. Kinematics accessible for an electron beam energy of 35 GeV.

Fig. 4 shows the kinematic region accessible for an incident beam energy of 35 GeV. We demand a minimum Q^2 of 1 (GeV/c)^2 to allow interpretation of the data in terms of the quark parton model. An upper cut at $y = 0.85$ removes the region of large radiative corrections and high hadronic background and a lower cut at $y = 0.15$ suppresses events from elastic scattering and resonance production. With these cuts the accessible x range extends from 0.02 to 0.8. It can be nearly completely covered by a spectrometer with an angular acceptance of $40 < \theta < 200$ mrad, where θ is the electron scattering angle.

Fig. 5 shows the planned detector configuration for HERMES. The spectrometer is very similar in design to that used by the UA6 group at CERN. To suppress low energy charged background, to discriminate electrons from positrons and to improve pion rejection a substantial magnetic field is required. This is accomplished by the use of a 1.5 Tm dipole magnet which is divided into two symmetrical parts by a horizontal iron plate. Both electron and proton beams, which are at the same height and horizontally separated by 88 cm, traverse this plate. The detector accepts particles with angles between 40 and 140 mrad vertically and 40 to 200 mrad horizontally. Charged particles will be tracked by silicon strip detectors at the target region, drift chambers in front of the magnet, proportional chambers in the magnet and drift chambers behind the magnet. The electron energy will be measured by a lead glass wall possibly including a preshower part of the same material, which provides good energy resolution ($\approx \frac{5\%}{\sqrt{E}}$), good pion discrimination, as well as coarse position measurement. This position measurement will form part of the first level trigger in conjunction with a scintillator hodoscope. To reject pions a transition radiation detector with a rejection factor of at least 100 will be used.

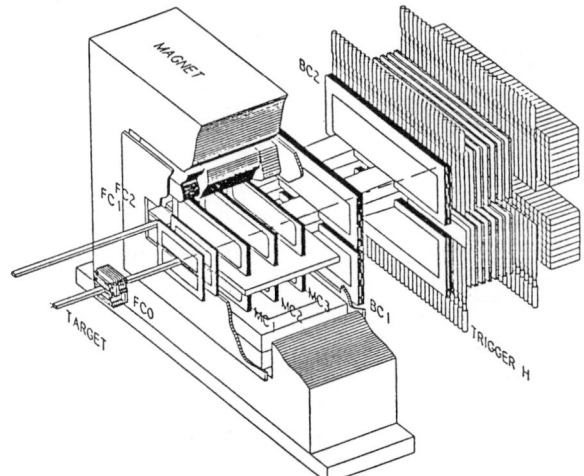

Figure 5. Schematic view of the HERMES electron spectrometer.

With this detector configuration, the energy resolution of the spectrometer is about 2% and the angular resolution is about 1 mrad. The resolution in Q^2 and x is about 5% over most of the kinematic plane. The rates of scattered particles in the detector have been studied extensively. The deep inelastic electron rate is about 10 Hz. The charged hadron rates were based on measurements from SLAC[28] and were cross-checked with the Lund code. The total charged hadron rates seen by the detector are relatively low and are easily handled using standard techniques. For example, the total direct charged particle rate in the annulus corresponding to 2° - 3° is about 700 Hz, or about 1 every 15,000 bunches. The ratio of pions to electrons is modest - a maximum of about 250 for the 7° - 8° at the lowest useful final energy. This can be easily handled by the TRD and electromagnetic shower counters. Electrons from Moller scattering constitute the highest background rate for the experiment. For an incident energy of 35 GeV, the cross-section is essentially independent of angle between 2° and 15° and gives a total rate in the detector of about 2 MHz in the case of the ^{3}He target. The energy of the electrons is low, however, varying from about 500 MeV at 2° to about 15 MeV at 15°.

V. EXPECTED ACCURACY OF HERMES

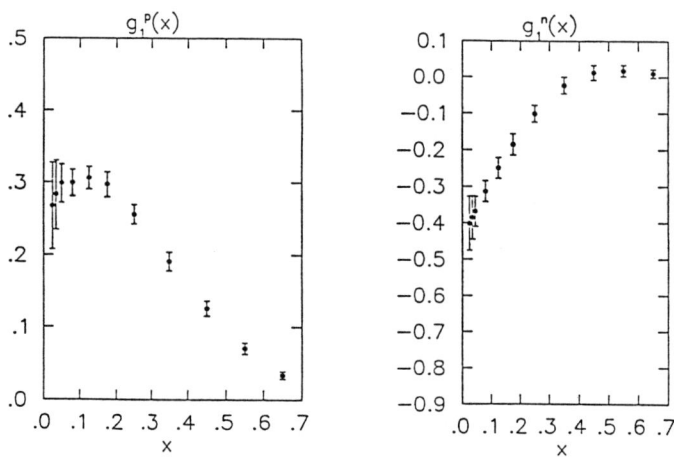

Figure 6. This figure shows the expected precision in a 400 hour run at HERA on (a) a 80% polarized hydrogen internal target with luminosity 3.5×10^{31} cm^{-2}s^{-1} and (b) on a 50% polarized ^{3}He internal target with luminosity 10^{32} atomscm^{-2}s^{-1}. The beam polarization has been assumed to be 50%.

Figures 6a and 6b show the precision as a function of x attainable in a 400 hour run by HERMES in a measurement of $g_1^p(x)$ and $g_1^n(x)$ on polarized proton and ^{3}He targets. The region in x extends from 0.02 to 0.8 and in Q^2 from 1 to 20 (GeV/c)^2. The low x limit is determined by elastic radiative tails and backgrounds at high $y = \frac{\nu}{E}$. In the case of ^{3}He a total systematic error of $\pm 15\%$ has been included. The value of $A_1^p(x)$ used was the measured value from the recent EMC data. The solid line in figure 6b is the prediction of a modified Carlitz-Kaur model which is in good agreement with the new EMC data and is constrained to obey the Bjorken sum rule[29].

We see that the statistical precision attainable at low x, where the main contribution to the sum rule arises, is very good. Systematic errors in HERMES will be reduced to a level comparable to the statistical error. The goal is to reach 3-5% for beam polarization and 2-3% for target polarization. It should be possible to perform a precise test of the fundamental Bjorken sum rule, deduce the fraction of nucleon spin carried by the quarks for both isospin states of the nucleon, and contribute greatly to our understanding of the structure of the nucleon. It is important that these measurements be carried out, especially in the light of recent data, which indicate that our understanding of the spin-structure of the nucleon is very incomplete.

VI. TIMESCALE

The HERMES collaboration has been occupied with the design of the experiment for over three years. In September 1990, HERMES was approved by DESY for about 10% of dedicated beam time at HERA conditional upon about 60% transverse electron polarization being observed in HERA. The first measurements of electron polarization will commence in 1991. The collaboration believes it will take approximately two years to construct the experimental apparatus. Thus, it is anticipated that data taking will commence in 1993.

This work is a collaborative effort of groups from Alberta, Argonne, Caltech, Heidelberg, Illinois, Leningrad, Los Alamos, Madison, Marburg, MIT, New Mexico State, Munchen, Torino, TRIUMF, and William and Mary. The author's research was supported in part by the U.S. Department of Energy, Nuclear Physics Division, under contract W-31-109-ENG-38, by the National Science Foundation (PYI) and by the MIT Sloan Fund.

REFERENCES

[1] R.L. Jaffe, CTP No. 1798, November 1989, submitted to Comments in Nuclear and Particle Physics
[2] EMC, J. Ashman et al.; Phys. Lett. B206 (1988) 364.
[3] EMC, J. Ashman et al.; Nucl. Phys. B328, 1 (1989)
[4] M.J. Alguard et al.; Phys. Rev. Lett. 37 (1978) 1261; 41 (1978) 70.
[5] G. Baum et al.; Phys. Rev. Lett. 51 (1981) 1135.
[6] P. Hoodbhoy, R.L. Jaffe and A. Manohar, Nucl. Phys. B312(1989)571.
[7] R.L. Jaffe and Aneesh Manohar, Phys. Lett., B223, 218 (1989)
[8] J. Ellis and R.L. Jaffe, Phys. Rev. D9 (1974), 1444
[9] R.L. Jaffe and A. Manohar; preprint CTP#1706 (1989), CTP#1712
[10] J. Kodaira, Nucl. Phys. B165 (1980) 129.
[11] J. Kodaira et al, Nucl. Phys. B159(1979)99.

[12] R. Carlitz and J. Kaur; Phys. Rev. Lett. 38 (1977) 673, 1102.
[13] S.J. Brodsky, J. Ellis and M. Karliner; Phys. Lett. B206 (1988) 309.
[14] J. Ellis and M. Karliner; Phys. Lett. B213 (1988) 73.
[15] M. Glueck and E. Reya; Z. Phys. C39 (1988) 569.
[16] L.A. Ahrens et al.;Phys. Rev. D35 (1987) 785.
[17] HERMES proposal to DESY, January 1990
[18] R. Woloshyn, Nucl. Phys. B (in press)
[19] A.A. Sokolov and I.M. Ternov, Sov. Phys. Doklady, 8, 1203 (1964)
[20] E.R. Kinney, 'Simulations of the Atomic Polarization and Denzity in the HERMES Polarized Internal Target', HERMES report 5-90
[21] Proceedings of the Workshop on Polarized Targets in Storage Rings, ed. R.J. Holt, Argonne Illinois (1984); Proceedings of the Workshop on Polarized Internal Targets, Minneapolis, Minnesota, ed. W. Haeberli and E. Steffens (1988)
[22] W. Haeberli, AIP Conference Proceedings No. 128 *Nuclear Physics with Stored, Cooled Beams*, ed. P. Schwandt and H.O. Meyer, p. 251 (1985)
[23] G. Graw et al., Proceedings of the Fourth Lear Workshop, Vol. 14 Nuclear Science Research Conference Series (Harwood Academic Publishers) ed. C. Amsler, p. 221 (1988)
[24] W. Haeberli, T. Wise, and A. Converse, Proceedings of the Fourth Lear Workshop, Vol. 14 Nuclear Science Research Conference Series (Harwood Academic Publishers) ed. C. Amsler, p. 217 (1988)
[25] R.A. Gilman et al., Phys. Rev. Lett. 65, 1733 (1990)
[26] R.G. Milner et al., Nucl. Instr. and Meth. A274, 56 (1989); C.E. Woodward et al., to be published
[27] S.I. Mishnev et al. , Proceedings of the Eighth International Symposium on High Energy Spin Physics, Minneapolis, Minnesota, 8-17 Sept. 1988
[28] A. Boyarski et al., Phys. Rev. Lett. 18, 363 (1967)
[29] A. Schafer, Phys. Lett. 208B, 175 (1988)
[30] O. Hausser, contribution to Paris Spin Conference; A. Rahav et al., TRIUMF preprint December 1990, submitted to Phys. Rev. Lett.
[31] C.E. Woodward et al., Phys. Rev. Lett., 65, 698 (1990)

PROPOSAL TO MEASURE THE SPIN-DEPENDENT NEUTRON STRUCTURE FUNCTION AT SLAC

P. A. Souder
Syracuse University, Syracuse, NY 13210

ABSTRACT

We are planning to measure the spin-dependent structure function of the neutron, g_1^n, in the kinematic range $0.04 < x < 0.6$ for $Q^2 > 1 (\text{GeV}/c)^2$. The experiment will use the 22.7 GeV polarized electron beam at SLAC scattering from a polarized ^{3}He gas target in End Station A. The scattered electrons will be detected by two fixed arm spectrometers. We expect to obtain our data in 1992.

INTRODUCTION

Measurements of asymmetries in the spin-dependence of deep inelastic scattering of polarized electrons from polarized nucleons determine how the spins of the constituent quarks are aligned relative to the spin of the nucleon. The simple SU(6) model predicts large asymmetries on the order of 55% for the proton. Indeed, this value is observed experimentally[1] at the kinematics (Bjorken $x \approx 1/3$) where this picture is most likely to have some validity. The same model gives zero asymmetry for the neutron, a lackluster prediction that has done little to inspire any experimental work.

More recently, results from the EMC collaboration[2], which cover a larger range of x, have stimulated a great deal of new interest in the subject. The reason is that the rigorous QCD prediction, the Bjorken sum rule[3]

$$\int_0^1 \left(g_1^p(x) - g_1^n(x) \right) dx = \frac{1}{6} \frac{g_A}{g_V} (1 - \alpha_s(Q^2)/\pi) = 0.191 \pm 0.002, \quad (1)$$

where $g_1^p(g_1^n)$ are the spin dependent structure functions of the proton(neutron), appears to be violated if $g_1^n = 0$. Here we use the notation of Jaffe[4]. In particular, the data yield

$$\Delta p \equiv \int_0^1 g_1^p(x) dx = 0.114 \pm 0.012(stat.) \pm 0.026(syst.), \quad (2)$$

which implies with the Bjorken sum rule that

$$\Delta n \equiv \int_0^1 g_1^n(x) dx = -0.077 \pm 0.029, \quad (3)$$

in striking contrast to the naive predictions.

An enormous number of papers have been published recently about this result and its various implications. I will not dwell on the theory, which has been discussed in other talks at this meeting. There is now great interest in measuring g_1^n, both to complete the test of the Bjorken sum rule and to obtain information on this fundamental quantity. New experiments are planned at several laboratories, including CERN, HERA, and SLAC. This paper will describe the plans for the SLAC experiment, which will be performed by a collaboration of physicists from American University, Harvard, Princeton, SLAC, Stanford, Syracuse, and Wisconsin.

EXPERIMENTAL METHOD

The goal of the experiment at SLAC is to obtain data on g_1^n covering $0.04 < x < 0.6$ for $Q^2 > 1(\text{GeV}/c)^2$. In order to obtain the data as soon as possible, we are designing the apparatus to be similar to the proven design of the proton experiment at SLAC, E130[1]. However, there will be many important changes to accommodate a new target and to take advantage of new technonogy.

The basic idea of the experiment is illustrated in Figure 1. A beam of 23 GeV polarized electrons impinges upon a polarized ^{3}He target. Two independent spectrometers, one at 4.5° and one at 7°, detect the scattered electrons. An experimental asymmetry

$$A_{exp} = \frac{N \uparrow\downarrow - N \uparrow\uparrow}{N \uparrow\downarrow + N \uparrow\uparrow}, \qquad (4)$$

where $N \uparrow\downarrow$ ($N \uparrow\uparrow$) is the number of electrons detected by one spectrometer when the spins of the electrons and ^{3}He are antiparallel(parallel), is then obtained.

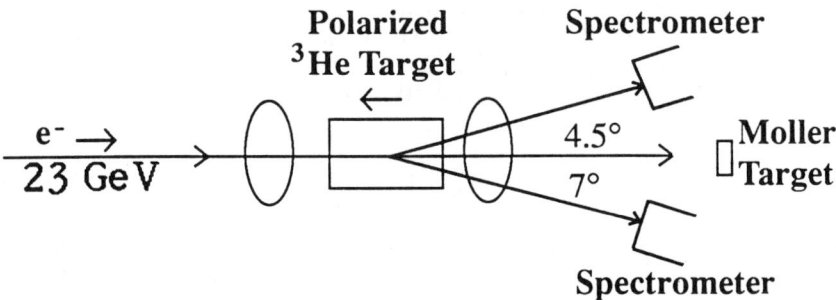

Fig. 1. Schematic diagram of the apparatus.

The beam polarization P_B is measured by Møller scattering from a thin magnetized foil, and the target polarization P_T is measured with NMR techniques. Then the neutron asymmetry A_1^n may be obtained from the formula

$$A_{exp} = A_1^n f P_T P_B D, \qquad (5)$$

where

$$f = \frac{\text{\# of polarized nucleons}}{\text{total \# of nucleons}} \qquad (6)$$

and D is a virtual photon depolarization factor determined by the kinematics. Finally, $g_1^n(x)$ is given by

$$g_1^n(x) = \frac{A_1^n(x) F_2^n(x)}{2x(1 + R(x))}, \quad (7)$$

where $F_2^n(x)$ is the usual unpolarized structure function and $R(x)$ is the ratio of cross sections for longitudinal versus transverse virtual photons.

APPARATUS

A high density polarized ^{3}He target[5] will provide the polarized neutrons. It is based on spin transfer from optically pumped Rb vapor, a technique that has been pioneered at Princeton[6] and applied recently to experiments at TRIUMF, LAMPF and Bates. Our target, which will be 30 cm long and operate at 10 atm, will be the largest of this type ever built. The recent development of the Ti-sapphire laser used for the optical pumping is an important factor in making this size target practical.

The target, shown in Figure 2, consists of two cells. The top cell, which is heated, contains the Rb vapor. Polarized laser light incident on this cell polarizes the Rb which in turn polarizes the He. Since the ionization of the Rb due to the beam is a significant depolarizing mechanism, we will provide a cooled lower cell without Rb vapor through which the beam passes. However, the He, which has a polarization lifetime of tens of hours, diffuses from one cell to the other, maintaining a uniform polarization on the order of 45%. The depolarizing effect of the beam in the lower cell is acceptable.[7] Also shown in the diagram are coils used to measure the polarization by the NMR method of adiabadic fast passage. Another possible method for polarimetry is to measure a frequency shift.[8]

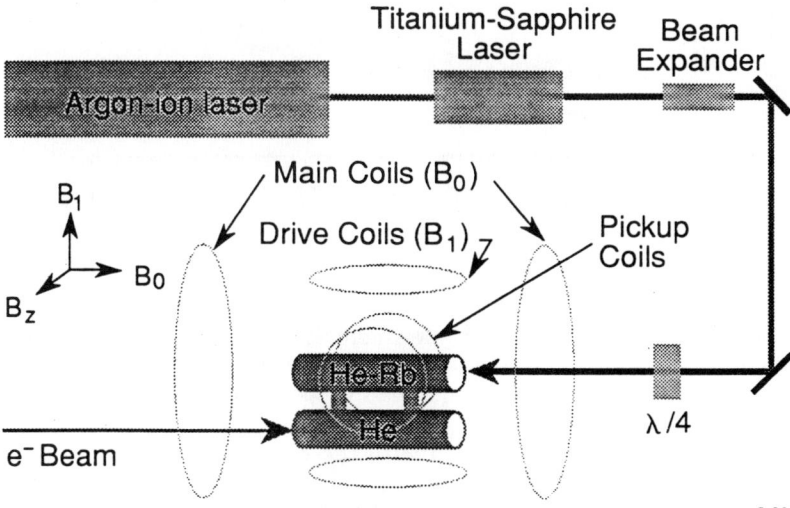

Fig. 2. Diagram of the polarized ^{3}He target. Shown are the argon-ion pump laser, the tunable titanuim-sapphire laser, laser optics, connected pumping and target cells, and NMR coils. The target is polarized along the B_0 direction.

318 Spin-Dependent Neutron Structure Function at SLAC

The polarized beam will be provided by a GaAs source being developed for the SLC. The major modification for this experiment will be providing a longer pulse length of 1.5 μsec. The polarization of the beam, which will be on the order of 40%, will be measured by Møller scattering to an accuracy of ±5%.

The experiment will use two independent spectrometers, one at 4.5° and the other at 7°. As shown in figure 3, they will use dipole magnets borrowed from the SLAC 8 GeV and 20 GeV spectrometers. They are designed to have a moderate solid angle acceptance over a large momentum range. This will enable data to be obtained simultaneously over the range $0.05 < x < 0.6$. Complementary data obtained with a lower central momentum will reduce the minimum x measured to 0.04.

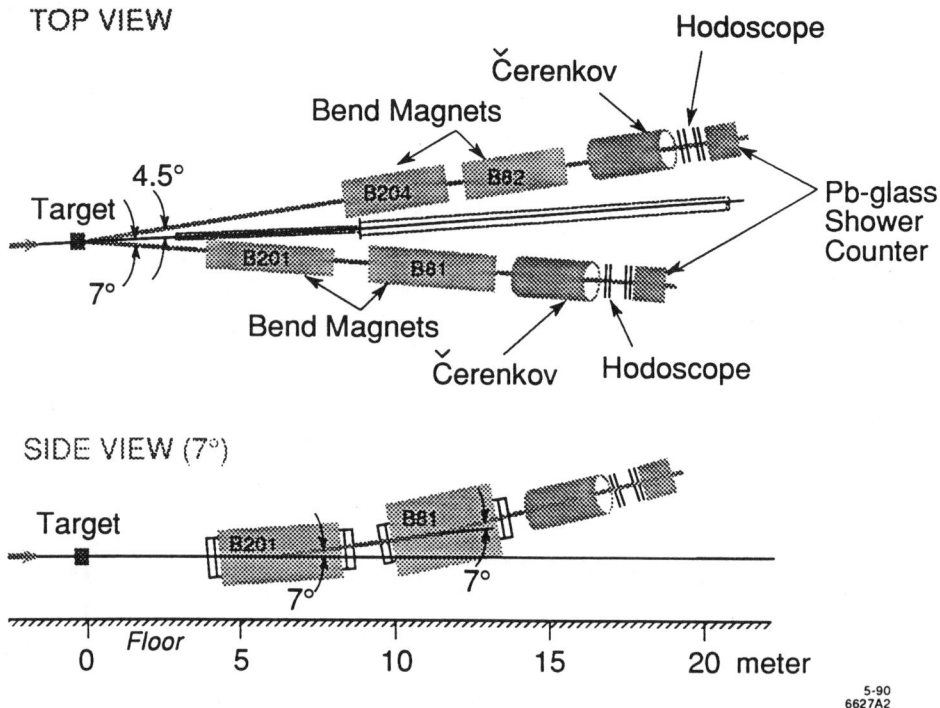

Fig. 3. Spectrometers for SLAC Experiment E142. Top and side views of the spectrometers for E142. Each spectrometer uses two dipoles. The detector package consists of gas Čerenkov counters, lead glass, and scintillation hodoscopes.

The detector package is designed with the requirement that more than one useful electron be detected during a SLAC beam pulse. Successful operation at these rates has not previously been achieved at SLAC. Therefore, there are some unique features in our design. There will be three primary types of particle detectors, gas Čerenkov counters, lead glass Čerenkov total absorption counters, and scintillation hodoscopes. The gas Čerenkovs will identify the electrons, the segmented lead glass counters will determine the energy and direction of the electron, and the hodoscopes will provide tracking information to verify the performance of the lead glass. Wire chambers will also be installed, but will be limited to recording one event per pulse.

One novel feature will be the use of pipeline TDC's to record the timing information for all the devices without electronic dead time. This feature will enable us to operate at higher event rates and also carefully evaluate any errors caused by the high rates.

PROJECTED SENSITIVITY

The quality of data that we plan to obtain (assuming 90 hours of good data recorded on tape) is shown in figure 4. Also shown are the results from previous experiments on the proton. Our statistical errors are quite competitive. They will contribute 0.004 to the error in Δn.

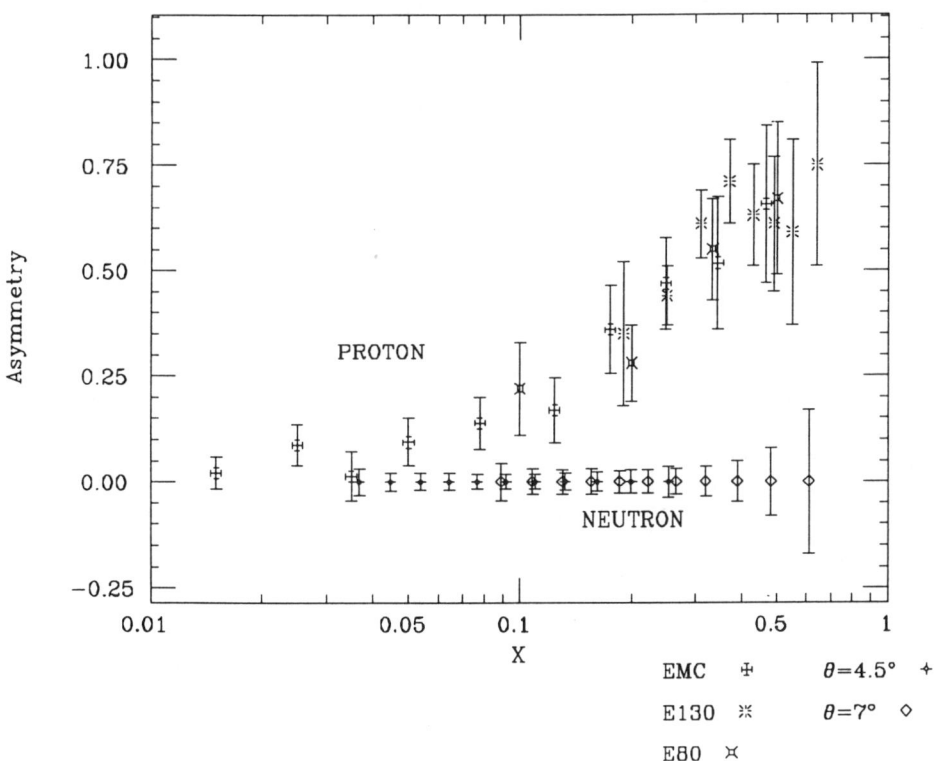

Fig. 4. Spin Dependent Structure Functions. Projected errors for E142 (solid points) together with the proton data from SLAC and EMC.

Since we can flip the beam polarization rapidly and can also change the target polarization, we expect $A_{exp} \approx 0.001$ to be dominated by the statistical errors. Indeed, the requirements for controlling spurious asymmetries are trivial compared to those required for some experiments using polarized electrons.[9]

Our systematic errors arise from the various factors of equations (5-7). Significant contributions will come from the P_B (5%), P_T (5%), f (4%), $F_2^n(x)$ and $R(x)$ (5%), and radiative corrections (2%). These estimates, which are quite conservative, are based on experience with similar experiments completed in the past. The errors combined amount to 10% of the measured asymmetry. They will dominate the statistical errors in g_1^n, contributing 0.007 to the error in Δn, if the asymmetry is as large as predicted by equation 3.

A second class of error arises from the fact that we are using ^{3}He instead of free neutrons.[10] To first order, polarized ^{3}He is just a polarized neutron together with a spinless proton pair. Corrections due to small components of the ^{3}He wavefunction, final state meson exchange currents, and possible three body forces, are expected to be on the order of 10%. There is controversy over what the error on these corrections should be; some experts might argue that they are as small as a few percent. At present, we use a very conservative estimate of 20% or 0.015 to Δn. This should be the dominant uncertainty in our experiment. However, future theoretical and experimental work might result in a significant decrease in this error.

Our final error in Δn will be 0.015 if we add the above errors in quadrature. With the present proton data (error on Δp of 0.024), we will be able to test the Bjorken sum rule to 0.024 or 12% of the expected value.

ACKNOWLEDGEMENT

This work is supported in part by the U. S. Department of Energy.

REFERENCES

1. G. Baum et al., Phys. Rev. Lett. **26**, 1135 (1983).
2. J. Ashman et al., Phys. Lett B **206**, 364 (1988); Nucl. Phys **B328**, 1, (1989).
3. J. D. Bjorken, Phys. Rev. **148**, 1467 (1966); Phys. Rev. D **1**, 1376 (1970).
4. R. L. Jaffe, Comments Nucl. Part. Phys **19**, 239 (1990).
5. T. E. Chupp et al., Phys. Rev. C **36**, 2244 (1987).
6. W. Happer et al., Phys. Rev. A **29**, 3092 (1984)
7. K. P. Coulter et al., Nucl. Instrum. Methods Phys. Res., Sect A **276**, 29 (1989).
8. S. R. Schaefer et al., Phys. Rev. A **39**, 5613 (1989).
9. P. A. Souder et al., Phys. Rev. Lett. **65**, 694 (1990).
10. R. M. Woloshyn, Nucl. Phys. **496A**, 749 (1989).

ADVANTAGES OF POLARIZATION EXPERIMENTS AT RHIC*

David G. Underwood
High Energy Physics Division, Argonne National Laboratory
Argonne, Illinois 60439 USA

ABSTRACT

We point out various spin experiments that could be done if the polarized beam option is pursued at RHIC. The advantages of RHIC for investigating several current and future physics problems are discussed. In particular, the gluon spin dependent structure function of the nucleon could be measured cleanly and systematically. Relevant experience developed in conjunction with the Fermilab Polarized Beam program is also presented.

Some people may think that the primary advantage of polarized colliding beams is to eliminate the unpolarized background which is present with solid fixed targets. The newer polarized target materials such as Li^6D make this less of an advantage than it used to be, but in any case, the biggest advantage is in the cm energy and luminosity available to do the apprporiate experiments for structure functions and particle searches.

For structure functions, RHIC seems to have a unique advantage in the root-S and luminosity. These allow one to do 2 to 2 physics with observation of the final two jets or gamma plus jet for applying kinematic constraints to determine x1 and x2. The use of such kinematic constraints is problematic for fixed target, even at 2 or 3 TeV. Also, the root-S at RHIC is lower than at SSC so that the moderate xBJ values are attainable, not just very low x.

Having the kinematic constraints is important for several reasons. In a fixed target experiment such as a direct gamma experiment experiment one generally does not measure the away side jet well if at all. At Fermilab fixed target polarized beam energies and luminosities an experiment would then measure the spin structure function over a region around x of .2 or .3, weighted in x by the usual structure functions and cross section. The same x coverage is true of experiments which depend on gluon-gluon fusion to produce χ_2. One depends on the physics process to choose the gluon as opposed to the quark. With kinematic constraints from well measured jets, one has an extra handle as well as being able to measure as a function of x. The extra handle comes from the fact that the valance quark densities are higher than the gluon densities at large x and smaller at low x. Furthermore, the polarization of valance quarks has been measured to increase with x and is almost .4 at x of 0.2.

So at RHIC,

1) One measures the structure function as a function of x,
2) In some kinematic regions one is allowed a kinematic choice of gluon vs. quark since x1 and x2 are determined and quark and gluon densities have different x dependence, and
3) In some regions a valance quark of known high polarization can be chosen to investigate the low x gluon region where some of the physics questions reside at present.

A simple and illuminating way to compare physics coverage of different facilities is to use a kinematics program for 90 degree parton-parton scattering with event weighting using structure functions and parton cross sections, and to plot contours of constant event probability. We have done this for Fermilab and RHIC energies and luminosities and have used EHLQ structure functions. As an example we find that a Fermilab fixed target direct gamma experiment could get 5% error bars for a two spin measurement at x1 approx = x2 approx 0.2, pt =5, lumonisity 5 E 29. At RHIC we can measure as a function of x from x1 = x2 =.1, pt = 25, η = 0 to x1 =.5, x2 = .02, pt = 25, η = 2.3 with about 5% error bars on direct gamma plus jet at lumonisity of about 5 E 31. As pointed out by others at this workshop, one could get very small error bars at pt = 20 and lumonisity of 1E 32.

Another point is that at large rapidity, small x of gluon, where it may be difficult to do a direct gamma experiment, it is probably not necessary to use direct gammas. Due to the distributions of quarks and gluons two jet events with both jets at large rapidity are mainly from a quark at large x and a gluon at small x and the use of the physics process to select gluons is not necessary. Also, one may do a poor-man's jet experiment with π°s as the leading particles of jets.

Apparently, protons on deuterons is not possible for the machine, but perhaps one could do deuterons on dueterons after the nucleon-nucleon is understood.

Having polarized protons on polarized deuterons would allow comparison of gluon spin distributions for protons vs. neutrons similar to the comparisons of quarks for checking the Bjorken sum rule, etc. According to some people, one should think about such comparisons for the distributions of polarization of strange sea quarks also.

A separate issue discussed at this workshop was measurement of the beam polarization. In our experiments we have established that three physics processes are suitable for polarimeters in this energy range. These are Coulomb-Nuclear interference, inverse Primakoff effect, and pion inclusive production at large x_F. All of these involve very forward particles, and some probably are best done from a gas jet target, although this should be studied.

The analyzing power of Coulomb-Nuclear interference[6] is calculated from scattering amplitudes. It involves the dominant imaginary strong amplitude and the real part of the EM spin flip amplitude. The peak analyzing power is about 5% and in a real experiment with resolution and acceptance, the effective analyzing power may be 3%. The advantage of C-N for RHIC is that the scattered particle would go down the beam pipe to Roman Pot detectors and the recoil from a gas jet target could be measured with silicon detectors. It may be that beam-beam collisions are also suitable, as elastic events are sufficiently clean at small angle to have been measured in SPS.

The Primakoff polarimeter[7] has a very high analyzing power which is related to measured asymmetries in low energy π° photoproduction. It may be a mismatch to RHIC unless one has close to 50 meters in a straight section for measuring forward protons outside the beam pipe. Also, a lead target must be used, possibly lead evaporated onto a carbon filament which would pass through the beam. Can one do proton-lead collisions in RHIC?

The large x pion inclusive polarimeter[8] ($x > .5$) might be viable with forward electromagnetic detectors near a standard collider detector. The analyzing power grows with x and is on the order of 10%.

---

* Work supported by the U.S. Department of Energy, Division of High Energy Physics, Contract W-31-109-ENG-38.

Table I

A TABLE OF INTERESTING POLARIZED COLLIDING PHYSICS

NUCLEON SPIN STRUCTURE FUNCTIONS:
    GLUON
        2-SPIN EXPERIMENTS:
            DIRECT GAMMA + JET
            JET-JET
            DRELL-YAN + JET
            (ALSO, P-D BEAMS)

    ORBITAL ANGULAR MOMENTUM
        1-SPIN AND 2 SPIN EXPERIMENTS:
            JET OR PI-ZERO
            NEW IDEAS

    STRANGE QUARK
        ?

SEARCHES
    SUPER SYMMETRY
        2-SPIN, 2 TO 2 PROCESSES:
        JET-JET
        LEPTON?

    RIGHT HANDED W, ETC.

OTHER PHYSICS:
    ELASTIC SCATTERING
        ODDERON WITH POMERON ?

    TOTAL CROSS SECTION
        RELATE $\Delta\sigma_L$ TO RISING CROSS SECTION ?

    PARITY VIOLATION
        LARGE DUE TO NON-VALANCE QUARK CONTRIBUTION ?

    SPIN OF JETS
        STUDY FRAGMENTATION
        LOOK AT ACAUSAL CORRELATIONS A LA BELL'S THEOREM

All of these topics have been treated in papers or conferences Refs. 1-5.

TABLE II

Comparison of Some Polarized Facilities

Laboratory	Momentum	Beam	Root-S	L Pol.	Target Junk	2-Spin	Direct gam. pt for 5% A_N	A_{LL}	Typical x Coverage
AGS	20 + fix	PP	6		YES	YES			
FNAL	200 fix	PP	19	2E29	YES	YES			
	200 fix	PP-	19	1E28	YES	YES			
	500 fix	PP	30	6E29	YES	YES	5	4-5	.2 ± .1
KEK			38						
CERN/SPS	315 gas	PP	24	>4E29	NO	NO	4-7	---	.2 ± .1
		PP-	24	>2E29	NO	NO			
UNK/POLEX	2000 fix	PP	61	>1E30	YES	YES	7	4-7	.1 ± ?
	700	PP-		5E28	YES	YES			
RHIC	250 + 250	PP	500	2E32	NO	YES	25	5-25	.02 to .1
FNAL	900 + 900	PP-	1800	1E31	NO	NO	30	---	
SSC	20T + 20T	PP	40T	1E33	NO	YES?	?	?	Very Low

Note that RHIC is the only place to do a clean 2-spin experiment as a function of x1 and x1 in the relevant x region for the gluon spin distribution function.

REFERENCES

1) Proc. of VIII Int. Symp. on High Energy Spin Physics, AIP Conf. Proc. 187, Particles and Fields 37, (1989). See also the VII and IX Int. Symp.
2) Polarized beams at SSC and Polarized Antiprotons AIP Conf. Proc. 145, Particles and Fields 34, (1985)
3) Physics at Fermilab in the 1990 s published by World Scientific.
4) Prediction of a Large Parity-Violating Total Cross-Section at High Energies, T. Goldman and D. Preston, Physics Letters, 168B, p. 4515 (1986).
5) Spin Correlations in Parton-Parton Scattering, I.G.Knowles, preprint Cavendish-HEP-88/5, or NSF-ITP-88-36 (Bell's theorem, etc).
6) Analzing-Power Measurements of Coulomb-Nuclear Interference with Polarized-Proton and -Antiproton Beams at 185 GeV/c, N. Akchurin et al., Phys. Lett. B229, 299 (1989).
7) Measurement of the Analyzing Power in the Primakoff Process with a High-Energy Polarized Proton Beam, D. C. Carey et al., Phys. Rev. Lett. 64, 357 (1990).
8) B. E. Bonner et al., Phys. Rev. Lett. 61, 1918 (1988) and a new paper with much more data to be published in early 1991.

The stimulated Stern-Gerlach effect in charged particle storage rings

Ya. S. Derbenev
Randall Laboratory of Physics, The University of Michigan
Ann Arbor, Michigan 48109-1120 USA

Abstract

The results of the study of possibilities to use spin-orbital force in order to split a circulating beam to two polarized beams are presented in this report. It is shown that the original spin-splitter[1,2] idea which is based on using intrinsic spin-orbital resonance is not sufficient for splitting, in principle, because the spin's long lasting effect on particle betatron oscillations is reduced just to a small tune shift. The theorem on the conservation of the sum or difference of quantum orbital and spin numbers, i.e. the combined spin-orbital invariance, is established for this case. The resonant RF magnetic field parallel to the plane of splitting is introduced in order to stabilize spin in the plane of its precession and remove the combined invariancy. The new double-resonance invariants are established, which describe the spin dynamics and the splitting process.

In addition, a method of spin-splitting a beam is considered using gradient RF magnetic field resonance to particle betatron oscillations. The field has a component along the axis of the equilibrium polarization in the storage ring. No resonance with spin precession around this axis is required.

The necessary conditions of beam splitting are discussed.

1. The neo-classical RF Stern-Gerlach method

We describe the principal aspects of this possibility. The spin-dependent part of the Hamiltonian corresponding to the resonant spin-orbit force can be written in the following form:

$$H_{sp} = S_n \vec{n} \cdot \vec{w}(\vec{r}, \vec{v}, t), \tag{1}$$

where S_n is the spin projection on $\vec{n}$, and vector $\vec{w}$ as a function of RF field can be written as follows:

$$\vec{w} \approx -(\frac{1}{\gamma} + G)\frac{\vec{B}(\vec{r}, t)}{B_o}, \tag{2}$$

where B_o is the average vertical field of the ring, and $\vec{B}(\vec{r}, t)$ is the gradient

RF field:
$$\vec{B}(\vec{r},t) = \frac{1}{2}\vec{B}_\omega(\vec{r})e^{-i\omega t} + \frac{1}{2}\vec{B}^*_\omega(\vec{r})e^{i\omega t}.$$

We assume that $\vec{B}_\omega$ linearly depends on x. Averaging of the Hamiltonian (1) for fast-oscillating terms, we get an effective Hamiltonian in the form:

$$H_{\text{eff}} = \frac{1}{2}S_n(g_\omega a \cdot e^{i\epsilon\theta} + g^*_\omega a^* e^{-i\epsilon\theta}), \qquad (3)$$

where θ is the generalized azimuth of the particle,

$$g_\omega = \frac{1}{2B_0}(\frac{1}{\gamma} + G) \cdot < \vec{n}\frac{\partial \vec{B}_\omega(\vec{r})}{\partial x} \cdot f_x \cdot e^{-i(k_x + \nu_x)\theta} >_{x,y=0} \qquad (4)$$

$$\epsilon = k_x + \nu_x - \frac{\omega}{\omega_0},$$

ω_0 is the revolution frequency, and a is the complex amplitude of the betatron x-oscillations:

$$x = \frac{1}{2}\left[af_x(\theta) + a^*f^*_x(\theta)\right].$$

Using the Hamiltonian equations we get:

$$a' = i\frac{R}{p}g^*_\omega e^{i\epsilon\theta} \cdot S_n, \qquad (5)$$
$$S'_n = 0;$$

here p is the particle total momentum and $2\pi R$ is the circumference. So, the spin projection on $\vec{n}$ is constant, and we have the resonance spin-splitting of the betatron oscillations:

$$a_+ - a_- = i\frac{R}{p} \cdot \hbar g^*_\omega \int_0^\theta e^{i\epsilon\theta} d\theta = -\frac{\hbar R}{p}g^*_\omega \frac{e^{-i\epsilon\omega_0 t} - 1}{\epsilon}. \qquad (6)$$

To conclude this section, we note that the method described above can be considered as an immediate extension of the classical Stern-Gerlach method to the case of circulating charged particles: the spin is stabilized by the basic conservative field, and the stable spin component is responding to the regular spin-orbit force; but, in order to enhance the splitting effect during many particle revolutions, we must apply the RF alternating gradient field with the frequency which is resonant to particle free oscillations.

From the practical point of view, one disadvantage of this method is the smallness of the value of the RF field strength compared to that of the intrinsic gradient field.

2. The RF-driven Stern-Gerlach effect near an intrinsic resonance

Now we add a dipole or solenoidal RF field to the intrinsic constant field of the ring:

$$\tilde{\vec{B}}(\vec{r}, t) = \text{Re}\left[\vec{B}^{\circ}(\vec{r}) \cdot e^{-i\omega t}\right].$$

The frequency ω of this field must be in resonance with particle betatron ν_x-value and with spin precession frequency ν_{sp}. So, we have the double resonance condition:

$$\frac{\omega}{\omega_o} \approx \nu_{\text{sp}} + k_{\text{sp}}, \qquad \frac{\omega}{\omega_o} \approx \nu_x + k_x, \qquad (7)$$

where k_{sp} and k_x are integers; note that $k_x - k_{\text{sp}} = k$, where k is the integer in resonance condition $\nu_{\text{sp}} \approx \nu_x + k$.

Let us observe the spin motions with respect to the system of base vectors

$$\vec{n}(\theta), \vec{n}_1(\theta), \vec{n}_2(\theta),$$

where

$$\vec{n}_1 + i\vec{n}_2 = \vec{e} \cdot e^{-i(\frac{\omega}{\omega_o} - k_{\text{sp}})\theta} \equiv \hat{\vec{n}}. \qquad (8)$$

Complex vector $\vec{e}$ is defined by the relation

$$\vec{e} = \vec{\eta} \cdot e^{i\nu_{\text{sp}}\theta},$$

where $\vec{\eta}$ is the solution of spin equations on the closed orbit, with the feature $\vec{\eta}(\theta + 2\pi) = \vec{\eta}(\theta) \cdot e^{-2\pi i \nu_{\text{sp}}}$.

We should pay some attention to a peculiar effect which arises when using the RF field. In order to provide a small value for the phase difference between the particle revolution and RF field oscillation, we need a bunched beam, i.e. we have to supply a longitudinal electric RF field with the frequency ω_o or $q\omega_o$, where q is an integer. In such a regime particle energy and relative phase $(\theta - \omega_o t)$ will oscillate near equilibrium values with some low frequency $\nu_\gamma \omega_o$, $\nu_\gamma \ll 1$. Also the betatron and spin ν-values will oscillate due to chromaticity parameters $\partial \nu_x / \partial \gamma$ and $\partial \nu_{\text{sp}} / \partial \gamma$. It is not difficult to extend the consideration taking these synchrotron oscillations into account. But, for simplicity, we assume here that the amplitudes of all of these oscillations are small enough to neglect them and also that all of the frequencies are constant and equal to the averaged values. We also assume that we can neglect phase "mismatching" $(\theta - \omega_o t)$ since the beam is short enough. There is also an advantage of using bunched beams, in that the contribution of energy spread to betatron and spin frequency becomes a value of the second order. Note that the parameter $\frac{\partial \nu_x}{\partial \gamma}$ could be canceled by the introduction of sextupoles and the parameter $\partial \nu_{sp}/\partial \gamma$ vanishes by using Siberian Snakes.

Then we have the equations of motion as follows:

$$S'_n = i\hat{w}\hat{S}_- - i\hat{w}^*\hat{S}_+,$$
$$\hat{S}'_+ = i\hat{w}S_n - i\epsilon_{sp}\hat{S}_+, \qquad \epsilon_{sp} = \nu_{sp} + k_{sp} - \frac{\omega}{\omega_o} \qquad (9)$$
$$\hat{a}' - i\epsilon_x \hat{a} = i\frac{R}{p}g\hat{S}_+; \qquad \epsilon_x = \nu_x + k_x - \frac{\omega}{\omega_o}$$

we define here

$$\vec{w} \equiv \tilde{w} + ga, \qquad \tilde{w} = \frac{<\vec{W}_{RF} \cdot \hat{n}>}{\omega_o}, \qquad ga = \frac{<\vec{W}_i \cdot \hat{n}>}{\omega_o},$$

where $\vec{W}_{RF}$ and $\vec{W}_i$ are the angle speed of the spin precession in the RF field, and in the intrinsic gradient field, respectively, according to the BMT equation.

Now we assume the relation

$$|\epsilon_x| << |\hat{w}|, \qquad (10)$$

which is not too precise a requirement from the practical point of view. With this condition, we can write the solution for spin motion in the base system (8) as:

$$\vec{S}(\theta) = \hat{S}_{||} \cdot \frac{\hat{\vec{w}}}{|\hat{\vec{w}}|} + S_n(\theta) \cdot \vec{n} + \frac{\vec{n} \times \hat{\vec{w}}}{|\hat{\vec{w}}|} \hat{S}_\perp(\theta), \qquad (11)$$

$$S_n(\theta) + i\hat{S}_\perp(\theta) = \left[S_n(0) + i\hat{S}_\perp(0)\right]\exp(-i\int |\hat{\vec{w}}|d\theta),$$

$$\hat{S}_{||} = \text{const}$$

where vector $\hat{\vec{w}}$ has the components:

$$\hat{\vec{w}} = (\hat{w}_1, \hat{w}_2, \epsilon_{sp}), \qquad \hat{w}_1 + i\hat{w}_2 = \hat{w}(\hat{a}).$$

Further, we assume for simplicity, that

$$|\epsilon_{sp}| << |\hat{w}|; \qquad (12)$$

then the axis of spin precession $|\hat{\vec{w}}|$ is close to the plane transverse to $\vec{n}$.

The evolution of amplitude a averaged for spin precession around $\hat{\vec{w}}$ is described by the equation:

$$a' + i\epsilon_x a = i\frac{R}{p}g^* \frac{\hat{\vec{w}}}{|\hat{\vec{w}}|}\hat{S}_{\|}. \tag{13}$$

With taking adiabatic invariancy of $\hat{S}_{\|}$ into account, the invariant of this equation is the Hamiltonian

$$\hat{H}_{\text{eff}} \to H_{\text{split}} = \frac{p}{2R}\int \epsilon_x d|a^2| + \hat{S}_{\|} \cdot |\tilde{\vec{w}} + g a|. \tag{14}$$

Thus, this Hamiltonian describes the splitting process, after using the eigenvalues of the operator $\hat{S}_{\|}$:

$$\hat{S}_{\|} \to \pm\frac{\hbar}{2}.$$

Let us assume that $\epsilon_x = $ const, i.e. x-oscillations are linear; then we obtain the trajectories of the splitting process in variables $|a|$ and ψ, i.e., $a = |a|e^{i\psi}$:

$$\frac{p}{2R}\epsilon_x|a|^2 \pm \frac{\hbar}{2}\sqrt{|\tilde{\vec{w}}|^2 + 2|\tilde{\vec{w}} g a|\cos\psi + |ga|^2} = \text{const}. \tag{15}$$

Let us focus our attention on the case of precise resonance, i.e. $\epsilon_x = 0$, when we can expect the biggest value of splitting. In this case, we have

$$|\hat{\vec{w}}| \equiv |\tilde{\vec{w}} + ga| = \text{const}, \tag{16}$$

i.e. total effective field $\hat{\vec{w}}$, which is the vector sum of $\tilde{\vec{w}}$ and the intrinsic field ga, just rotates in the plane transverse to $\vec{n}$. We can get the angle speed $\Delta\omega$ of this rotation (with respect to system $\vec{n} = \vec{n}_1, \vec{n}_2$) using the equation (13):

$$\hat{\vec{w}}' = \pm i\frac{R\hbar}{2p}|g|^2 \frac{\hat{\vec{w}}}{|\hat{\vec{w}}|}, \tag{17}$$

and then we have

$$\Delta\omega = \pm\frac{R\hbar}{2p}\frac{|g|^2}{|\tilde{\vec{w}} + ga_o|}\omega_o, \tag{18}$$

where a_o is the initial value of a.

The time progress of the splitting process depends on the evolution of $\hat{w}$:

$$\hat{w} = \hat{w}(0) \cdot \exp(i\Delta\omega \cdot t), \qquad (19)$$

or

$$\tilde{w} + ga = (\tilde{w} + ga_o)\exp\left\{\pm iR\hbar|g|^2\omega_o t/2p|\tilde{w} + ga_o|\right\}. \qquad (20)$$

Particularly, at $a_o = 0$ (i.e., $|ga_o| \ll |\tilde{w}|$) we have

$$a(t) = \frac{\tilde{w}}{g}\left[\exp(\pm i\frac{R\hbar}{2p}\frac{|g|^2}{|\tilde{w}|}\omega_o t) - 1\right]. \qquad (21)$$

Thus, we have a process with a beat period of $2\pi/\Delta\omega$. The maximum splitting arises at the moments

$$t_n = (n + \frac{1}{2})\frac{\pi}{\Delta\omega}, \quad n = 0, 1, 2. \ldots \qquad (22)$$

when it is equal to

$$|a_+ - a_-|_{\max} = 2|\frac{\tilde{w}}{g}|. \qquad (23)$$

It is also important to note that for the condition $|ga_o| \ll |\tilde{w}|$, the value $|a|$ does not depend on $|\tilde{w}|$ in the initial stage of the splitting process, when $t \ll 1/\Delta\omega$:

$$|a_+ - a_-| \approx \frac{R\hbar|g|}{p}\omega_o t, \quad t \ll \frac{1}{\Delta\omega}, \qquad (24)$$

but the phase of $a \pm$ depends on the phase of $\tilde{w}$, which gives rise to a spin resonance motion:

$$a_\pm \approx \pm i\frac{R\hbar g^*}{2p}\frac{\tilde{w}}{|\tilde{w}|}\omega_o t, \quad t \ll 1/\Delta\omega. \qquad (25)$$

3. The conditions of beam splitting

Now, we have to discuss and formulate the conditions which must be satisfied in order to realize the perfect splitting of a beam. First of all, note that after splitting the beam to the value of $|a_+ - a_-|$ above the beam size σ, we can frequently accelerate the process by engaging in some kind of instability. It could be quadrupole instability as a result of the beam interaction with some dissipative elements of the chamber, but, possibly, the simplest and most efficient way to accelerate the splitting process is to create the parametric instability by switching on the RF gradient magnetic field with the frequency

$$\omega = (k \pm 2\nu_x)\omega_o;$$

the field must be switched off when the amplitudes a_+ and a_- achieve the perfect value that is necessary for separation of two polarized parts of the beam.

So, in practice, the time of splitting would be limited by the value

$$\tau_{sp\ell} \sim \frac{\gamma M \sigma}{\hbar |g|} \sqrt{\nu_x}. \tag{26}$$

Correspondingly, the necessary strength of the spin-driving RF field has the order of value:

$$|\tilde{w}| \gtrsim |g|\sigma\sqrt{\nu_x}. \tag{27}$$

The betatron tune spread $\Delta\nu_x$ must be small enough to satisfy the condition

$$\Delta\nu_x < 1/\omega_o \tau_{sp\ell}. \tag{28}$$

The spin-driving RF field must be parallel to the plane of splitting, in order not to excite the dipole x-oscillation:

$$\vec{B}_{RF} \parallel (\vec{x}, \vec{v}). \tag{29}$$

Taking into account the admissible value of the dipole amplitude x_d, we obtain the limitation on the deviation angle α of the field $\vec{B}_{RF}$:

$$\alpha \lesssim \frac{x_d}{\beta} \cdot \frac{<B>}{<B_{RF}>} \cdot \Delta\nu_d, \tag{30}$$

where β is the β-function, B is the bending field, B_{RF} is the amplitude of the RF field, the brackets $<\ldots>$ mean averaging along the closed orbit, and $\Delta\nu_d$ is the detune between frequencies of dipole and quadrupole beam oscillations.

Note also that $\vec{B}_{RF}$ should not be parallel to the direction of the periodic polarization $\vec{n}$ in the storage ring. At low energies this condition can be satisfied together with (29) by using the RF solenoid. At high energies we should use the transverse RF field; in this case, the directions of $\vec{n}$ and x must be different.

Conclusion

Thus, the conclusions are as follows:
- the Stern-Gerlach method can work in storage rings, in principle, only with application of resonant RF magnetic field;

- the necessary time for splitting is defined by the beam size; therefore, it is especially important to provide a beam size as small as possible, using some cooling technique;
- the question of how to monochromize the betatron particle motion during the splitting process is still the most crucial problem from a practical point of view.

References

1. R. Rossmanith, Proceedings of the 8th International Symposium on High Energy Spin Physics, Minneapolis, Minnesota (September 1988), AIP Conf. Proc., V. 2, p. 1085-1092.
2. T. O. Niinikoski and R. Rossmanith, AIP Conf. Proc. **N145**, Particles and Fields Series 34.

Parity violating e^- deuteron scattering as a probe of the strangeness content of the nucleon.

S.J. Pollock

Institute for Nuclear Theory
University of Washington
Seattle, WA 98195

Parity violation in elastic (polarized) electron- (unpolarized) deuteron scattering is considered in some detail, as a means to extract information on both vector and axial-vector currents, and thus the strange quark content of the deuteron. Numerical estimates are made, using a simple model when required, to find the sensitivity to s-quark components of the form factors.

I. INTRODUCTION AND PRELIMINARIES

A complete set of electroweak form factor measurements on the nucleon provides an excellent means to search for heavy quark contributions to ground state nucleon structure. By comparing weak with electromagnetic form factors, and using the good $SU(2)$ isospin symmetry for the nucleon, one can in principle separate the various $SU(3)_f$ matrix elements of quark current operators between nucleon states. Since free neutron targets are unavailable, we consider means to extract form factor data from the deuteron. (see also ref ([1]), and references therein.) This nucleus offers many advantages: it is abundant, weakly bound (minimizing nuclear medium modifications) and excellent nuclear wave functions are known. Since the deuteron is isoscalar, non-zero heavy quark contributions, which are manifestly isoscalar, should appear quite directly in the axial form factor. In addition, the isoscalar magnetic moment of the nucleon is small, so extra s-quark current contributions should show up more clearly than in the free nucleons themselves. The fact that the deuteron spin is 1 means magnetic form factors contribute at backward angles, an advantage over spin-0 nuclei which are primarily sensitive only to electric form factors.

The matrix elements of weak currents between deuteron states can be written in terms of six a priori unknown body form factors, $G_1(q^2)$ through $G_6(q^2)$, as discussed for example in Hwang and Henley[2] (see also [1,3]). The three odd-parity, or axial, form factors $G_{4,5,6}$ are usually completely discarded, because the deuteron is an isoscalar system, and there are no axial isoscalar currents in the standard model if one ignores strange (or heavier) quarks. In that case, one would expect exact relations[4] between weak and electromagnetic deuteron form factors:

$$G_{1,2,3}(q^2) = -2\sin^2\theta_W \, G^\gamma_{1,2,3}(q^2),$$
$$G_{4,5,6}(q^2) = 0. \tag{1}$$

We are interested in the contributions of possibly non-zero $G_{4,5,6}$ to observables, as well as deviations from this simple relation for $G_{1,2,3}$.

The presence of strange quarks in the nucleon modifies the *weak* isoscalar form factors. For example, recent neutrino scattering experiments[6] give $F_A(0) \approx -0.15 \pm .08$, with F_A the nucleon isoscalar axial form factor. The vector form factors may also receive contributions from strange quarks.[7] Forming the isoscalar combination immediately gives the weak nucleon (vector) isoscalar form factors

$$F_i = F_i^\gamma(-2\sin^2\theta_W) - \frac{1}{2}F_i^{st}. \tag{2}$$

where we define s-quark form factors in exact analogy with the usual electromagnetic ones:

$$<p'|\bar{s}\gamma_\mu s|p> \equiv \bar{u}(p')[\gamma_\mu F_1^{st}(q^2) + i\sigma_{\mu\nu}q^\nu F_2^{st}(q^2)]u(p). \tag{3}$$

The presence of s-quarks thus violates the simple relation of eqn (1), and a measurement of such weak vector form factors could then be directly connected via eqn (3) to the strange quark distribution in the nucleon.

II. RESULTS AND DISCUSSION

In order to look for sensitivities to the possible s-quark contributions to nucleon structure, we focus on the asymmetry, defined as the (normalized) difference in cross section for right and left-hand longitudinally polarized e^-'s scattering off unpolarized deuterons. To connect deuteron measurements to the nucleon's properties, one needs a consistent model of the deuteron. In this note, for simplicity, only non-relativistic wave functions[5] are used. This yields acceptable predictions for electromagnetic form factors for $0 \leq -q^2 < 0.5 - 1.0$ GeV^2 (below the first diffraction minima), and thus should be a reasonable first step. Relativity and meson-exchange corrections, among others, should not seriously modify any conclusions made about *general* sensitivity of deuteron cross sections to s-quark content of the nucleon, although they may certainly alter some quantitative details.

To examine the effects of possible s-quarks, we have calculated the asymmetry in several ways[1]. First, we ignore all strange contributions, and use the standard model relations between isoscalar form factors, eqn (1). We then recalculate allowing an extra axial isoscalar contribution by giving $F_A(0)$ a non-zero value. We also allow for the possibility that s-quarks might appear via the magnetic vector form factor, i.e. $F_M^{st}(0) \equiv \mu^{st} \neq 0$. We arbitrarily choose an extremely large value of $\mu^{st} = 0.8$, just to see the effects. The deviation scales linearly with μ^{st}. This quantity is currently essentially unknown. In all cases, we assume a simple (dipole) q^2 dependence for all form factors. This is not theoretically required, but serves as a rough guide, without adding too many extra parameters.

In general, parity violation measurements on the deuteron in the kinematic regime considered here would be very difficult. Systematic error sources have *not* been investigated here. Many possible modifications of the asymmetry, due e.g. to higher order diagrams, isospin breaking, or odd parity admixtures in the deuteron wave function, have also not yet been considered here. (See e.g. ref 8)

The sensitivity to possible strange quark contributions are shown in the figures where we plot the predicted asymmetry in $d(\vec{e}, e')d$ versus scattering angle for fixed momentum transfers, and versus q^2 for fixed angle. We also show, in fig (2), the scale of possible s-quark effects compared to rough statistical error predictions. Some of the more important features are the following:

(1) The asymmetry at backward angles is reasonably sensitive to a significant s-quark component in the magnetic form factor. This sensitivity is somewhat deuteron-model dependent, but because the asymmetry involves a ratio of cross sections, this model dependence appears to be quite weak. A large assumed s-quark contribution to the magnetic nucleon form factor of $\mu^{st} = 0.8$ could result in modifications in the asymmetry of more than 100%. Figure of merit considerations based only on statistics imply that moderate $q^2 \approx 0.1$ GeV^2 and backward angles optimize the sensitivity to magnetic contributions. (cf. fig 3.)

(2) The asymmetry at forward angles is reasonably sensitive to both magnetic and electric form factors, with the electric slightly dominating. Moderate $q^2 \approx 0.2$ GeV^2 (which corresponds to much higher beam energy than the backward angle case) optimizes the sensitivity. Uncertainties in the asymmetry coming from, e.g. errors in the weak mixing angle become comparable to our assumed s-quark contributions at lower q^2.

(3) Because of the "accidental" value of $\sin^2\theta_W \approx \frac{1}{4}$, the axial contributions due to a possible non-zero isoscalar current are strongly suppressed. Even at backward angles, the axial form factor modifies the asymmetry by amounts less than our estimated reasonable statistical errors.

III. SUMMARY

The deuteron is an interesting nuclear target for parity violating electron scattering measurements. We see that, if there turns out to be a fairly large s-quark component in the nucleon, the parity violating asymmetry in electron – deuteron scattering would be significantly modified from its value when strangeness is ignored. Such modifications, while slightly deuteron-model dependent, could provide a useful means to measure and separate the various pieces of the strange currents in nucleons. This would of course be further aided by combining data from various angles, as well as with complementary direct data from free nucleons, and other nuclei.

The author would like to thank F. Gross, R. Ent, and T.W. Donnelly for their useful comments, suggestions, and help. This work was supported by the Foundation for Fundamental Research (FOM) and the Netherlands Organization of Scientific Research (NWO) at NIKHEF-K.

---

[1] S.J. Pollock, *Phys. Rev.* **D42**, 3010 (1990)

[2] W-Y.P. Hwang, E.M. Henley, *Annals of Physics* **129**, 47 (1980)

[3] F. Gross, *Phys. Rev.* **136**, B140 (1964)

[4] S.J. Pollock, Stanford PhD Thesis, 1987, unpublished

[5] R. Reid, *Annals of Physics* **50**, 411 (1968), R. Machleidt, K. Holinde, Ch. Elster, *Physics Reports* **149**, 1 (1987)

[6] L.A. Ahrens et al., *Phys. Rev.* **D35**, 785 (1987)

[7] D.B. Kaplan, A. Manohar, *Nucl. Phys.* **B310**, 527 (1988)

[8] M. Musolf, Princeton PhD Thesis, 1989, unpublished

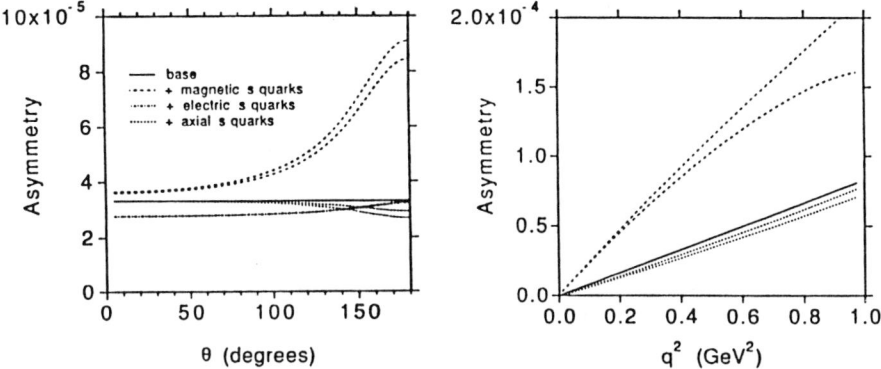

FIG. 1a. Asymmetry vs. θ_{e^-}, at $q^2 = -0.4$ GeV^2. The solid curve has no s-quark contributions. The two long-dashed curves include a strange magnetic moment contribution of $+0.8$. They differ in assumed nucleon-nucleon interaction potential. The dot-dashed curves include only an extra electric s-quark form factor. The short-dashed curves include only an axial isoscalar form factor of -0.15 at $q^2 = 0$.

FIG. 1b. Asymmetry vs. q^2, at $\theta_{e^-} = 180°$. The curve descriptions match those in figure 1a.

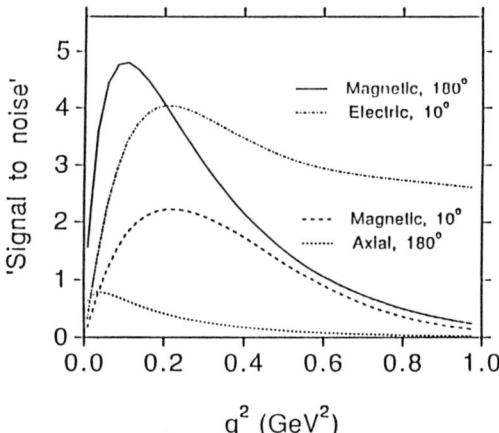

FIG. 2. Ratio of the deviation in A_D caused by s-quark contributions, to an estimated reasonable statistical error. The solid curve assumes a magnetic s-quark contribution at 180° ($\mu^{st} = 0.8$), the long-dash curve is 10°. The dot-dash curve has electric contributions at 10°, the short-dash curve assumes just an axial contribution at 180°.

Additional Contributions

SELF POLARIZATION OF STORED (ANTI-)PROTONS
STATUS OF THE SPIN SPLITTER EXPERIMENT AT IUCF

R. Rossmanith

CEBAF, 12000 Jefferson Avenue, Newport News, Virginia, USA

representing the Spin-Splitter Collaboration *

Abstract

Several years ago a selfpolarization effect for stored (anti-)protons and ions was investigated theoretically. The effect is based on the well-known Stern-Gerlach effect in gradient fields. The aim of the ongoing measurements at IUCF is to verify experimentally the various assumptions on which this effect is based. The final goal is to demonstrate this new polarization effect. The proposed effect could be a powerful tool to produce polarized stored hadron beams both in the low energy range and at SSC and LHC energies.

Introduction

During the last years a new polarization effect for stored protons or antiprotons was investigated theoretically[1,2]. The idea is very simple: a storage ring consists of strong nonlinear fields, the quadrupoles. The nonlinear fields act on the spin of the particle by the Stern-Gerlach effect: the particles are kicked by a very small amount. This small kick produces betatron oscillations. During the next revolution the gradient of the quadrupole is (naturally) the same but the spin direction can be altered. As a result the strength and the direction of the kick can be different from revolution to revolution. When this time dependent kick follows a betatron oscillation the kicks can add up over many revolutions and a macroscopic oscillation can be excited which can be detected by an outside monitor. The principle is the same as in the one used for measuring the Q-value: small kicks add up over many revolutions and produce an oscillation with a significant amplitude.

Aim of the ongoing measurements is to establish this effect. Although the above mentioned concept is very simple and staightforward and somewhat an analogy of well established techniques it differs in two points from these techniques.

---

* N. Akchurin (Iowa), Laura Badano (Genova), M. Conte (Genova), R. Giacomich (Trieste), M. Giorgi (Trieste), J. Hall (New Mexico), S. Hsueh (Fermilab), H. Kreiser (Hamburg), J. McPherson (Iowa), Jane Nachtman (Iowa), Y. Onel (Iowa), G. Pauletta (Udine), A.Penzo (Trieste), A. Pisent (CERN), M. Pusterla (Padova), R. Rossmanith (CEBAF), U. Strohbusch (Hamburg), A. Yegneswaran (CEBAF)

First of all the excitation strength is small and therefore a long time is needed to see a macroscopic effect. Since the strength of the kick is comparable with the quantum mechanical uncertainty a long discussion took place, if the laws of quantum mechanics are not imposing a principle limit on this effect. In addition the long accumulation time rises questions concerning the stability of the field. Long term drifts may smear out the effect.

And, secondly, the driving force is related to the spin direction. The spin direction cannot be measured in a simple way, especially when the spin direction changes from revolution to revolution.

It was amusing to learn that similar problems arise when gravitational wave detectors are designed [6,7]. Although different in the final goal the underlying principles and limitations are nearly identical.

The Principal Concept

When a proton passes a quadrupole the acquired kick depends on the momentum of the particle:

$$\eta' = p_t/p$$

where $p = \beta\gamma m_p c$, $\gamma = (1 - \beta^2)^{-1/2}$ and m_p is the (anti)proton mass. p_t depends on the force F and the length of the interaction τ. When L is the length of the interaction and βc is the velocity, η' becomes:

$$\eta' = \frac{FL}{\beta^2 \gamma m_p c^2}$$

The Stern-Gerlach force is $F = \nabla(\mu.B)$ with $\mu = 1,41.10^{-10} JT^{-1}$. For low energies $\gamma \approx 1, \beta \approx 0.1$, a field of 20 Tm^{-1} the kick is 9.4x10^{-14} rad. In order to calculate a displacement in the machine the kicks have to be multiplied with the square roots of the beta function. Assuming that the betafunction both at the point of observation and the point where the kick is applied is 100 m the betatron amplitude is $\approx 10^{-8}$mm. 10^8 revolutions are required to accumulate a betatron amplitude of 1 mm.

Since most of the low energy storage rings have revolution times in between 10^{-6} and 10^{-7} seconds a measurable amplitude should be accumulated within 10 to 100 seconds. During this time the spin motion and the betatron motion must be in phase. The resonance condition is

$$Q_{spin} = Q \pm n \qquad (1)$$

where Q is the Q-value of the machine, n is any integer and Q_{spin} is the number of spin precessions per revolution. In a machine without any spin manipulations Q_{spin} is $G\gamma$, where G is the anomalous magnetic moment of the proton (G=1.793).

It should be mentioned that the superposition of Stern-Gerlach kicks on single charged particles is a well established effect[3]. The difference to the single particle technique is only that an ensemble of particles and an ensemble of spins have to be considered. In order to avoid depolarization in such an ensemble the spin motion of all particles must be (within certain limits) the same. The same is valid for the betatron motions. In addition both motions must fulfill equation (1).

Coherent Spin Motions

Equation (1) defines the range of the spin tune. When Q is an integer or a half integer the storage ring will not work. As a consequence the spin tune Q_{spin} must differ from an integer or half an integer. In other words the spin direction has to change from revolution to revolution. Repetitions after each or each second revolution are not allowed. One of the aims of our experiments is to verify that such a spin oscillation is stable and does not lead to depolarization.

In principle several different techniques can be applied to rotate the spin around the stable spin direction (so called n-axis). One is the Spin-Splitter idea[8], where the rotation is performed by a solenoid which rotates the spin by less than 180 degree. Fig. 1 shows the whole ensemble which is more or less similar to a Siberian Snake with a solenoid. The skew quadrupoles have two functions: they compensate the influence of the solenoid on the particle trajectories and they act as Stern-Gerlach magnets. The magnets have opposite polarity but due to the 180 degree rotation of the solenoid the Stern-Gerlach kicks add.

The other proposed technique is based on the application of time dependent fields[9,10] which manipulate the spin in a similar way as the solenoid does.

When we considered this experiment for the first time an Ann Arbor-IUCF group[4] discussed a Siberian-Snake experiment with a strong solenoid at IUCF in Bloomington, Indiana. IUCF is a proton storage ring with an electron cooler operating in the 100 MeV to several 100 MeV range. After the Siberian Snake group have finished parts of their shifts we asked for a few shifts to test some of our ideas.

The experiment was planned in three steps:

a.) The polarization is injected along the n-axis: the spin direction repeats from revolution to revolution (fig. 2). The maximum beam current is accumulated. The polarization of the stored beam is measured. Since Q_{spin} is an integer an accumulation of Stern-Gerlach kicks is impossible. This measurement allows to study the behaviour of the polarimeter and the efficiency of the accumulation process (stacking) in the presence of a Siberian Snake.

b.) The spin is injected 90 degree relative to the n-axis. The spin is supposed to oscillate around the stable condition. Since the ideal Siberian Snake rotates the spin by 180 degree the spin direction repeats after each second revolution (fig. 3). A Stern-Gerlach accumulation can only be observed on a half integer Q-value (which is is not a stable condition). Nevertheless, this experiment allows to study some technical details of the concept. Since the polarization must be the same after each second revolution, the polarimeter only measures a non-zero polarization when the input is gated with a subharmonic of the revolution frequency. The aim of this experiment is to study the stability of a polarization vector oscillating around the n-axis.

c.) The 180 degree rotation of the Siberian Snake is changed into a 180-Δ degree rotation. The spin does not repeat every second revolution (fig. 4) and the resonance condition (1) can be fulfilled.

So far we had only time to finish the measurements on point a.).

ad a.) We injected the beam with a polarization along the n-axis. The n-axis is parallel to the field axis of the solenoid. We were able to accumulate a polarized beam of up to $2\mu A$ at 107 MeV ($G\gamma=2$). The solenoid was not compensated during these shifts by the skew quadrupoles. The polarimeter shows a polarization of typically

$$P_H = 86 \pm 10\%, P_V = 4 \pm 10\%$$

We were able to store up to $20\mu A$ polarized beam when, in a next step, the solenoid was completely compensated by skew quadrupoles. The compensation was verified by measuring the coupling between the horizontal and vertical plane.

ad b.) We tried to inject a beam with a horizontal polarization of 90 degree relative to the n-axis (see fig. 3). The polarization should flip from revolution to revolution by 180 degree. By timing the injection with a subharmonic of the revolution frequency we tried to inject only beam with the correct polarization relative to the stored beam. Unfortunatly during accumulation the stored beam has to be debunched. With the used electronic scheme we were not able to keep the debunching time short enough. The beams are debunched during a period of 10^6 revolutions. A particle with an energy deviation of 10^{-3} performs 1000 revolutions more or less than a particle with nominal energy and the spin flips thousand times more or less often leading to a depolarization of the stacked beam. A remedy could be an additional cavity keeping the bunches together or a shorter time in which the beam is debunched. Another, even simpler remedy is to inject the beam polarized along the n-axis as in a.), accumulate the beam and rotate the polarization after accumulation with a time dependent field into the required position. During the next shifts such a time dependent field will be available.

In the meantime extensive computer simulations of the spin stability were performed and will be published soon[11]. The main results are the following: with a Siberian Snake and a $G\gamma$ of 2 the oscillations around the stable solution do not lead to depolarization. Outside this regime the stability is only granted when so-called spin matching conditions are applied. These conditions are described in detail in [12].

Coherent Betatron Motions

A rather short time was spent to get more experimental information on Landau effects. Equation (1) assumes that the Q-value is the same for all particles. In reality the Q-value depends on γ, the phase and the amplitude of the betatron motion. The chromaticity (dependance of Q on γ) is not compensated at IUCF. One would assume that with a single kick the at the beginning coherent oscillation is converted into an incoherent oscillation in a short time. This was verified experimentally[5].

But even with a fully compensated chromaticity the Q-spread is big enough to reduce the excitation of the Stern-Gerlach kicks by Landau-damping. A more detailed description of Landau-damping can be found in [13]. Here only the basic elements are given.

The focussing strength of a nonideal quadrupole (quadrupole with sextupole, octupole, etc., components) depends on the amplitude of the particle trajectory. Amplitude means the distance between the center of the quadrupole and the actual trajectory. The amplitude changes from revolution to revolution (see fig. 5) due to the fact that the Q-value has to be non-integer. As a result the focussing strength is slightly different from revolution to revolution. Taking many revolutions into account a distribution of Q-values around a central Q-value is obtained (fig. 6).

Consider an ensemble of particles. Each particle has at each revolution a slightly different Q-value, even when the central Q-value for all particles is the same. When such an ensemble is excited by a monochromatic frequency from outside only parts of the ensemble react with the maximum amplitude. Others are only partly or not excited due to the fact that their oscillation frequency is different from the excitation frequency. The particles being fully excited change position with the partly excited and viceversa. The behaviour of the ensemble can be described in a formal way as a damping, the so-called Landau damping. In order to produce a significant amplitude the excitation has to overcome the Landau damping constant.

During the last year the effect of Landau damping on the Spin-Splitter experiment was discussed by several authors[5,10,11]. H. Kreiser[11] pointed out, that this effect can reduce the achievable degree of polarization. Derbenev[10] argued that a coupling between the betatron oscillations can reduce the Landau-damping effects. This is well known and described in[13] in more detail. Here the effect is explained by two pendulums with slightly different frequencies.

When these two pendulums are independent from each other and excited by one source the sum of the amplitudes can be described by the above mentioned Landau-damping picture. If the two pendulums are coupled by a (relatively weak) additional spring, the situation does not significantly change except for the fact that the whole assembly has an additional resonance frequency. This frequency is sharp. Derbenev argued that a quadrupole type betatron oscillation is generated when particles with different spins get separated. The two parts of the quadrupole oscillations are coupled by strong electromagnetic forces when the particles are not relativistic. As a result a new sharp frequency is generated and the Landau damping does no longer reduce the separation.

Part of the oncoming measurements in 1991 will be used to get a more detailed information on these effects.

Acknowledgements

The Spin Splitter group wishes to thank Prof. Cameron from IUCF for his support. We also want to thank the IUCF machine group for their help during the measurements. Finally one of the authors (RR) wants to thank Prof. Leeman from CEBAF for his continuous interest and encouragement.

Literature

[1] T. Niinikoski and R. Rossmanith, Nucl. Instr. Meth. A 255,460 (1987)

[2] M. Conte and M. Pusterla, Il Nuovo Cimento, 103A, 1087 (1990)

[3] R. van Dyck, P. Schwinberg and H.Dehmelt, Phys. Rev. D, 722 (1986)
   H. Dehmelt, 1989 Nobel prize talk, Rev. Mod. Phys., 62, 525 (1990)

[4] A. D. Krisch et al., Phys. Rev. Lett., 63, 1137 (1989)
[5] T. Ellison, private communication, IUCF 1990
[6] N. F. Ramsey, IEEE Trans. Instr. Meas. IM-36, 155(1987)
[7] V. P. Chebotayev, Appl. Phys. B 51, 303 (1990)
[8] Y. Onel, A. Penzo and R. Rossmanith, AIP Conf. Proc 150, 1229 (1986)
[9] M. Conte, A. Penzo, A. Pisent and M. Pusterla, INFN report INFN/TC-88/25
[10] Ya. Derbenev, University of Michigan Report, UM HE 90-23
    Ya. Derbenev, in proceedings of this workshop
[11] H. Kreiser, PhD thesis, Univ. of Hamburg, to be published
[12] R. Rossmanith and R. Schmidt, Nucl. Instr. Meth., A236, 231(1985)
[13] R. Talman, in AIP Conference Proc. 153, New York 1987, p789

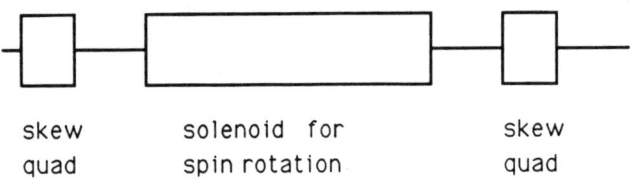

skew quad   solenoid for spin rotation   skew quad

Fig. 1 Magnet Arrangment of the Spin Splitter

Fig. 2 The stable spin direction in the presence of a 180 degree spin rotator

Fig. 3 The injected polarization is oscillating around the stable spin direction

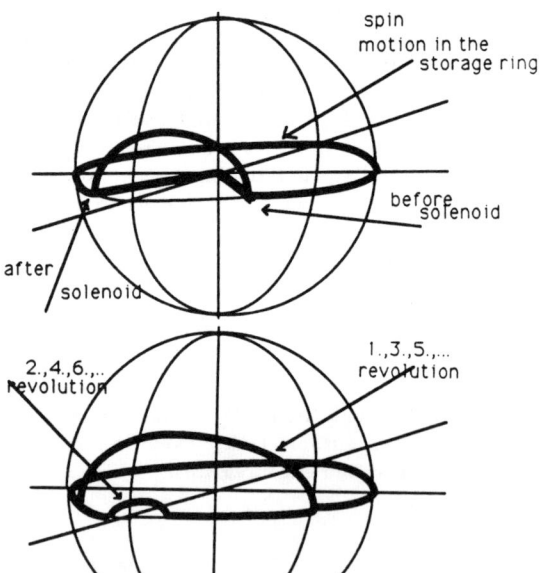

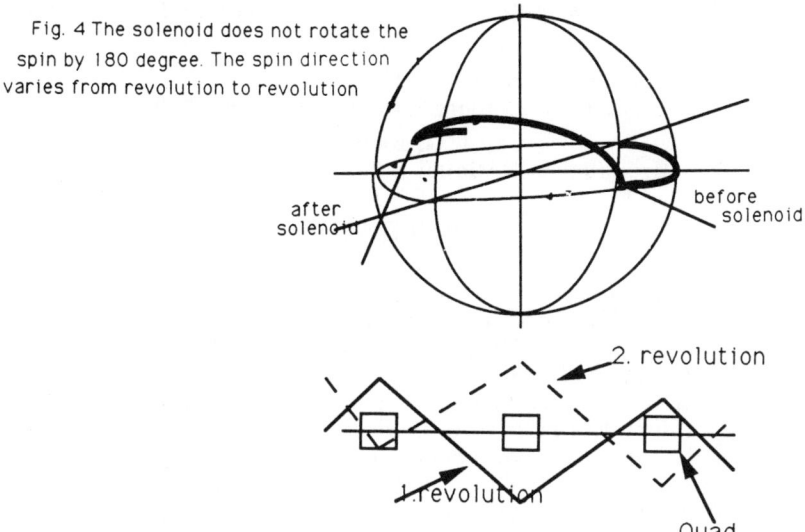

Fig. 4 The solenoid does not rotate the spin by 180 degree. The spin direction varies from revolution to revolution

Fig. 5 The betatron amplitude in a quad changes from revolution to revolution. As a result the nonlinearities in the quad change the Q-value from revolution to revolution.

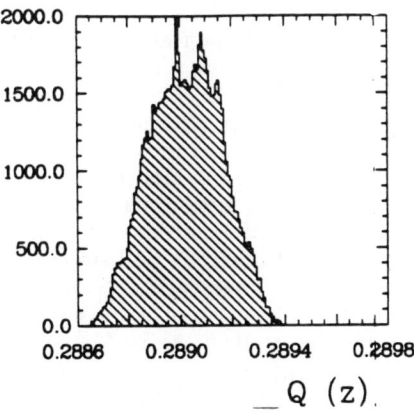

Fig. 6 Calculated Q-value distribution (in relative units) for IUCF

Feasibility of a Polarized Deuteron Beam in the AGS and RHIC

S.Y. Lee
Department of Physics
Indiana University, Bloomington, IN 47405

L.G. Ratner
Brookhaven National Laboratory, N.Y. 11973

ABSTRACT

The Possibility of accelerating polarized deuterons in the AGS and RHIC is discussed. In the AGS, the polarized deuteron can be accelerated up to 15 GeV/c without depolarization. In RHIC, the snake designed for use in accelerating polarized protons are applicable as partial snake in inducing spin flip to overcome the imperfection resonances. Gradient error resonances in RHIC are found to be negligible. Difficulties arise from intrinsic resonances in RHIC, which limit the polarized deuteron collision energy to less than $\gamma \leq 29$ or to energy less than 27.2 GeV/u, if corrections are not made. A type 3 snake is found to be impractical for a spin tune jump. Betatron tune jump by using quadrupoles or induced betatron oscillations may be used to jump through intrinsic resonances. The achievable luminosity is found to about $2 \cdot 10^{30}$ cm^{-2}s^{-1} at $\gamma \approx 30$. If intrinsic resonances can be overcome, one expects to obtain higher energy and higher luminosity.

1. Introduction

Spin is a fundamental property of the basic constituents of elementary particles. Studies in spin physics lead to the important understanding of fundamental interactions between elementary particles[1,2]. Polarized beams are used very often as a research tool in studying the interactions.

Along with polarized proton collisions, scattering data with polarized neutrons may offer supplementary information on the fundamental interactions. Unfortunately the acceleration of a polarized neutron beam is difficult. The next best source of neutrons is a beam of loosely bound deuterons. Experiments using polarized deuteron collisions may offer interesting information regarding polarized neutron scattering.

Deuterons consist of a loosely bounded proton and neutron system with the properties that mass is $M_d \approx 2m_p$ and the magnetic dipole moment is $\vec{\mu}_d = 0.8573\vec{S}$[nM] $\approx \vec{\mu}_p + \vec{\mu}_n$. The equivalent anomalous magnetic g-factor is therefore $G = -0.1427$, which is about 13 times smaller than the corresponding anomalous g-factor for the proton. Because of the small magnitude of G, the deuteron spin precession tune is small. There are fewer resonances in the circular accelerator. Since the resonance strength depends essentially on the value of $G\gamma$, the strength seen by the deuteron beam is therefore small. However, the problem is that snakes used for polarized proton beams are too weak for the acceleration of polarized deuterons. A 100% snake for protons is equivalent to only about 0.5% snake for deuterons.

In the past, polarized deuterons have been accelerated to 10 GeV at Dubna[3] and 12 GeV at the Argonne ZGS[4]. In spite of our knowledge of the deuteron nucleus and of deuteron sources[3], there has been no systematic study on the possibility of accelerating polarized deuteron beams. In this paper, we analyze the feasibility of polarized deuteron beams in the AGS and RHIC. We shall study possible depolarization resonances, the achievable energy and luminosity and the difficulties involved. Section 2 calculates the expected and tolerable resonance strengths in AGS and RHIC. Section 3 discusses the luminosity and the conclusions are given in section 4.

2. Spin Resonance Strength

The most important information needed in accelerating polarized beam is the depolarizing resonance strength. Since spin motion in circular accelerators follows the Thomas-BMT equation[5], the spin depolarizing resonance strength[6] is given by

$$\epsilon_K = \frac{1+G\gamma}{2\pi} \int_0^{2\pi R} \frac{\partial B_z/\partial x}{B\rho} z e^{iK\theta} ds \;, \tag{1}$$

where θ is the orbital bending angle, z is the vertical displacement from the center of a quadrupole, $\partial B_z/\partial x$ is the gradient of the quadrupole and $B\rho = 6.26\beta\gamma$ [T-m] is the momentum rigidity of the deuteron beam. Using basic accelerator physics[7], we have $z = z_{co} + z_\beta$, with closed orbit displacement, z_{co}, and betatron displacement, z_β. Then

$$z_{co}(s) = \sqrt{\beta_z(s)} \sum_{k=integer} \frac{\nu_z^2 f_k e^{ik\phi(s)}}{\nu_z^2 - k^2} \;, \tag{2}$$

$$z_\beta(s) = \sqrt{\frac{\beta_z(s)\epsilon_N}{\pi\gamma}} \cos(\nu_z\phi + \chi) \;, \tag{3}$$

where ν_z, ϵ_N are the vertical betatron tune of the machine and the normalized emittance of the beam respectively, $\beta_z(s)$ is the vertical betatron amplitude function, and χ is an arbitrary phase factor depending on the bunch distribution in the betatron phase space. The betatron phase, $\phi(s)$, and the Fourier amplitude, f_k, are given by

$$\phi(s) = \int^s \frac{ds}{\nu_z\beta_z(s)} \;;\; f_k = \frac{1}{2\pi\nu_z} \int_0^{2\pi R} \sqrt{\beta_z} \frac{\Delta B}{B\rho} e^{-ik\phi} ds.$$

where ΔB is the dipole field error in the accelerator. Because of the periodicity of circular accelerators, there are three types of spin resonances, i.e. intrinsic, imperfection and gradient error resonances, to be discussed below.

1) The imperfection resonances, due to z_{co} of the closed orbit errors, are located at $K = integer$.
2) The intrinsic resonances, due to the betatron motion z_β, are located at $K = nP \pm \nu_z$, where n is integer and P is the superperiodicity of the accelerator.
3) The gradient error resonances arise from the error in the quadrupole gradients, i.e. $\Delta(\frac{\partial B_z/\partial x}{B\rho})$. They are located at $K = n \pm \nu_z$.

Since $G = -0.1427$ for the deuteron is a small number, we expect that the

$|G\gamma|$ value is in the interval of $(0,2)$ in AGS and $(1,20)$ in RHIC. Experience in the acceleration of polarized protons in the AGS[8] indicates that resonances $G\gamma \leq 6$ do not require corrections. Therefore there are no important deuteron depolarization resonances in the AGS, where the available deuteron energy is below 15 GeV (Z/A=1/2). The polarized deuteron however will loose about 5% polarization in the AGS to RHIC transfer line if there are no spin correctors there. This however can be compensated by some spin rotators located in the AGS to RHIC transfer line.

2.1a Intrinsic Resonance Strength

In RHIC, the intrinsic resonances are located at $K = nP \pm \nu_y$. Since $\nu_y = 28.825$ and $P = 3$, the resonance location are, at

$$K = 1.175, 1.825, 4.175, 4.825, 7.175, 7.825, \cdots, 19.175 \tag{4}$$

The resonance strength can be expressed as[9]

$$\varepsilon_K = -\frac{1+G\gamma}{4\pi}\sqrt{\frac{\varepsilon_N}{\pi\gamma}}\{[F_c^+ E_M^+ + X_{ins}^+]E_P^+ + [F_c^- E_M^- + X_{ins}^-]E_P^-\}, \tag{5}$$

where $F_c^{\pm}$ are contributions from a single FODO cell in the arc, i.e.

$$F_c^{\pm} = g_F\sqrt{\beta_z(F)} - g_D\sqrt{\beta_z(D)}e^{i(K\pm\nu_B)\pi/MP}.$$

Here g_F, g_D are focusing and defocusing quadrupole strengths. $\beta_z(F), \beta_z(D)$ are respectively the vertical betatron amplitude functions at horizontal focusing and defocusing quadrupoles. ν_y is the vertical betatron tune for the accelerator and $2\pi\nu_B$ is the vertical betatron phase advance accumulated through all the dipole cells, where the spin also precesses at a rate of $G\gamma$. $X_{ins}^{\pm}$ are strength contributions from insertions. The enhancement factors, $E_P^{\pm}, E_M^{\pm}$, due to the P superperiods in an accelerator and M FODO cells in each superperiod, are respectively,

$$E_P^{\pm} = \frac{1-e^{i2\pi(K\pm\nu_y)}}{1-e^{i2\pi(K\pm\nu_y)/P}} \; ; \; E_M^{\pm} = \frac{1-e^{i2\pi(K\pm\nu_B)/P}}{1-e^{i2\pi(K\pm\nu_B)/PM}}.$$

Thus the resonance strength is enhanced by a factor of P at $K = nP \pm \nu_y$ and by a factor of M at $K = nPM \pm \nu_B$. In RHIC, $P = 3, M = 24, \nu_y = 28.825, \nu_B \approx 22.825$. At the resonances K of Eq.(4) except $K = 19.175$, we have $|E_P^{\pm}| = 3, |E_M^{\pm}| \approx 1$. We thus expect the intrinsic resonance strength for the deuteron to be,

$$|\varepsilon_K^{intrinsic}| \approx 2 \times 10^{-4}(1+G\gamma)\sqrt{\frac{\varepsilon_N[\pi\mu m]}{\gamma}} \approx 5 \cdot 10^{-4}. \tag{6}$$

At $K = 19.175$, we obtain $|E_P^{\pm}| = 3, |E_M^{\pm}| \approx 4$. Whence the resonance strength is increased by a factor of 4, i.e. $\varepsilon_{19.175}^{intrinsic} \approx 2.2 \cdot 10^{-3}$.

2.1b Imperfection Resonance Strength

The imperfection resonance can also be estimated similarly[9] as,

$$|\varepsilon_K^{imperfection}| \approx 7 \cdot 10^{-4}(1 + G\gamma)\sigma_{co}[\text{mm}] ,\qquad(7)$$

where σ_{co} is the vertical RMS closed orbit displacement. After closed orbit corrections with $\sigma_{co} \leq 0.3$mm, we expect $|\varepsilon_K^{imperfection}| \leq 4 \cdot 10^{-3}$. Thus the snake strength[10] needed for the imperfection resonances in deuteron acceleration is less than 0.5%.

2.1c Gradient Error Resonance Strength

For the gradient error resonances at $K = n \pm \nu_y$, we can also estimate the resonance strength as following:

$$|\varepsilon_K^{gradient\ error}| \leq 2 \cdot 10^{-3}(1 + G\gamma)\sqrt{\frac{\varepsilon_N[\pi\mu\text{m}]}{\gamma}}\sigma_{\frac{\Delta g}{g}} ,\qquad(8)$$

where $\sigma_{\frac{\Delta g}{g}}$ is the gradient error. Thus we expect the strength of gradient error resonances to be less than 10^{-6} provided that $\sigma_{\frac{\Delta g}{g}} \leq 10^{-4}$ is achieved.

2.2 Tolerable Resonance Strength

Let us now analyze the tolerable resonance strength for deuteron acceleration in RHIC. The acceleration rate for deuteron in RHIC is $\dot{\gamma} = 2.0$ s^{-1}. Whence the Froissart-Stora[11] accelerating rate, α, is given by

$$\alpha \equiv dG\gamma/d\theta = 5.8 \cdot 10^{-7} .$$

Using the Froissart-Stora formula, $P_f/P_i = 2e^{-\pi|\varepsilon|^2/2\alpha} - 1$, we find that the resonance strength must be less than $4.3 \cdot 10^{-5}$ to maintain a 1% depolarization. A resonance strength of $5 \cdot 10^{-4}$ will fully depolarize the beam. A resonance strength of $1.4 \cdot 10^{-3}$ is needed to achieve a 99% spin flip.

The above analysis indicates that gradient error spin resonances are not important. The imperfection resonances can induce spin flip without loosing polarization. Alternately, the partial snake for polarized deuterons (100% snake for proton) in RHIC can also be used to overcome imperfection resonances. The intrinsic resonance will depolarize the beam. The first depolarization resonance for the deuteron beam is located at $G\gamma = 4.175$, which corresponds to $\gamma = 29.2$. Unless a clever correction method is discovered, the energy for polarized deuteron operation in RHIC will be limited by the intrinsic resonance. The task of correcting intrinsic resonances can be accomplished by jumping either the spin tune or the resonance tune. We shall discuss resonance correction in the following sections.

2.3a Spin Tune Jump with Type 3 Snake

Recently, a type 3 snake, consisting of vertical orbit bump magnets and solenoid magnets, was found to give rise to a spin precession tune shift[13]. The type 3 snake is composed of the magnets: (-V $\frac{1}{2}$S$_c$ V -S -V $\frac{1}{2}$S$_c$ V), where V is the vertical orbit bumper (horizontal field) and S_c, S are solenoid field magnets. The spin tune is given by

$$\cos(\pi\nu_s) = \cos(G\gamma\pi)(\cos\frac{\Phi}{2}\cos\frac{\Phi_c}{2} + \sin\frac{\Phi}{2}\sin\frac{\Phi_c}{2}\sin\psi) \\ + \sin(G\gamma\pi)(\cos\frac{\Phi}{2}\sin\frac{\Phi_c}{2}\sin\psi + \frac{1}{2}\sin\frac{\Phi}{2}\sin^2\frac{\Phi_c}{2}\sin(2\psi)),$$
(9)

where Φ, Φ_c are the spin rotation angles of the solenoid S and S_c respectively, and ψ is the spin rotation angle of the V magnet. Expanding Eq.(9) in power series of the spin rotation angles, we obtain

$$\Delta\nu_s \approx -\frac{1}{2\pi}\psi\Phi,$$
(10)

where we have also assumed that the solenoid fields are properly compensated, i.e. $\Phi = \Phi_c$. The bilinear dependence of the spin tune shift on the spin rotation angles in Eq.(10) indicates that the procedure is rather impractical.

Assuming the vertical orbital kicker magnets, V, to be 2 Tesla-meters each, i.e. $\psi = \pm 46$ mrad, we obtain that the required solenoid field will be $\Phi = (1+G)\frac{B_s\ell}{B\rho} = 68\Delta\nu_s$, where $B_s\ell$ is the integrated solenoid field strength. To correct intrinsic resonances[9], we need $\Delta\nu_s \geq 6 \cdot \varepsilon^{intrinsic} \approx 3 \cdot 10^{-3}$. Therefore the solenoid field strength required is $B_s\ell \approx 40$ Tesla-meters at $\gamma = 30$. The required field strength increases linearly with energy. Therefore the type 3 snake is not practical for the intrinsic resonance correction.

2.3b Induced Betatron Oscillations

The betatron tune jump method had been commonly used to overcome intrinsic resonances in the AGS and the ZGS. Since the tune jump needed in RHIC for deuteron acceleration is small, i.e $\Delta\nu \geq 6 \cdot \varepsilon^{intrinsic} \approx 3 \cdot 10^{-3}$, one can either use special quadrupoles or the induced betatron oscillations to achieve the tune jump. The tune jump with quadrupole has been studied extensively in the polarized proton operation at the AGS. The rise time for the fast quadrupole should be less than 1000 ns by taking advantage of the abort gap in RHIC. The available tune space should stay within the limited tune space of about 0.033 to avoid nonlinear resonances. Therefore, a tune jump of 0.003 is achievable. On the other hand, the betatron tunes normally depend quadratically on the amplitudes of the induced betatron oscillations. This effect may also be useful in jumping through the spin resonance. However the bunch motion will be detuned and the emittance will be increased unless the bunch is kicked onto stable fixed points of a betatron resonance. The time needed to stay in the stable fixed point is about 10 ms for $\Delta G\gamma = 3 \cdot 10^{-3}$. Such a method has never been tested.

Another method to correct the intrinsic resonance is to kick the beam into large amplitude betatron oscillations. The resonance strength is proportional to the amplitudes of betatron oscillation. Thus the polarization may flip without depolarization. Since the beam size at $\gamma = 30$ is 0.7 times that at the injection $\gamma \approx 15$, one can use the available dynamical aperture for the betatron oscillations. Due to limited available dynamical apertures, the resulting resonance strength is too small to induce spin flip in RHIC except the resonance at $G\gamma = 19.175$.

2.3c Requirement of the Pulsed Quadrupole Strength

Based on our experience at the AGS, the betatron tune jump can easily be accomplished by the pulsed quadrupoles. We consider using 3 fast ferrite quadrupoles (one for each superperiod). The highest energy intrinsic resonance for the polarized deuteron beam in RHIC is at $|G\gamma| = 19.125$, where the resonance strength is $\varepsilon \approx 2.2 \cdot 10^{-3}$. This resonance may be too large to be corrected by the tune jump method in the storage ring. The next intrinsic resonance correctable by the betatron tune jump method, with $\Delta\nu_z \leq 0.003$, is located at $G\gamma = 16.825$. The corresponding $B\rho$ value is 738 T-m. To correct this resonance, we need a quadrupole strength of

$$B'\ell = \frac{4\pi\Delta\nu_z}{\beta} B\rho \approx 0.556 \text{T}$$

where $\beta = 50$ m, $\Delta\nu_z = 3 \cdot 10^{-3}$ and $B\rho = 738$ T-m have been used. Using three quadrupoles with length $\ell = 0.25$ m each, we obtain $B' = 0.742$ T/m. Let the radius of the quadrupole be $R = 0.04$ m. The required current becomes

$$NI = \frac{B'R^2}{2\mu_0} = 399 \text{ Ampere} - \text{turns.}$$

A one turn coil then requires 399 amperes. The inductance of such a quadrupole is given by

$$L = 8\mu_0 N^2 y(y + \frac{2}{3}w)\frac{\ell}{R^2} = 4.48 \cdot 10^{-6} \text{H} ,$$

where y is the distance from the quadrupole center to the coil, and w is the coil width. We choose $y = 0.05$m and $w = 0.01$m. Taking the advantage of the bunch spacing for the RHIC abort system, the kicker rise time can be chosen to be 1000 ns. The pulsed quadrupoles are then fired in sequence with 4 µs interval. Whence the voltage requirement becomes (1000 ns rise time),

$$V = L\frac{dI}{dt} = 1.8 \text{ kV}$$

In comparison, the AGS tune jump system requires 2000 Amperes, 15 kV at a rise time of 1600 ns. The suggested RHIC tune jump system requires 399 Amperes, 1.8 kV at a rise time of 1000 ns.

3. Luminosity

The polarized deuteron source has been studied previously[3,4]. Assuming the same intensity as the polarized proton source, we can expect $N_B = 10^{11}$ particles per bunch through accumulation of 20 LINAC pulses in the AGS Booster. The beam is accelerated in the Booster, AGS and RHIC to the desired energy. Based on the basic properties of the RHIC insertion design, we have

$$\beta^* = 6.6\frac{\epsilon_N[\pi\mu\text{m}]}{\gamma} \text{ [m]}.$$

The luminosity is then given by

$$\mathcal{L} = 2.9 \cdot 10^{30} \frac{\frac{B}{57}(\frac{\gamma}{30})^2(\frac{N_B}{10^{11}})^2}{(\frac{\epsilon_N[\pi\mu m]}{10})^2} \text{ cm}^{-2}\text{s}^{-1},$$

where B is the number of bunches in the accelerator and N_B is the number of particles per bunch. The luminosity depends quadratically on energy provided that the beam normalized emittance is constant. If the intrinsic resonances are corrected by the betatron tune jump, we expect a luminosity of $5 \cdot 10^{31}$ cm^{-2}s^{-1} at the full RHIC energy. Another way to increase luminosity is to increase the number of bunches, B, or to increase the number of particles per bunch, N_B, where effort in source development is needed.

4. Conclusion

In conclusion, we have analyzed the feasibility of deuteron acceleration in the AGS and RHIC. Our findings are summarized below.

1) The deuteron polarization will be preserved in the AGS during the acceleration from $\gamma \approx 1$ to 15.
2) The AGS to RHIC transfer line will lose about 5% polarization. The injection energy has to be $G\gamma \geq \nu_y^{RHIC} - 27$ in order to avoid an intrinsic resonance in RHIC. To minimize loss of polarization in the transfer line[12], the $\frac{G\gamma}{12}$ value should chosen to be near an integer unless proper spin rotators are found to eliminate polarization loss.
3) The imperfection resonance strength after the closed orbit correction in RHIC is expected to be less than $4 \cdot 10^{-3}$ in this energy range. They can induce spin flip without polarization loss. The snake designed for the polarized proton can also be used as a partial snake for the deuteron to overcome imperfection resonances.
4) The split snake configuration for polarized proton discussed in reference 14 does not have enough strength to achieve deuteron helicity spin state.
5) The gradient error resonances in RHIC are not important.
6) The essential problem arises from the intrinsic resonance at $G\gamma = 33 - \nu_y^{RHIC}$ or $\gamma = 29$. The resonance strength is expected to be $|\varepsilon| \approx 5 \cdot 10^{-4}$, which would depolarize the beam due to the slow acceleration rate in RHIC.
7) The luminosity within this limited energy range is about $2 \cdot 10^{30}$cm^{-2}s^{-1} at $\gamma = 30$.
8) When the betatron tune jump is successfully used to jump through intrinsic resonances, the luminosity will be $5 \cdot 10^{31}$cm^{-2}s^{-1} at top energy of about 126 GeV/u.

Although we may encounter difficulty at $\gamma > 29$ in RHIC, the implementation of polarized deuterons in the AGS and RHIC complex is straight forward. To achieve high energy polarized deuteron collisions, spin tune jump and betatron tune jump were studied. The solenoid field needed in the type 3 snake[13] is too large to be practical in jumping through these intrinsic resonances. Induced betatron oscillations may increase the intrinsic resonance strength. But the strength is still too small to induce spin flip except the resonance at $G\gamma = 48 - \nu_y^{RHIC} = 19.175$. Possible betatron tune jump with the magnitude $\Delta\nu \approx 3 \cdot 10^{-3}$ can be accomplished through either excitation of fast quadrupoles or induced betatron oscillations by a fast orbit kicker. The resulting betatron tune jump can be used to overcome the intrinsic spin resonances. By taking advantage of the abort gap in RHIC, the quadrupole rise time is about 1000 ns, which is independent of the number of bunches, B. Our calculation indicates that there is no major difficulty in the voltage and current requirements of the fast pulsed quadrupoles. Such a method has never been tested in a storage ring. If no unforseen difficulties arise, however, the polarized deuteron can be accelerated to reach full energy in RHIC without losing polarization by using the pulsed quadrupole technique. Further studies are indeed important to realize polarized deuteron collisions at high energies.

References

1. K.J. Heller ed., High Energy Spin Physics, AIP Conf. Proc. No.187, (1988).
2. A.D. Krisch ed., Proc. of Workshop on Polarized Beams at SSC, AIP Conf. Proc. No.145, (1985).
3. A.A. Belushkina et al.,"Acceleration of polarized deuterons at the synchrophastron", Proc. of the VII International Synposium on High Energy Spin Physics, edited by L.D. Soloviev et al.,Vol. 2, p215(1986).
   T.B. Clegg, p.1227 in Ref. 1(1989).
4. E.F. Parker et al., IEEE Trans. Nucl. Sci. NS26, 3200 (1979).
5. L.H. Thomas, Phil. Mag. 3, 1 (1927).
   V.Bargmann, L. Michel, and V.L. Telegdi, Phys. Rev. Lett. 2, 435 (1959).
6. E.D. Courant and R. Ruth, BNL report, BNL-51270 (1980).
7. E.D. Courant and H.S. Snyder, Ann. Phys. 3,1(1958).
8. F.Z. Khiari et al., Phys. Rev. D39, 45 (1989).
   L. Ahren, p. 1068 in Ref. 1 (1988).
9. S.Y. Lee, p. 1105 of Ref. 1 (1988).
   S.Y. Lee, p. 189 in Ref.2, (1985).
10. T. Roser, "Properties of Partially Excited Siberian Snakes", AIP Conf. Proc. No. 187, p. 1442 (1988).
11. M. Froissart, and R. Stora, Nucl. Inst. Meth. 7, 297 (1960).
12. S.Y. Lee and E.D. Courant, "Effect of the Vertical Bends in ATR Transfer Line on the Polarized Proton Operation", AD/RHIC-63 (1990).
13. M.G. Minty et al.,"Spin Tune Shift due to Type III Snake in the Electron Cooling System", to be published.
14. S.Y. Lee, "Snakes and Spin Rotators", BNL 52248/UC-414 (1990).

EFFICIENT CALCULATION OF ONE-LOOP POLARIZED QCD AMPLITUDES

Zvi Bern

Department of Physics, University of Pittsburgh, Pittsburgh, PA 15260

David A. Kosower

Fermi National Accelerator Laboratory, Batavia, IL 60510

ABSTRACT

We present the one-loop pure glue correction to polarized $gg \to gg$ scattering. The computation is performed using a new and efficient technique for calculating loop amplitudes in a gauge theory. The technique is based on the technology of four-dimensional heterotic superstrings.

An obvious difficulty in the analysis of data from polarized beam experiments is the amount of algebra involved in computing the relevant QCD helicity amplitudes.[1] Even at tree level the two to six gluon scattering amplitude would involve nearly 35000 Feynman diagrams and a half billion terms. At loop level the situation is even worse; for example, the $\mathcal{O}(\alpha_s^3)$ corrections to polarized $gg \to gg$ scattering have not been computed previously. Here we present the pure glue contributions to these corrections which we obtained by using a new and efficient technique[2] for computing QCD one-loop amplitudes. We will also present an outline of our new technique which is based on the technology of four-dimensional superstrings. The details of the method as well as a discussion of certain technical issues will be given elsewhere.[3] As a non-trivial check on our method we have compared our results[2] for unpolarized $gg \to gg$ scattering, to those of Ellis and Sexton;[4] we find complete agreement for this quantity.

There are several aspects of the method which indicate its advantages over the conventional diagrammatic technology. In essence, the starting point — the superstring amplitude for n-gluon scattering — already sums up all Feynman diagrams, organizes them in a color decomposition, and performs all momentum integrals, leaving only integrals analogous to Feynman parameter integrals to be done. This bypasses all algebra associated with the large number of terms generated by the nonabelian gauge vertex factors. Furthermore, as the initial expression is a function solely of the external momenta, polarization vectors, and color indices (as well as Feynman parameters), it is very well suited to use of the spinor-helicity basis;[5] which reduces the complexity of the amplitude enormously.

The color decomposition, which emerges from the string amplitude, organizes the full amplitude into a sum over certain permutations of color factors times partial amplitudes. Each partial amplitude is gauge-invariant (under on-shell gauge transformations) and contains contributions from many Feynman diagrams eliminating most of the large cancellations typical of Feynman diagram computations. Another aspect of the reorganization is the absence of extra Faddeev-Popov ghost diagrams, even though the string formalism is completely covariant.

The new formalism also lends itself to a richer set of consistency checks than does the conventional one. One has the usual checks: on gauge invariance (for example, by repeating the computation with a different choice of reference momenta in the spinor helicity basis); on unitarity (via the optical theorem); and on cancellation of infrared divergences against the soft and collinear divergences of $(n+1)$-point tree cross sections. In addition, the various gauge theory partial amplitudes are related via decoupling equations;[6] these are the one-loop version of the tree-level 'twist' or 'subcyclic' identities.[7]

Although the use of a string-like reorganization of the amplitude is by now standard in tree-level gauge theory computations, a number of technical complications might appear to impede the application of such a formalism to loop computations. The most obvious issue is that at loop level control of the massless string spectrum is required. Other difficulties are factors of 0/0 appearing in the amplitudes[8], the IR and UV divergences and the connection of the regularization and renormalizations schemes to standard field theory schemes.

Although we will not go into details here, we have overcome these difficulties.[2,3,9] We have, for example, contructed consistent strings whose field theory limits contain pure Yang-Mills[10] using the technology of four-dimensional string theories;[11,12] however, for practical calculations in the field theory limit a fully consistent string is not required. We have also developed a string version of dimensional regularization,[13] based on the work of Brink, Green, and Schwarz.[14] While knowledge of string theory is important in resolving the technical issues outlined above, for practical computations one can rely on a set of rules presupposing ignorance of string theory.

The starting point of the computation of $gg \to gg$ is the one-loop N-gluon string amplitude in the formalism of Kawai, Lewellen, and Tye[12] (KLT), given in refs. [9]. Schematically, it has the form

$$\mathcal{N} \begin{pmatrix} \text{color} \\ \text{charges} \end{pmatrix} \int \frac{d^2\tau}{(\text{Im}\,\tau)^2} \int \left(\prod_{\substack{i=1\ldots N \\ m=1\ldots 4}} d\theta_{i,m} \right) \int \left(\prod_{l=1}^{N} d^2\nu_l \right) \sum_{\vec{\alpha},\vec{\beta}} C^{\vec{\alpha}}_{\vec{\beta}} Z^{\vec{\alpha}}_{\vec{\beta}}(\tau)$$

$$\times \prod_{i<j} \exp \left[\lambda k_i \cdot k_j \, G_B(\nu_{ij}) - \theta^2_{\bullet,\bullet} \delta \begin{pmatrix} \text{color} \\ \text{indices} \end{pmatrix} G_F(\nu_{ij}) \right.$$

$$\left. - \theta^2_{\bullet,\bullet} \left\{ \lambda k_i \cdot k_j, i\sqrt{\lambda} k_i \cdot \varepsilon_j, \varepsilon_i \cdot \varepsilon_j \right\} \left\{ G_F(\bar\nu_{ij}), \dot{G}_B(\bar\nu_{ij}) \right\} + \theta^4_{\bullet,\bullet} \varepsilon_i \cdot \varepsilon_j \, \ddot{G}_B(\bar\nu_{ij}) \right] \tag{1}$$

where τ is the complex modular parameter describing the torus which is the one-loop world-sheet of the string, the ν_i are the Koba-Nielsen variables[15] describing the locations of the external gluon vertex operators on the world sheet ($\nu_{ij} = \nu_i - \nu_j$), and the ε_i are the ordinary gluon polarization vectors. The vectors of rational numbers $\vec{\alpha}$ and $\vec{\beta}$ describe the choices of world-sheet boundary conditions for the world-sheet fermions. One must sum over these boundary conditions with the KLT coefficients $C_{\vec{\beta}}^{\vec{\alpha}}$. The fermionic Green functions G_F (and fermionic contributions to the partition function $\mathcal{Z}_{\vec{\beta}}^{\vec{\alpha}}(\tau)$) depend on the choices of these world-sheet boundary conditions, while the bosonic Green functions G_B (and bosonic contributions to the partition function) are independent of the the boundary conditions (in the fermionic formulation of superstrings[12]).

Performing the integrations over the Grassmann parameters $\theta_{i,m}$ leads to a result which is multilinear in the polarization vectors. It immediately gives a color-decomposed form[6] a sum of terms, where each term consists of three pieces: a color factor — one or two traces of products of color charge matrices T^a (times left-mover Green functions), a kinematic tensor — a product of dot products of polarization vectors and momenta (times right-mover Green functions), and a kinematic core (consisting of the exponentiated bosonic Green functions $\exp(\sum \lambda k_i \cdot k_j G_B(\nu_{ij}))$). It is convenient, and possible, to integrate by parts with respect to the $\bar{\nu}_i$ so as to remove all appearances of double derivatives of the bosonic Green's functions. The amplitude then has a uniform positive power of the inverse string tension λ sitting in front.

The field theory limit is simply the limit $\lambda \to 0$. In the case of the four-point function, the amplitude contains an over-all factor of λ^2; thus in order to extract a non-vanishing contribution we must extract two poles in λ. There are two sources of such poles. One is a pinch of Koba-Nielsen variables $\nu_i - \nu_j \to 0$. In this limit we obtain contributions of the form

$$\int d^2\nu |\nu|^{-2-\lambda k_i \cdot k_j/\pi} \sim -\frac{2\pi^2}{\lambda k_i \cdot k_j}. \tag{2}$$

The other is the large $\operatorname{Im}\tau$ region where we obtain

$$\int d\operatorname{Im}\tau \, (\operatorname{Im}\tau)^p e^{-\lambda A \operatorname{Im}\tau} \sim \frac{\Gamma(p+1)}{(\lambda A)^{p+1}} \tag{3}$$

where the power of p depends on how many unpinched Koba-Nielsen variables remain. It is only in these limits that we need the expansions of the Green function and partition function. Here we shall not display the explicit form of these expansions, but will instead give the behavior of typical combinations of Green functions which occur in the amplitude.

Consider first the color-charge factors (described by the left-movers). For each ordering of the $\operatorname{Im}\nu_i$ there is a distinct contribution, each one of which is trivial. For example, with the ordering $\operatorname{Im}\nu_1 \leq \operatorname{Im}\nu_2 \leq \operatorname{Im}\nu_3 \leq \operatorname{Im}\nu_4 = \operatorname{Im}\tau$ one finds in the gauge theory limit

$$\begin{aligned} Z_L G_F(\nu_{21}) G_F(\nu_{32}) G_F(\nu_{43}) G_F(\nu_{14}) &\longrightarrow -N_c \\ Z_L G_F(\nu_{21}) G_F(\nu_{31}) G_F(\nu_{43}) G_F(\nu_{24}) &\longrightarrow 0 \end{aligned} \tag{4}$$

where N_c is the number of colors and Z_L is the left-mover partition function.

The kinematic tensor (described by the right-movers) has a slightly richer structure but also simplifies in the gauge theory limit. For example, with the same ordering of Im ν's given above one finds that

$$Z_R G_B(\bar{\nu}_{21}) G_B(\bar{\nu}_{12}) G_F(\bar{\nu}_{43}) G_F(\bar{\nu}_{34}) \longrightarrow \frac{1}{2}(1 - 2x_{21})^2$$
$$Z_R G_B(\bar{\nu}_{21}) G_B(\bar{\nu}_{32}) G_B(\bar{\nu}_{43}) G_B(\bar{\nu}_{14}) \tag{5}$$
$$\longrightarrow \frac{1}{2^4}(2 - \epsilon)(1 - 2x_{21})(1 - 2x_{32})(1 - 2x_{43})(1 - 2x_{41})$$

where $\epsilon = 4 - D$. (The sum over world-sheet boundary conditions with appropriate coefficients is included in these simplifications.) With $x_i \equiv \mathrm{Im}\, \nu_i / \mathrm{Im}\, \tau$, the $x_{ij} \equiv x_i - x_j$ are standard Feynman parameters.

The kinematic core results in an expression of the form

$$\int d\,\mathrm{Im}\,\tau\,(\mathrm{Im}\,\tau)^{1-\epsilon/2} \exp\left[\frac{\lambda}{2}\mathrm{Im}\,\tau\Big(s(G_B(\nu_{12})+G_B(\nu_{34}))+t(G_B(\nu_{14})\right.$$
$$\left.+G_B(\nu_{23}))+u(G_B(\nu_{13})+G_B(\nu_{24}))\Big)\right] \tag{6}$$
$$\longrightarrow \Gamma(2-\epsilon/2)\Big(-\lambda(sx_1x_2 + tx_2x_3 + ux_1x_3 + tx_1 - tx_2)\Big)^{-2+\epsilon/2}$$

and is the same in both dimensional regularization and reduction.

Combining the color factor (4), the kinematic tensor factors (5), the kinematic core (6), and summing over the various terms then yields a Feynman parameterized form of the full amplitude. The evaluation of these final Feynman parameter integrals can then be done by standard methods.

In the four-point amplitude, there are 43 formally independent terms, the number of multilinear functions of the polarization vectors. Physically, however, they are redundant, because they are related by constraints of gauge invariance. The entire physical content is contained in the three helicity amplitudes $\mathcal{A}(+++\,+)$, $\mathcal{A}(-++\,+)$, and $\mathcal{A}(--++\,)$. (Note that we use the convention that all momenta are outgoing, that is $k^0_{1,2} < 0$.) The spinor-helicity basis[5] chooses $\varepsilon^{(+)}_\mu(k;q) = \langle q_-|\gamma_\mu|k_-\rangle/\sqrt{2}\langle q_-|k_+\rangle$ and $\varepsilon^-_\mu(k,q) = \langle q_+|\gamma_\mu|k_+\rangle/\sqrt{2}\langle k_+|q_-\rangle$, where k is the gluon momentum, q is an arbitrary reference momentum such that $q^2 = 0$, $k \cdot q \neq 0$, and $|k_\pm\rangle$ is a Weyl spinor. A judicious choice of the reference momenta allows us to extract the physical information efficiently, by forcing many terms to vanish and by combining others. In the new formalism, the spinor helicity basis can be used immediately in the starting formula, equation (1).

Using the results for the partial amplitudes presented in ref. [2] the result for the squared matrix element through $\mathcal{O}(\alpha_s^3)$ (summed over the two final state helicities) is

$$\sum_{\text{colors}} |\mathcal{A}_4(1^+, 2^+)|^2 = \sum_{\text{colors}} |\mathcal{A}_4(1^-, 2^-)|^2$$

$$= g^4(\mu^2)(\mu^2)^\epsilon \Bigg\{ 2N_c^2(N_c^2 - 1)\frac{s^2}{t^2 u^2}(s^2 + t^2 + u^2)$$

$$+ \frac{\alpha_s}{2\pi} N_c^3(N_c^2 - 1) C(\epsilon) \Bigg[\left(-\frac{8}{\epsilon^2} - \frac{22}{3\epsilon} + \frac{11}{6}l_Q(\mu^2)\right)\frac{s^2}{t^2 u^2}(s^2 + t^2 + u^2)$$

$$+ \frac{2}{\epsilon}\frac{s^2}{t^2 u^2}\Big((s^2 + t^2)l_Q(u) + (u^2 + s^2)l_Q(t) + (u^2 + t^2)l_Q(s)\Big)$$

$$- \frac{(s^2 + t^2 + u^2)^2}{4t^2 u^2}l_Q(t)l_Q(u) - \frac{s^2}{t^2}l_Q(s)l_Q(t) - \frac{s^2}{u^2}l_Q(s)l_Q(u)$$

$$- \frac{2u^2 + 3tu + 2t^2}{2tu}(l_Q^2(u) + l_Q^2(t)) - \frac{s(3t^2 - st + 7s^2)}{3tu^2}l_Q(u) - \frac{s(3u^2 - su + 7s^2)}{3t^2 u}l_Q(t)$$

$$- \frac{s^2}{2tu} + \left(\frac{\pi^2}{2} - \frac{32}{9} - \frac{1}{6}\right)\frac{s^2}{t^2 u^2}(s^2 + t^2 + u^2) - \frac{\pi^2}{2}\frac{2s^2 - ut}{tu} \Bigg] \Bigg\} \tag{7}$$

$$\sum_{\text{colors}} |\mathcal{A}_4(1^+, 2^-)|^2 = \sum_{\text{colors}} |\mathcal{A}_4(1^-, 2^+)|^2$$

$$= g^4(\mu^2)(\mu^2)^\epsilon \Bigg\{ 2N_c^2(N_c^2 - 1)\frac{(u^4 + t^4)}{s^2 t^2 u^2}(s^2 + t^2 + u^2)$$

$$+ \frac{\alpha_s}{2\pi} N_c^3(N_c^2 - 1) C(\epsilon) \Bigg[\left(-\frac{8}{\epsilon^2} - \frac{22}{3\epsilon} + \frac{11}{6}l_Q(\mu^2)\right)\frac{(u^4 + t^4)}{s^2 t^2 u^2}(s^2 + t^2 + u^2)$$

$$+ \frac{2}{\epsilon}\frac{u^4 + t^4}{s^2 t^2 u^2}\Big((t^2 + s^2)l_Q(u) + (u^2 + s^2)l_Q(t) + (u^2 + t^2)l_Q(s)\Big)$$

$$- \frac{2u^2 + 3su + 2s^2}{2su}l_Q^2(u) - \frac{2t^2 + 3st + 2s^2}{2st}l_Q^2(t) + \frac{u^2 + t^2}{tu}l_Q^2(s) \tag{8}$$

$$- \frac{(u^2 + s^2)(u^2 + t^2)}{s^2 u^2}l_Q(s)l_Q(u) - \frac{(t^2 + s^2)(u^2 + t^2)}{s^2 t^2}l_Q(s)l_Q(t)$$

$$- \frac{u^4 + t^4}{t^2 u^2}l_Q(t)l_Q(u) - \frac{t(3s^2 - st + 7t^2)}{3su^2}l_Q(u) - \frac{u(3s^2 - su + 7u^2)}{3st^2}l_Q(t)$$

$$- \frac{(7u^2 - 8tu + 7t^2)(s^2 + t^2 + u^2)}{6tus^2}l_Q(s)$$

$$- \frac{u^2}{2st} - \frac{t^2}{2su} + \left(\frac{\pi^2}{2} - \frac{32}{9} - \frac{1}{6}\right)\frac{t^4 + u^4}{s^2 t^2 u^2}(s^2 + t^2 + u^2) \Bigg] \Bigg\}$$

where

$$C(\epsilon) = 4\left(\frac{4\pi\mu^2}{Q^2}\right)^{\epsilon/2} \frac{\Gamma(1 + \epsilon/2)\Gamma^2(1 - \epsilon/2)}{\Gamma(1 - \epsilon)}, \tag{9}$$

s, t and u are the usual Mandelstam invariants, μ^2 is the renormalization scale, Q^2 is the factorization scale, $l_Q(x) = \ln|x/Q^2|$, and we have used the $\overline{\text{MS}}$ renormalization prescription. Here we are retaining all the observable external polarization vectors in four dimensions. Unitarity requires that in this scheme unobservable external states be continued to $(4 - \epsilon)$-dimensions.

One point that could be checked using the squared matrix elements (7) and (8) is whether the conclusion that the pure QCD double polarization asymmetry is small[16] continues to hold at one loop; a small QCD asymmetry would imply that in principle one can search for new physics by experimentally measuring the asymmetry.

Computations of higher point one-loop amplitudes including massless fermions is straightforward using our string based technology. This formalism allows for the computation of one-loop corrections to the three-jet cross section from which the value of α_s can be extracted. Computations involving massive fermions and higher loops should also be possible. We expect refinements of our string based technique to lead to a deeper undertanding of perturbative QCD.

We thank R.K. Ellis and A.H. Mueller for helpful discussions on QCD, and D.C. Dunbar and K. Roland for helpful discussions on some of the string aspects of this work. This work was supported in part by the National Science Foundation, grant PHY-87-20221.

REFERENCES

1. See e.g., J. Guillet, in these proceedings.
2. Z. Bern and D.A. Kosower, preprint Fermilab-Pub-90/225-T, Pitt-90-21.
3. Z. Bern and D.A. Kosower, in preparation.
4. R.K. Ellis and J.C. Sexton, Nucl. Phys. **B269**, 445 (1986).
5. F.A. Berends, R. Kleiss, P. De Causmaecker, R. Gastmans, and T.T. Wu, Phys. Lett. **103B**, 124 (1981); P. De Causmaeker, R. Gastmans, W. Troost, and T.T. Wu, Nucl. Phys. **B206**, 53 (1982); Z. Xu, D.-H. Zhang, L. Chang, Tsinghua University preprint TUTP-84/3 (1984), unpublished; R. Kleiss and W.J. Stirling, Nucl. Phys. **B262**, 235 (1985); J.F. Gunion and Z. Kunszt, Phys. Lett. **161B**, 333 (1985); Z. Xu, D.-H. Zhang, and L. Chang, Nucl. Phys. **B291**, 392 (1987).
6. Z. Bern and D.A. Kosower, preprint Fermilab-Pub-90/115-T, LA-UR-90-2548.
7. M. Mangano, S. Parke, and Z. Xu, Nucl. Phys. **B298**, 653 (1988); D.A. Kosower, B.-H. Lee, and V.P. Nair, Phys. Lett. **201B**, 85 (1988); M. Mangano, Nucl. Phys. **B309**, 461 (1988); D.A. Kosower, Nucl. Phys. **B315**, 391 (1989); F.A. Berends and W.T. Giele, Nucl. Phys. **B306**, 759 (1988) ; D.A. Kosower, Nucl. Phys. **B335**, 23 (1990).
8. J. Minahan, Nucl. Phys. **B298**, 36 (1988).
9. Z. Bern and D.A. Kosower, Nucl. Phys. **B321**, 605 (1989); Z. Bern, D.A. Kosower, and K. Roland, Nucl. Phys. **B334**, 309 (1990).
10. Z. Bern and D.A. Kosower, Phys. Rev. **D38**, 1888 (1988); Z. Bern and D.A. Kosower, in *Perspectives in String Theory*, eds. P. Di Vecchia and J. L. Petersen (World Scientific, 1988).
11. L. Dixon, J. Harvey, C. Vafa, and E. Witten, Nucl. Phys. **B261**, 678 (1985); Nucl. Phys. **B274** 285 (1986); K.S. Narain, Phys. Lett. **169B**, 41 (1986) 41; K.S. Narain, M.H. Sarmadi and C. Vafa, Nucl. Phys. **B288**, 551 (1987); W. Lerche, D. Lust and A. N. Schellekens, Nucl. Phys. **B287**, (1987) 477.
12. H. Kawai, D.C. Lewellen and S.-H.H. Tye, Phys. Rev. Lett. 57, 1832 (1986); Nucl. Phys. **B288**, 477 (1987) ; I. Antoniadis, C.P. Bachas and C. Kounnas, Nucl. Phys. **B289**, 87 (1987).
13. G. 't Hooft and M. Veltman, Nucl. Phys. **B44**, 189 (1972).
14. M. B. Green, J. H. Schwarz and L. Brink, Nucl. Phys. **B198**, 472 (1982).
15. Z. Koba and H.B. Nielsen, Nucl. Phys. **B12**, 517 (1969).
16. C. Bourrely, J. Soffer, F. M. Renard and P. Taxil, Phys. Rep. **177**, 319 (1989).

Is There a Relation Between the Distributions of Transversely and Longitudinally Polarized Quarks?

John Collins

Physics Department, Pennsylvania State University,
University Park, PA 16802, U.S.A.

Abstract

The Wandzura-Wilczek and Burkhardt-Cottingham sum rules have been used to suggest a relation between the distributions of transversely and longitudinally polarized quarks, and to deduce that there is little transverse quark polarization in the valence region, contrary to the situation for longitudinal polarization. It is shown that these deductions are unjustified.

1 Introduction

An interesting possibility for a polarized hadron collider is to measure the distribution of transversely polarized quarks in a transversely polarized hadron, and to investigate hard scattering processes (e.g., jet production) with transversely polarized beams. The energy would be high enough that the hard scattering would be unambiguously in the perturbative domain. To date, the only measurements of the polarization of partons have been of the helicity, or longitudinal polarization, in deeply inelastic electron scattering.

What predictions for the distributions can be made in advance of experiment?

The simplest idea is that the transverse distributions should be the same as the helicity distributions, since different spin states of a particle should be related by rotation invariance. In fact, this is a false argument: The concept of a parton density is appropriate to a measurement on a fast moving hadron. A rotation in the rest frame of the hadron also rotates the measuring apparatus, which is moving ultra-relativistically with respect to the hadron. Indeed, QCD predicts that the gluon helicity is nonzero, but[1] angular momentum conservation implies that the distribution of linearly polarized gluons, which have spin-1, in a transversely polarized spin-half hadron is zero.

Even if exact equality between transverse and longitudinal distributions does not hold, one can still conjecture, reasonably, that the transverse and helicity distributions are similar: The polarization should be up to 100% in the valence region and should decrease at small x. This simply reflects the fact that the valence quarks carry the quantum numbers of the hadron. (The whole of the EMC controversy is about the small x region. Nothing has touched the basic idea that the quarks at large x and moderate Q correspond fairly directly to the constituent quarks.)

There has been some work in the literature that contradicts this idea, and it was brought to the attention of this workshop by Sivers at the second round table discussion.[2] In their pioneering work on high p_T hadron production with a transversely polarized beam and target, Hidaka, Monsay and Sivers[3] attempted to estimate the transverse quark density from the helicity density by using the Wandzura-Wilczek sum rule.[4] A related sum rule is due to Burkhardt and Cottingham.[5] In this note I will demonstrate that this collection of arguments is fallacious.

The experimental importance of these discussions is that they concern the kinematic regions where the spin asymmetries are large. In contrast to the simple ideas on valence quarks, the Wandzura-Wilczek sum rule applied to the quark densities implies that the transverse spin asymmetry is largest at small x.

Let us define $q(x)$ to be the number density of quarks of some particular flavor, and let use define $q_T(x)$ and $q_L(x)$ to be their density weighted by transverse and longitudinal polarization, when the parent hadron is respectively 100% transversely or longitudinally polarized. (Then $q_T(x) = q(x)$ corresponds to 100% polarization parallel to the hadron's polarization.)

In Sec. 2, I will give a physical argument that rotational invariance alone, or even Lorentz invariance, does not provide a relation between q_L and q_T. In Sec. 3, I will summarize the sum rules that are claimed to relate q_T and q_L, and in Sec. 4, I will show that there are false, by an explicit counter example. Then, in Sec. 4, I will dispose of one other argument, that a boost aligns spin with momentum. My conclusions are in Sec. 5.

2 Rotational invariance does not help

If we apply a 90° rotation to a longitudinally polarized particle of definite helicity, its spin vector is certainly rotated through 90°. But the particle remains longitudinally polarized, with the same helicity, because its direction of motion is rotated through exactly the same angle. Similarly, a rotation preserves a state of transverse polarization.

Now we generally think of a parton density as being a property of a fast moving hadron. So we see that rotation invariance does not manifestly imply a relation between q_T and q_L.

However, physics is invariant under the whole Lorentz group and not just under rotations. So let us apply a boost to the particle to bring it to rest. The states of longitudinal and transverse polarization are now related by a simple 90° rotation. Surely we can now obtain a relation between q_T and q_L.

But such a relation still cannot be derived. To see this, we must consider what a parton density means physically. For example, consider deeply inelastic lepton scattering, in the rest frame of the incoming hadron. The electron measures some integral of the hadron's wave function along an almost light-like line; the line becomes exactly light like in the Bjorken limit.

Alternatively, consider the Drell-Yan process in the rest frame of one of the initial hadrons. In lowest order, the process goes by quark-antiquark annihilation. Each quark (or antiquark) in the moving hadron probes the distribution of its antiparton in the stationary hadron, again along an almost light-like line.

These cases are generic. Whenever we have a hard scattering, the parton distribution is something that is measured along a specific line, the collision axis. To change the hadron's polarization state from longitudinal to transverse, we must apply a 90° rotation to the hadron, while keeping the collision axis fixed. Since the collision axis is at a different angle to the spin axis, there is no guarantee that the polarization of the parton density is unchanged. Rather the opposite is likely to be true.

The best one can hope for is that in some approximation there is no coupling between the spin and rotational degrees of freedom of the partons. In that case the parton spin is being probed at a point. The direction of the line of measurement is irrelevant. Then q_T and q_L are equal. Altarelli-Parisi evolution will certainly upset that relation, because the Altarelli-Parisi kernel does have some spin-orbit coupling, as we will see in Sec. 4.

A very simple example where there is no spin-orbit coupling in the wave function, is where we treat the hadron as a gas of free quarks. In that case the relation $q_T = q_L$ follows from its truth for non-interacting quarks.

3 Sum Rules

The Wandzura-Wilczek sum rule relates the structure functions $g_1(x, Q)$ and $g_2(x, Q)$ in polarized deeply inelastic scattering:

$$g_1(x) + g_2(x) \approx \int_x^1 \frac{dy}{y} g_1(x). \tag{1}$$

It is only supposed to be approximate, and follows from a hypothesis about the proton matrix elements of certain operators in the operator product expansion. The Burkhardt-Cottingham sum rule[5] is supposed to be exact and independent of QCD:

$$\int_0^1 dx\, g_2(x) = 0. \tag{2}$$

Hidaka et al.[3] deduce results for the polarized parton densities:

$$q_T(x) = \int_x^1 \frac{dy}{y} q_L(x), \tag{3}$$

$$\int_0^1 dx\, q_T(x) = \int_0^1 dx\, q_L(x). \tag{4}$$

Their deduction relies on a naive parton model relation between g_2 and q_T that is, in fact, false, since the transverse spin dependence of deeply inelastic scattering is higher twist. Feynman[6] also derived eqs. (2) and (4) on the basis of such parton model ideas. As we will see, these equations cannot be true for all values of the scale used to define the parton densities.

One cannot avoid the objections to the parton model derivations by an appeal to the ambiguities in the definition of the parton densities. The definition is pretty unambiguous physically[7] — the ambiguities lie in the renormalization scheme dependence, which is a very minor issue for our purposes. The statement that the transverse spin dependence of deeply inelastic lepton scattering is higher twist has the physical implication that the hard scattering involves significant effects from parton virtuality and transverse momentum, and from correlations between different partons. This is not an appropriate situation in which to measure parton densities.

Both the Drell-Yan process and jet production in doubly transversely polarized hadron-hadron collisions are leading twist processes and directly probe the transverse parton densities, with the Ralston-Soper[7] definition.

4 Counterexample

The anomalous dimensions of the parton densities are used to show how they evolve with scale Q. Let us define moments by

$$\tilde{f}(j, Q) \equiv \int_0^1 dx\, x^{j-1} f(x, Q). \tag{5}$$

Then the evolution equation for the transverse quark density has the form

$$\frac{d}{d\ln Q^2} \tilde{q}_T(j, Q) = \gamma_T(j, \alpha_s(Q))\, \tilde{q}_T(j, Q), \tag{6}$$

while for the longitudinal density we have

$$\frac{d}{d\ln Q^2}\tilde{q}_L(j,Q) = \gamma_L(j,\alpha_s(Q))\, q_L(j,Q) + \text{gluon term}. \qquad (7)$$

An easy calculation shows that[1]

$$\gamma_T - \gamma_L = \frac{\alpha_s}{2\pi} C_F \frac{1}{j(j+1)}. \qquad (8)$$

Since the anomalous dimensions differ at $j = 1$, the first moments of the transverse and longitudinal quark densities evolve differently with Q. Therefore, the sum rule Eq. (4), that claims that $\tilde{q}_T(j=1)$ and $\tilde{q}_L(j=1)$ are equal, cannot be maintained for all Q. This is the sum rule that was supposed to follow from the Burkhardt-Cottingham sum rule. Changes in the definitions of the parton densities cannot upset this argument, since the anomalous dimensions at order α_s are scheme independent.

5   Does a boost align spin with momentum?

It has been claimed that the transverse quark density should be small at large x, since a boost lines up the spin with the momentum[3,4]. This argument rests on the correct observation that when one boosts a polarized particle from rest to a high momentum, its spin vector, s^μ, acquires a longitudinal component that is proportional to the momentum, whereas the transverse component of s^μ is invariant.

However, what is physically relevant is not the longitudinal component of the spin vector, but the helicity. Indeed when the energy of a particle is very large, one can write the spin vector in terms of a helicity h and a transverse spin vector s_T^μ:

$$s^\mu = hp^\mu/m + s_\perp + O(m/E). \qquad (9)$$

This is divergent in the zero mass limit. But the concept of helicity is perfectly well-defined for a massless particle, e.g., a photon, and it is in fact Lorentz invariant. The correction term in Eq. (9) is unimportant in a hard collision.

In a hard collision, it is the helicity and transverse spin that are the correct concepts to consider. Note that for a massive particle, the transverse spin is only defined when one specifies an axis: This axis is the direction of the collision, and is the same axis as appears in the definition of the parton density.

6   Conclusions

The fact that the transverse spin asymmetries in deeply inelastic lepton scattering are higher twist has given rise to a number of misconceptions. Even in the face of correct results for QCD, the misconceptions have persisted because one is prejudiced that parton model results always represent the correct leading order physics in QCD. This prejudice fails for the case of transverse polarization.

A particular case is the approximate Wandzura-Wilczek sum rule, Eq. (1), between g_1 and g_2. One might hope to use the parton model relation between structure functions and parton densities to deduce a corresponding sum rule, Eq. (3), that relates q_T and q_L. This simply does not work: The parton model relation between the structure functions and the distribution of transversely polarized quarks is false.

An explicit demonstration that the sum rule cannot be a good QCD result is that it is violated by Altarelli-Parisi evolution, as shown in Sec. 4.

These issues are rather important when planning experiments. The Wandzura-Wilczek sum rule, when applied to the quark densities, indicates[3] that the highest transverse polarization occurs at small x. This disagrees with naive bound state ideas, which suggest that the highest polarizations occur in the valence region, at large x. These naive ideas are correct to a reasonable degree of approximation for all other quantum numbers, including helicity. There is no reason to abandon them for transverse polarization.

Ultimately, of course, parton distributions are nonperturbative quantities. Our present state of understanding of nonperturbative QCD is not so strong that one be totally sure of the predictions.

The field is ripe for further study. It belongs as much in the domain of nuclear physics as of particle physics. It is surely true that the comparison of q_T and q_L should shed some light on the amount of spin-orbit coupling in the hadron wave function.

Acknowledgements

This work was supported in part by the U.S. D.O.E. under grant DE-FG02-90ER-40577. I would like to thank many of the participants at the workshop for useful conversations.

References

1. X. Artru and M. Mekhfi, *Z. Phys. C* **45** (1990) 669; X. Artru, these proceedings.

2. Collider Round Table, these proceedings.

3. K. Hidaka, E. Monsay, and D. Sivers, *Phys. Rev.* **19** (1979) 1503.

4. S. Wandzura and F. Wilczek, *Phys. Lett.* **72B** (1977) 195; S. Wandzura, *Nucl. Phys.* **B122** (1977) 412.

5. H. Burkhardt and W.N. Cottingham, Ann. Phys. **56** (1970) 453.

6. R.P. Feynman, "Photon-Hadron Interaction", (Benjamin, Reading, 1972).

7. J. Ralston and D.E. Soper, *Nucl. Phys.* **B152** (1979) 109; J. Ralston, these proceedings.

Spin Asymmetries: From Fixed Target to Colliders

S. Heppelmann

Pennsylvania State University, University Park, PA 16802

ABSTRACT

While the thrust of this workshop has been toward the physics of spin dependent hard scattering of partons at collider energies, there has been an attempt to connect the discussion of a new program with the rich tradition of fixed target experiments at lower energies and transverse momenta. Several contributions to this workshop are discussed in this context.

In recent years we have come to understand more about the kinds of experimental variables that can be unambigously predicted within the Standard Model using today's computational methods. We have learned that, in the laboratory of the hadron collider, high p_t observables are predictable without the need for additional unproven assumptions. Indeed, the single strong message which comes out of this workshop is that the high p_t spin asymmetry measurements at a polarized hadron collider will provide a large and varied quantity of data which is quantitatively predictable without model dependence. These measurements serve as tests of the standard model or equivalently as searches for physics beyond the standard model.

In lower energy experiments, the strong interaction phenonema are not yet directly calculable in a model free manner. QCD can provide insight into how effective models can be constructed, but direct application of QCD is not currently possible.

The existing data from polarized hadron interactions have all been obtained at fixed target facilities with low energy and/or luminosity. Experimenters have been frustrated with the inability of theory to make low energy predictions that could stand the test of time. Polarization asymmetry experiments have historically revealed large asymmetries when effective models of QCD have predicted small asymmetries. With this background, I believe that many have been led to two incorrect impressions.

First, there is a vague fear that the interpretation of spin measurements will always be inconclusive. Experimentalists are concerned about making measurements which, even though contrary to expectation, seem to have little or no impact. One important mission of this workshop is to assure experimentalists that things are different now, that if deviations from standard model predictions are observed in large p_t collider physics, such discoveries will be significant. In collider physics, we can now expect firm and nontrivial predictions from the Standard Model.

The other impression is that the lower energy measurements, with their large observed asymmetries, are not important because they cannot directly test QCD. It is often argued that it is unnecessary to make measurements of quantities which are not currently calculable in the theory. Many of us believe, however, that there are some observables which are so intuitively intrinsic to a physical picture of a process that their measurement must be endorsed in spite of our current limitations in calculation methods. We may never be able to test QCD with these low energy inclusive and exclusive measurements, but inspired by QCD, we can build working models that help us characterize the lower energy and perhaps even bound state phenomena of strong interactions. Whether such studies fall within the scope of Particle Physics or Intermediate Energy/Nuclear Physics is open to discussion. In either case, we believe that these data are important. The characterization of these phenonema in models may lead to new insights into QCD that, in turn, could redirect computational methods.

Several contributions to this workshop deal with progress that has been made in understanding exclusive and inclusive hadronic scattering in the p_t range from 1 to 5 GeV/c, including polarization effects. There are a number of approaches to understanding these semi-hard processes. One can, for example, try to extend the formal perturbative techniques to higher twist (Qui) for purposes of extending the predictablity to somewhat lower p_t.

Analysis of the perturbative QCD calculation (Botts) for exclusive scattering with special attention to the Landshoff singularities provides a mathematical model for some of the long distance effects which dominate the lower p_t exclusive scattering data. These and other long distance effects relate to issues of Color Transparency or nuclear filtering as well as providing mechanisms for producing transverse spin asymmetries which are prohibited in the leading twist approximations to perturbative QCD (Frankfurt, Sivers, Ralston, Ratcliffe).

Clearly the exciting technical developments which make polarized hadron colliders possible will inspire experimental efforts in hard scattering experiments with jets, direct photons, Drell Yan, etc. It should be pointed out that beyond this it is possible to consider new kinds of experiments as well. The light cone parton distributions, with longitudinal and transvers spin, can be fully unfolded in such a facility. The possibility of p-nucleus collisions in the nucleon-nucleon center of mass provides a new kind of nuclear experiment.

Color transparency results indicate that it will be possible to study nuclear light cone distributions with proton-nucleon quasielastic scattering. It seems that the general problem of the relation between bound state and light cone distributions can be studied in both nucleon and nuclear systems with a machine like RHIC. The ideas of color transparency and the filtering of spin effects at large energy seem to provide a new handle on the connection between transverse spin assymmetries and long distance or confinement effects in the proton.

Finally, it is important to realize that within the current design of RHIC, there are unique opportunities [1] which should be exploited. The surprisingly minor modifications of the RHIC design necessary to produce polarized beams is one such example. As another example, the BC1 magnets, located after every intersection region in RHIC, may provide a ready-made spectrometer for light cone momentum analysis of nuclear fragments produced in quasielastic p-nucleus collisions. Such measurements have never been done and provide a new approach to the study of nuclear structure.

In summary, a polarized p-p collider facility, with the capability for p-A and A-A collisions, can have two novel applications. The first is to complement the ongoing high energy collider program by providing spin dependent parton distributions and testing Standard Model predictions for spin dependent interactions. The second is the movement into a new high luminosity, high energy frontier in QCD inspired phenomenology, a frontier which may represent a mariage of particle physics and intermediate energy nuclear physics. This would include extensions of intermediate p_t spin dependent inclusive and exclusive scattering measurements, experiments in nuclear filtering of spin observables, and the determination of light cone nuclear wave functions.

References

[1] "Fourth Workshop on Experiments and Detectors for a Relativistic Heavy Ion Collider, ed. M. Fatyga and B. Boskawitz, Brookhaven National Laboratory (1990)".

Roundtable Summaries

THEORETICAL INTERPRETATION OF THE EMC RESULTS: SUMMARY OF THE ROUND-TABLE DISCUSSION

R.D. Carlitz

Department of Physics and Astronomy, University of Pittsburgh
Pittsburgh, PA 15260

Aneesh V. Manohar*

Department of Physics, University of California, San Diego,
9500 Gilman Drive, La Jolla, CA 92093-0319.

ABSTRACT

In this article on the EMC results, we would like to note the points on which the various groups agree.

There have been many papers about the theoretical implications of the EMC result, and in particular, about the role of gluons. There was a round-table discussion of these issues at the conference. The panelists were R.D. Carlitz, J.C. Collins, A.V. Efremov, J. Mandula, A.V. Manohar, and P.A. Souder. Rather than summarize the discussion, we felt it would be more useful if we noted the points which were agreed on by all the participants. These are stated below, followed by a brief discussion of the points on which there was disagreement. One topic of the round-table discussion which is not mentioned here is the lattice computation of Δg and its implications for the EMC experiment. This is covered in J. Mandula's contribution to the conference proceedings.[1]

a) The sum rule for the g_1 structure function can be derived using the operator product expansion (OPE),[2]

$$\int_0^1 dx\, g_1^p(x, Q^2) = \frac{1}{18} \left[4\Delta u(Q^2) + \Delta d(Q^2) + \Delta s(Q^2) \right], \qquad (1)$$

and the quark distributions are given by

$$2\Delta q(\mu^2) s_\alpha = \langle p, s | \overline{q} \gamma_\alpha \gamma_5 q |_{\mu^2} | p, s \rangle, \qquad (2)$$

where $|p, s\rangle$ is the proton state of momentum p and spin s. The label μ refers to the mass scale at which the axial vector current operator is renormalized. In this simplified version, we have omitted higher order corrections and heavy quark effects. There is no gluon operator which contributes to the right hand side of the sum rule, because there is no twist two gauge invariant local gluon operator which can contribute.

b) The absence of a local gluon operator for the first moment of $g_1(x)$ in the OPE does not imply that the OPE definition of $\Delta g(x)$ has zero first moment. It only means that the coefficient of Δg in the sum rule (1) vanishes. Thus

$$\Delta g = \int_0^1 \Delta g(x) dx \qquad (3)$$

---

* On leave from the Department of Physics, Massachusetts Institute of Technology, Cambridge, MA 02139.

is non-zero, but cannot be determined in a fully inclusive deep inelastic scattering experiment. It can, however, be measured in other processes such as semi-inclusive deep inelastic scattering (which distinguishes events with one or two recoiling jets) or in Drell-Yan production. These other processes cannot be analyzed using the OPE, but they can be analyzed in the framework of perturbative QCD using factorization.

c) The first moment of Δg can be expressed as the proton matrix element of a gauge invariant, but non-local gluon operator. This operator differs from K^μ in light-cone gauge by a surface term. In the Schwinger model, which is an exactly solvable two dimensional model with many of the qualitative features of QCD[3], the forward matrix element of K^μ is gauge dependent, and thus does not provide a definition of Δg.

d) The EMC measurement[4] that $\Delta u + \Delta d + \Delta s \simeq 0$ is not in conflict with the fact that the proton has spin-1/2. The total angular momentum of the proton has orbital and spin contributions from both quarks and gluons. The orbital terms, in particular, contribute to the angular momentum, but cannot be measured in high energy scattering experiments, because they correspond to operators which involve explicit factors of the coordinate $\vec{r}$.

e) At lowest order in α_s there is agreement between the quark and gluon distributions defined using the OPE or those defined using the parton model with a specific factorization scheme. At next order, there are factorization ambiguities, so the distributions can differ. A particular choice advocated by several groups is to use a redefined quark distribution[5]

$$\Delta q' = \Delta q + \frac{\alpha_s}{2\pi} \Delta g, \qquad (4)$$

where Δg is still the OPE definition (2). If one uses $\Delta q'$ as the definition of the polarized quark distribution, then there is a gluon contribution to (1) obtained by replacing all the Δq's by $\Delta q' - (\alpha_s/2\pi)\Delta g$. One must be careful to use the same definition of the quark distributions in all processes. Thus if one redefines the quark distribution using (4), then a similar redefinition must also be used in other processes such as neutral current scattering. Thus the Z^0 boson axial coupling becomes

$$\Delta u - \Delta d - \Delta s \to \Delta u' - \Delta d' - \Delta s' + \frac{\alpha_s}{2\pi} \Delta g. \qquad (5)$$

f) The EMC experiment combined with F and D extracted from hyperon semileptonic decay can be used to determine Δq for $q = (u, d, s)$, or equivalently, to determine $\Delta q' - (\alpha_s/2\pi)\Delta g$ for $q' = (u', d', s')$. Δg can be determined by studying other hadronic processes. [The OPE and parton model definitions of Δg are identical, because we only need the value of Δg to lowest order in α_s.] By using the measured values of Δg and Δq (or $\Delta q'$), one can predict the results of other experiments which depend on polarized quark and distributions. All groups would agree on these predictions, because (4) is merely a redefinition of the quark distributions.

g) The redefined distributions $\Delta q'$ are renormalization group invariant, i.e.

$$\mu \frac{d}{d\mu} \Delta q' = 0, \qquad (6)$$

in a minimal subtraction scheme.

The point of disagreement is whether there is any advantage to using $\Delta q'$ over Δq. The place where such an advantage might occur is where one attempts to construct models for the functional forms of the various parton distributions. Since the definition

$\Delta q'$ involves a cutoff on the partons' transverse momenta, it has been argued that such a definition is necessary for discussing any model which involves a Fock space picture of the parton wavefunctions. One of us (A.M.) believes there are no advantages in using $\Delta q'$ rather than Δq.[6] If one elects to avoid the predictions of any such models and concentrate solely upon the comparison of data from different hard scattering processes, then there is no substantial disagreement as to how this data should be analyzed. There remain only the usual ambiguities associated with renormalization and factorization scheme dependences of higher order calculations.

ACKNOWLEDGEMENTS

R.C. was supported in part by the National Science Foundation. A.M. was supported in part by a grant from the Alfred P. Sloan Foundation, by a National Science Foundation Presidential Young Investigator award #PHY-8958081, and by the Department of Energy under grant #DE-FG03-90ER40546.

REFERENCES

1. J. Mandula, this proceedings.
2. J. Kuti and V.F. Weisskopf, Phys. Rev. **D4**, 3418 (1971);
   J. Ellis and R.L. Jaffe, Phys. Rev. **D9**, 1444 (1974).
3. A.V. Manohar, UCSD/PTH 90-31.
4. J. Ashman, et al., Phys. Lett. **206B**, 364 (1988);
   J. Ashman, et al., Nucl. Phys. **B328**, 1 (1989).
5. The choice of primed and unprimed distributions used here is the opposite of that in the literature. The unprimed distribution in Eq. (4) is defined by Eq. (2).
6. A.V. Manohar, The g_1 Problem: Much Ado About Nothing, this proceedings.

Roundtable Discussion on Prospects for Polarized Collider Physics

Participants: J. Soffer (Moderator), S. Y. Lee, T. Ludlam, A. Yokosawa, D. Sivers, D. Underwood, G. Bunce, P. Taxil. (Comments from members of the audience are also included.)

Soffer opened the roundtable and asked Sivers to present some brief comments on polarized gluon distributions. Sivers noted that quite independently of interpretations of EMC results and analyses using the axial anomaly, one can argue for the possible largeness of polarized gluon distributions based on their QCD evolution equations. He noted that one can calculate the running of the unpolarized and polarized gluon distributions ($G(x, Q^2)$ and $\Delta G(x, Q^2)$) using the first order Altarelli-Parisi kernel (and using reasonably well-measured polarized valence u- and d-quark distributions as input). If the assumed initial polarized gluon distribution is relatively small (as in Fig. 1, curve (a), where the figure plots $\Delta G(x)/G(x)$ vs x), then for larger values of Q^2 it is driven by the QCD evolution to the straight line, $\Delta G(x)/G(x) = x$. (Additional structure at low x arising from the evolution equations is not included in the plot.) If the initial (low Q^2) distributions are already highly polarized (curve (c)), they are then driven to be almost completely polarized down to even lower values of x (curve (d)). (He also observed that the integro-differential equations for $\Delta G/G$ can be written down explicitly and the different boundary behaviour in the large Q^2 limit can be easily seen.) The curve approaching the straight line then gives

$$\int \Delta G(x)dx \approx \int xG(x)dx \approx 0.5$$

plus slowly varying logarithmic corrections, while the large curves can have $\int \Delta G(x)dx \approx 5-10$.) Upon questioning from Yokosawa he noted that these are quite general features of the QCD evolution of parton distributions. He thought the case of large initial gluon polarization might be confronted soon with data. He reiterated that this observation is a good argument to look for possible large gluon polarization independent of any EMC-type results. He also stressed that the higher order Altarelli-Parisi kernel for polarized distributions had not been calculated.

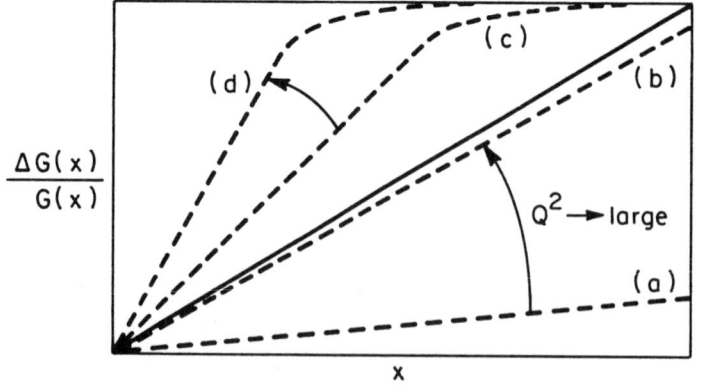

Fig. 1

Soffer commented that χ_2 production, which in principle provides a direct measurement of $\Delta G(x)/G(x)$, is then very useful in mapping out the polarized gluon distribution. He also noted that the x dependence of $\Delta G(x)$ is completely unknown and that $\Delta G(x)$ could even change sign, and the standardly accepted sign for $\Delta G(x)$ is, at the moment, just a theoretical prejudice. Yokosawa pointed out that since the observed χ_2 asymmetry is proportional to $(\Delta G(x)/G(x))^2$, no information on the sign of $\Delta G(x)$ is available this way.

Yokosawa asked if a similar phenomenon ($\Delta G(x,Q^2)/G(x,Q^2)$ increasing with increasing Q^2) occured with transverse polarization. Ratcliffe reminded the audience that the Altarelli-Parisi kernel for transversely polarized parton distributions had been written down but had never been integrated. Artru commented that a distribution of transversely polarized gluons (like linear polarization) cannot exist in protons/ neutrons but only in hadrons with spin greater than one so that the question of transversely polarized glue in the proton wasn't relevant but was of relevance for quark distributions.

Sivers reminded the audience of the Burkhardt-Cottingham (BC) sum rule[1] which states that

$$\int_0^1 \Delta q_L(x) dx = \int_0^1 \Delta q_T(x) dx \qquad (1)$$

where $\Delta q_L(x)$ ($\Delta q_T(x)$) are appropriate longitudinal (transverse) parton distributions. He argued that this was a quite general result using only Lorentz covariance and analyticity. (See, however, the contribution by Collins in these proceedings.) He also mentioned the stronger argument, due to Feynman[2],

$$\Delta q_T(x) \approx \Delta q_L(x). \qquad (2)$$

He reminded the audience that this assumes various factorization properties of spin and momentum degrees of freedom à la SU_6 and is argued only for the non-relativistic limit. He noted that these distributions were related to g_2 up to possible twist-3 operators. Artru argued that he has constructed models in which the proton is thought of as spin-0 diquark with a spin-1/2 quark in which the BC sum rule is violated. Sivers reiterated that he believed one could find consistent definitions of $\Delta q_T(x)$ and $\Delta q_L(x)$ which satisfied the sum rule, Eqn. 1. Leader pointed out the possible difficulty with convergence of the sum rule, assuming, as it does at one point, an interchange of integration limits. Qiu commented that in the operator product expansion picture, the twist 3 contributions did not automatically satisfy this sum rule, and Sivers responded that the BC sum rule was quite general, emphasizing again its simple assumptions (Lorentz covariance and analyticity.)

Changing the subject from technical details to more general questions, especially involving experimentation at RHIC, Soffer asked the experimentalists on the panel for their views. Ludlam, in turn, asked the panel if they foresaw a single experiment at a polarized RHIC facility or rather a program of measurements and what implications this had for the development of detectors. Sivers responded by suggesting that a program to measure polarized parton distributions and to study transverse spin effects would be appropriate. Yokosawa suggested that a study of such questions would be of fundamental importance and asked for others to comment on that statement. Bunce also answered Ludlam by arguing for some discussion of what the workshop could accomplish towards elucidating concensus on theoretical issues (such as interpretations of the EMC effect). In a similar vein, he also stressed the need for a discussion of a possible collaboration aimed at examining the prospects for experimentation at RHIC using polarized protons. Bunce then commented that there can be two differing views on experimentation in a new regime. One approach is to have clearly delineated theoretical predictions to confirm or shoot down, while another is to make use of such new 'windows' and to systematically examine and describe new phenomena. One example of both these issues is the examination of what the precise (and statistics and not systematics limited) measurement of possible small polarization asymmetries could probe in terms of standard model physics and beyond.

Soffer commented that many of the effects that he and collaborators have considered that have small (2 − 3%) asymmetries are now possibly measureable. (An example is the search for compositeness using single spin asymmetries in direct photon events.) Taxil noted that searches at HERA would be sensitive to large compositeness scales but would probe only eq interactions directly, while a polarized RHIC (or other polarized hadron collider) would be able to search for direct quark compositeness and hence be complementary.

Robinett asked Bunce to comment on the possible use of a general purpose 4π detector at RHIC noting that many of the possible processes discussed at the workshop as being sensitive probes of polarized distributions (direct photons, jets, W/Z, quarkonia, heavy quarks) have all been well-studied at UA1/UA2 using such detectors. Bunce responded that this was indeed a possibility but that the problem of the high luminosities available at a pp machine compared to those encountered at a $p\bar{p}$ facility would necessitate careful study of the properties of such standard collider detectors. Possible problems might include obtaining sufficiently short drift times and dealing with overlapping events. A specific problem of this kind was the search for new physics requiring a missing energy or p_T trigger. Ludlam asked the panel if this implied the 'borrowing' (and possible upgrade) of an existing collider detector.

Yokosawa commented that he could think of no strikingly new feature of strong interaction spin physics which has been revealed by a new 1 − 2% measurement. Tannenbaum noted that a similar statement could have been made in the early 1970's before the systematic study of large transverse momentum phenomena made the careful study of QCD possible. He argued for a program of systematic measurements, at first descriptive and then confronting theoretical work, to be carried out at a polarized proton collider. Leader also noted that in contrast to existing data on polarized hadron collisions, where some would argue that perturbative QCD (PQCD) might not even be applicable, in the collider setting envisaged during the workshop, PQCD would be highly relevant to the large values of Q^2 available. He also argued for a systematic program to measure the polarized distribution functions, $\Delta G(x)$, $\Delta u(x)$, $\Delta d(x)$, etc., looking for internal consistency among many such experiments.

Underwood was asked by the chairman on his views on the possible form and number of detectors for such a polarized collider. He mentioned the need for forward detection as well as central coverage. Bunce commented that since one requires two Siberian snakes to achieve polarization in the first place (at RHIC), the simplest scheme might be to have two intersection regions making use of polarized proton collisions, one of them 'coming for free', as it were. He then mentioned that one experimental region could indeed make use of general purpose 4π detector while the other could be more specialized in terms of, say, of forward detection. He thought that initially one collaboration could study the needs of various kinds of experiments which might then naturally lead to the formation of two experimental groups along the lines above.

Ludlam noted that the larger high energy physics community must be brought in on the project and convinced of its worthiness. He reiterated a point from an earlier workshop talk that 8-10 weeks of the year the RHIC facility would not be running heavy ions and therefore would be available for the high energy physics program. He added that the nuclear facility would want to run with protons for at least some of the time in any event (for calibration purposes in a sense), and if they were polarized, so much the better for other physics studies. He stressed the need for early studies to see if any addition to the RHIC machine lattice or new experimental areas would be required for the polarization option to remain viable. Sivers added that members of the nuclear physics community are interested in the study of polarization phenomena and should be enlisted. He stressed that the question of the proton spin content was at least as important to the nuclear community as to the HEP community.

Igo commented that perhaps the most well-studied mechanisms should be examined first for any new effects, and he suggested electromagnetic phenomena, especially Drell-Yan production, as an important process. Sivers noted that if you could study Drell-Yan processes, you could study almost all of the processes described in the workshop. Tannenbaum pointed out that at least one of the proposed RHIC heavy ion detectors might be able to study Drell-Yan production reasonably easily.

At that point, Soffer called the roundtable discussion to a close with the comment that the group had been charged with discussing the prospects for physics at a polarized collider facility and that the problem could benefit from more discussion and study and he suggested future meetings on the topic.

References

1. H. Burkhardt and W. H. Cottingham, Ann. Phys. <u>56</u>, 453 (1970).
2. R. Feynman, Photon Hadron Interactions (Benjamin, Reading, MA, 1972)

Participants

Anferov, Vladimir	University of Michigan
Anselmino, Mauro	University of Cagliari and INFN
Artru, Xavier	CNRS, Lyon
Bern, Zvi	University of Pittsburgh
Bilchak, Cynthia	Shippensburg University
Bodwin, Geoffrey	Argonne National Laboratory
Botts, J.	University of Washington
Bunce, Gerry	Brookhaven National Laboratory
Bystricky, Jiri	DPhPe, CEN-Saclay
Carlitz, Robert	University of Pittsburgh
Carroll, Alan	Brookhaven National Laboratory
Collins, John	Penn State University
Contopanagos, Harry	University of Michigan
Courant, Ernest	Brookhaven National Laboratory
Derbenev, Yaroslav	University of Michigan
Doncheski, Michael	University of Wisconsin
Dukes, Edmond	University of Virginia
Durrant, Simon	Penn State University
Efremov, Anatoli	JINR-Dubna
Frankfurt, Leonid	University of Illinois-Leningrad NPI
Grotch, Howard	Penn State University
Guillet, J. Ph.	CERN
Heppelmann, Steven	Penn State University
Hughes, Vernon	Yale University
Hyun, J.	Penn State University
Igo, George	UCLA
Imai, Kenichi	Kyoko University
Ji, X. -D.	MIT
Kaufmann, Russell	Argonne National Laboratory
Kaufman, William	University of Michigan
Krisch, Alan	University of Michigan
Ladinsky, Glenn	Penn State University
Leader, Elliot	Birkbeck College
Lee, S. -Y.	University of Indiana
Li, Hsiang Nan	SUNY-Stony Brook
Lopiano, David	Argonne National Laboratory
Ludlam, Thomas	Brookhaven National Laboratory
Luo, Ma	SUNY-Stony Brook
Makdisi, Yousef	Brookhaven National Laboratory
Mandula, Jeffrey	Department of Energy
Manohar, Aneesh	University of California, San Diego
Milner, Richard	MIT
Minor, Ellsworth	Penn State University
Nardulli, Giuseppe	University of Bari

Participants

Pollock, Steven	University of Washington
Qiu, Jianwei	SUNY-Stony Brook
Ralston, John	University of Kansas-Ecole Polytechnique
Ratcliffe, Philip	INFN-Milano
Ratner, Lazarus	Brookhaven National Laboratory
Robinett, Richard	Penn State University
Roser, Thomas	University of Michigan
Rossmanith, R.	CEBAF
Ryzak, Zbigniew	Harvard University
Scalise, Randall	Penn State University
Sivers, Dennis	Argonne National Laboratory
Soffer, Jacques	CNRS-Marseille
Souder, Paul	Syracuse University
Staiano, Amedeo	University of Torino
Tannenbaum, Michael	Brookhaven National Laboratory
Taxil, Pierre	CNRS-Marseille
Tornqvist, Nils	University of Helsinki
Tung, Wu-Ki	Illinois Institute of Technology
Underwood, David	Argonne National Laboratory
Vuaridel, Bertrand	University of Michigan
Weinkauf, Laura	Penn State University
Yokosawa, Akihiko	Argonne National Laboratory
Yuan, Chien-Peng	Johns Hopkins University

AIP Conference Proceedings

		L.C. Number	ISBN
No. 144	Magnetospheric Phenomena in Astrophysics (Los Alamos, NM, 1984)	86-71149	0-88318-343-9
No. 145	Polarized Beams at SSC & Polarized Antiprotons (Ann Arbor, MI & Bodega Bay, CA, 1985)	86-71343	0-88318-344-7
No. 146	Advances in Laser Science–I (Dallas, TX, 1985)	86-71536	0-88318-345-5
No. 147	Short Wavelength Coherent Radiation: Generation and Applications (Monterey, CA, 1986)	86-71674	0-88318-346-3
No. 148	Space Colonization: Technology and The Liberal Arts (Geneva, NY, 1985)	86-71675	0-88318-347-1
No. 149	Physics and Chemistry of Protective Coatings (Universal City, CA, 1985)	86-72019	0-88318-348-X
No. 150	Intersections Between Particle and Nuclear Physics (Lake Louise, Canada, 1986)	86-72018	0-88318-349-8
No. 151	Neural Networks for Computing (Snowbird, UT, 1986)	86-72481	0-88318-351-X
No. 152	Heavy Ion Inertial Fusion (Washington, DC, 1986)	86-73185	0-88318-352-8
No. 153	Physics of Particle Accelerators (SLAC Summer School, 1985) (Fermilab Summer School, 1984)	87-70103	0-88318-353-6
No. 154	Physics and Chemistry of Porous Media—II (Ridge Field, CT, 1986)	83-73640	0-88318-354-4
No. 155	The Galactic Center: Proceedings of the Symposium Honoring C. H. Townes (Berkeley, CA, 1986)	86-73186	0-88318-355-2
No. 156	Advanced Accelerator Concepts (Madison, WI, 1986)	87-70635	0-88318-358-0
No. 157	Stability of Amorphous Silicon Alloy Materials and Devices (Palo Alto, CA, 1987)	87-70990	0-88318-359-9
No. 158	Production and Neutralization of Negative Ions and Beams (Brookhaven, NY, 1986)	87-71695	0-88318-358-7
No. 159	Applications of Radio-Frequency Power to Plasma: Seventh Topical Conference (Kissimmee, FL, 1987)	87-71812	0-88318-359-5
No. 160	Advances in Laser Science–II (Seattle, WA, 1986)	87-71962	0-88318-360-9
No. 161	Electron Scattering in Nuclear and Particle Science: In Commemoration of the 35th Anniversary of the Lyman-Hanson-Scott Experiment (Urbana, IL, 1986)	87-72403	0-88318-361-7
No. 162	Few-Body Systems and Multiparticle Dynamics (Crystal City, VA, 1987)	87-72594	0-88318-362-5

No. 163	Pion–Nucleus Physics: Future Directions and New Facilities at LAMPF (Los Alamos, NM, 1987)	87-72961	0-88318-363-3
No. 164	Nuclei Far from Stability: Fifth International Conference (Rosseau Lake, ON, 1987)	87-73214	0-88318-364-1
No. 165	Thin Film Processing and Characterization of High-Temperature Superconductors (Anaheim, CA, 1987)	87-73420	0-88318-365-X
No. 166	Photovoltaic Safety (Denver, CO, 1988)	88-42854	0-88318-366-8
No. 167	Deposition and Growth: Limits for Microelectronics (Anaheim, CA, 1987)	88-71432	0-88318-367-6
No. 168	Atomic Processes in Plasmas (Santa Fe, NM, 1987)	88-71273	0-88318-368-4
No. 169	Modern Physics in America: A Michelson-Morley Centennial Symposium (Cleveland, OH, 1987)	88-71348	0-88318-369-2
No. 170	Nuclear Spectroscopy of Astrophysical Sources (Washington, DC, 1987)	88-71625	0-88318-370-6
No. 171	Vacuum Design of Advanced and Compact Synchrotron Light Sources (Upton, NY, 1988)	88-71824	0-88318-371-4
No. 172	Advances in Laser Science–III: Proceedings of the International Laser Science Conference (Atlantic City, NJ, 1987)	88-71879	0-88318-372-2
No. 173	Cooperative Networks in Physics Education (Oaxtepec, Mexico, 1987)	88-72091	0-88318-373-0
No. 174	Radio Wave Scattering in the Interstellar Medium (San Diego, CA, 1988)	88-72092	0-88318-374-9
No. 175	Non-neutral Plasma Physics (Washington, DC, 1988)	88-72275	0-88318-375-7
No. 176	Intersections Between Particle and Nuclear Physics (Third International Conference) (Rockport, ME, 1988)	88-62535	0-88318-376-5
No. 177	Linear Accelerator and Beam Optics Codes (La Jolla, CA, 1988)	88-46074	0-88318-377-3
No. 178	Nuclear Arms Technologies in the 1990s (Washington, DC, 1988)	88-83262	0-88318-378-1
No. 179	The Michelson Era in American Science: 1870–1930 (Cleveland, OH, 1987)	88-83369	0-88318-379-X
No. 180	Frontiers in Science: International Symposium (Urbana, IL, 1987)	88-83526	0-88318-380-3
No. 181	Muon-Catalyzed Fusion (Sanibel Island, FL, 1988)	88-83636	0-88318-381-1
No. 182	High T_c Superconducting Thin Films, Devices, and Application (Atlanta, GA, 1988)	88-03947	0-88318-382-X
No. 183	Cosmic Abundances of Matter (Minneapolis, MN, 1988)	89-80147	0-88318-383-8
No. 184	Physics of Particle Accelerators (Ithaca, NY, 1988)	89-83575	0-88318-384-6

No. 185	Glueballs, Hybrids, and Exotic Hadrons (Upton, NY, 1988)	89-83513	0-88318-385-4
No. 186	High-Energy Radiation Background in Space (Sanibel Island, FL, 1987)	89-83833	0-88318-386-2
No. 187	High-Energy Spin Physics (Minneapolis, MN, 1988)	89-83948	0-88318-387-0
No. 188	International Symposium on Electron Beam Ion Sources and their Applications (Upton, NY, 1988)	89-84343	0-88318-388-9
No. 189	Relativistic, Quantum Electrodynamic, and Weak Interaction Effects in Atoms (Santa Barbara, CA, 1988)	89-84431	0-88318-389-7
No. 190	Radio-frequency Power in Plasmas (Irvine, CA, 1989)	89-45805	0-88318-397-8
No. 191	Advances in Laser Science–IV (Atlanta, GA, 1988)	89-85595	0-88318-391-9
No. 192	Vacuum Mechatronics (First International Workshop) (Santa Barbara, CA, 1989)	89-45905	0-88318-394-3
No. 193	Advanced Accelerator Concepts (Lake Arrowhead, CA, 1989)	89-45914	0-88318-393-5
No. 194	Quantum Fluids and Solids—1989 (Gainesville, FL, 1989)	89-81079	0-88318-395-1
No. 195	Dense Z-Pinches (Laguna Beach, CA, 1989)	89-46212	0-88318-396-X
No. 196	Heavy Quark Physics (Ithaca, NY, 1989)	89-81583	0-88318-644-6
No. 197	Drops and Bubbles (Monterey, CA, 1988)	89-46360	0-88318-392-7
No. 198	Astrophysics in Antarctica (Newark, DE, 1989)	89-46421	0-88318-398-6
No. 199	Surface Conditioning of Vacuum Systems (Los Angeles, CA, 1989)	89-82542	0-88318-756-6
No. 200	High T_c Superconducting Thin Films: Processing, Characterization, and Applications (Boston, MA, 1989)	90-80006	0-88318-759-0
No. 201	QED Stucture Functions (Ann Arbor, MI, 1989)	90-80229	0-88318-671-3
No. 202	NASA Workshop on Physics From a Lunar Base (Stanford, CA, 1989)	90-55073	0-88318-646-2
No. 203	Particle Astrophysics: The NASA Cosmic Ray Program for the 1990s and Beyond (Greenbelt, MD, 1989)	90-55077	0-88318-763-9
No. 204	Aspects of Electron–Molecule Scattering and Photoionization (New Haven, CT, 1989)	90-55175	0-88318-764-7
No. 205	The Physics of Electronic and Atomic Collisions (XVI International Conference) (New York, NY, 1989)	90-53183	0-88318-390-0
No. 206	Atomic Processes in Plasmas (Gaithersburg, MD, 1989)	90-55265	0-88318-769-8

No. 207	Astrophysics from the Moon (Annapolis, MD, 1990)	90-55582	0-88318-770-1
No. 208	Current Topics in Shock Waves (Bethlehem, PA, 1989)	90-55617	0-88318-776-0
No. 209	Computing for High Luminosity and High Intensity Facilities (Santa Fe, NM, 1990)	90-55634	0-88318-786-8
No. 210	Production and Neutralization of Negative Ions and Beams (Brookhaven, NY, 1990)	90-55316	0-88318-786-8
No. 211	High-Energy Astrophysics in the 21st Century (Taos, NM, 1989)	90-55644	0-88318-803-1
No. 212	Accelerator Instrumentation (Brookhaven, NY, 1989)	90-55838	0-88318-645-4
No. 213	Frontiers in Condensed Matter Theory (New York, NY, 1989)	90-6421	0-88318-771-X 0-88318-772-8 (pbk.)
No. 214	Beam Dynamics Issues of High-Luminosity Asymmetric Collider Rings (Berkeley, CA, 1990)	90-55857	0-88318-767-1
No. 215	X-Ray and Inner-Shell Processes (Knoxville, TN, 1990)	90-84700	0-88318-790-6
No. 216	Spectral Line Shapes, Vol. 6 (Austin, TX, 1990)	90-06278	0-88318-791-4
No. 217	Space Nuclear Power Systems (Albuquerque, NM, 1991)	90-56220	0-88318-838-4
No. 218	Positron Beams for Solids and Surfaces (London, Canada, 1990)	90-56407	0-88318-842-2
No. 219	Superconductivity and Its Applications (Buffalo, NY, 1990)	91-55020	0-88318-835-X
No. 220	High Energy Gamma-Ray Astronomy (Ann Arbor, MI, 1990)	91-70876	0-88318-812-0
No. 221	Particle Production Near Threshold (Nashville, IN, 1990)	91-55134	0-88318-829-5
No. 222	After the First Three Minutes (College Park, MD, 1990)	91-55214	0-88318-828-7
No. 223	Polarized Collider Workshop (University Park, PA, 1990)	91-71303	0-88318-826-0
No. 224	LAMPF Workshop on (π, K) Physics (Los Alamos, NM, 1990)	91-71304	0-88318-825-2
No. 225	Half Collision Resonance Phenomena in Molecules (Caracus, Venezuela, 1990)	91-55210	0-88318-840-6
No. 226	The Living Cell in Four Dimensions (Gif sur Yvette, France, 1990)	91-55209	0-88318-794-9
No. 227	Advanced Processing and Characterization Technologies (Clearwater, FL, 1991)	91-55194	0-88318-910-0
No. 228	Anomalous Nuclear Effects in Deuterium/Solid Systems (Provo, UT, 1990)		0-88318-833-3